Food Security in India

Food Security in Asia

Food Security in Asia

Food Security in Asia

AMITAVA MUKHERJEE

$SAGE www.sagepublications.com
Los Angeles • London • New Delhi • Singapore • Washington DC

First published in 2012 by

 SAGE Publications India Pvt Ltd
B1/I-1 Mohan Cooperative Industrial Area
Mathura Road, New Delhi 110 044, India
www.sagepub.in

SAGE Publications Inc
2455 Teller Road
Thousand Oaks, California 91320, USA

SAGE Publications Ltd
1 Oliver's Yard, 55 City Road
London EC1Y 1SP, United Kingdom

SAGE Publications Asia-Pacific Pte Ltd
33 Pekin Street
#02-01 Far East Square
Singapore 048763

Published by Vivek Mehra for SAGE Publications India Pvt Ltd, typeset in 10/12pt Palatino by Star Compugraphics Private Limited, Delhi and printed at Chaman Enterprises, New Delhi.

Library of Congress Cataloging-in-Publication Data

Mukherjee, Amitava, 1951–
    Food security in Asia/Amitava Mukherjee.
        p. cm.
    Includes bibliographical references and index.
    1. Food security—Asia. 2. Food-supply—Asia. I. Title.

HD9016.A2M85   338.1'9.5—dc23   2012   2012027010

ISBN: 978-81-321-0906-8 (HB)

The SAGE Team: Sharel Simon, Dhurjjati Sarma, Vijay Sah and Rajinder Kaur

To

The Hallowed Memory of My Mother
Late Priti Kalyani Devi of Hazaribagh who turned many
winters of despair into glorious summers of hope for us!!

Contents

List of Tables

List of Figures

List of Boxes

List of Abbreviations

APDS	Alternative Public Distribution System
APIFA	Asia-Pacific International Food Agency
APIGB	Asia-Pacific International Grain Bank
BIMAS	Bimbingan Massal Swa Sembada Bahan Makanan
BRAC	Bangladesh Rural Advancement Committee
CARP	Comprehensive Agrarian Reform Programme
CBO	Community-based Organization
CGF	Community Grain Fund
CIP	Climate Insurance Pool
CPR	Common Property Resources
CSA	Community-Supported Agriculture
ETF	Exchange Traded Funds
FAO	Food and Agriculture Organization
FMIS	Farmer-Managed Irrigation Systems
FPS	Fair Price Shops
GDP	Gross Domestic Product
GGRC	Gujarat Green Revolution Company
GHI	Global Hunger Index
GM	Genetically Modified
HEIA	High External Input Agricutlure
HRS	Household Responsibility System
ICDS	Integrated Child Development Services
ICESCR	International Covenant on Economic, Social and Cultural Rights
IFAD	International Fund for Agricultural Development
IGNWPS	Indira Gandhi National Widow Pension Scheme
IGVGD	Income Generation for Vulnerable Groups Development
IPCC	Intergovernmental Panel on Climate Change
IPM	Integrated Pest Management
IPR	Intellectual Property Rights
ISFM	Integrated Soil Fertility Management

MCII	Munich Climate Insurance Initiative
MDG	Millennium Development Goals
MF	Microfinance
MFI	Microfinance Institution
MMR	Maternal Mortality Rate
NCEUS	National Commission on Enterprises in the Unorganized Sector
NGO	Non-governmental Organization
NREGS	National Rural Employment Guarantee Scheme
NTFP	Non-timber Forest Produce
OPK	Operasi Pasar Khusus
PDS	Public Distribution System
PPP	Purchasing Power Parity
R&D	Research and Development
RLCL	Rural Land Contract Law
SHG	Self-help Group
TFP	Total Factor Productivity
TFR	Total Fertility Rate
TRIPS	Trade-related Intellectual Property Rights
UDHR	Universal Declaration of Human Rights
UNEP	United Nations Environment Programme
UNESCAP	United Nations Economic and Social Commission for Asia and the Pacific
VAT	Value Added Tax
VFA	Vietnam Food Association
VGD	Vulnerable Group Development
WFP	World Food Programme
WTO	World Trade Organization

BANGLADESH BANK
(Central Bank of Bangladesh)

Dr Atiur Rahman
Governor

Foreword

I t is a matter of great pleasure for me to write this foreword for the treatise titled *Food Security in Asia* by Dr Amitava Mukherjee. I am happy to learn that the treatise is being published by SAGE Publications (Los Angeles, London, New Delhi, Singapore, and Washington DC). This comprehensive book covers major aspects of food security and provides rigorous analyses of the critical dimensions of food security, drawing upon cross-country experiences in the Asia Pacific.

The book commences with exploring the state of food security in the Asia-Pacific region and major regional and international responses to the current food security issues. The correlation between economic growth and poverty alleviation, particularly in the Asia-Pacific, is well analyzed in Chapter 2 where the author argues that despite having strong correlation between the two, food insecurity could not be eliminated completely unless the poor are ensured equal opportunities of advancement opportunities. Gender dimension of food insecurity and other major causes like poor land quality, exposure to environmental risks and isolation from labour and other markets are also analyzed in this chapter. Chapter 3 envisages the nature and character of food insecurities faced by women and girl children, and stresses on the need for treating food insecurities faced by women separately. It then goes on to suggesting measures that would require urgent attention to tackle gender dimension of food insecurity.

Chapter 4 analyzes numerous measures taken by governments in the Asia-Pacific region to tackle food insecurity including developing public distribution system, land reform, fiscal and trade policies, protecting disadvantaged people from food insecurity and so on. Chapter 5 captures innovations at the community level to prevent food insecurity such as increasing food availability and economic access to food, production and consumption responses, food storage and production responses and so forth. Based on these experiences, the author draws a set of policy conclusions that can facilitate community responses to the issues of food insecurity.

Chapter 6 focuses on the role of social protection in ensuring food security for all population segments in the society and highlights social policy challenges of ensuring access to food for all. The author also discusses the urgent need for reducing gender-based food insecurities and promoting policies targeting at ensuring more equitable resources for all. In the concluding chapter (Chapter 7), the author focuses on the contours of the ways and means to increase food production including management of idiosyncratic and covariate risks, social protection and managing climate change.

In Bangladesh, central bank refinance supported credit programs for bridging market failures and gaps in financing agricultural production and SME/self-employment initiatives directly contribute to food security. Adequate central bank refinance lines supporting local production of diverse range of food crops, e.g., agricultural credit at the rate of interest 2 per cent on pulse, oil seed, spices and maize; lending to sharecroppers owning little or no holding of farmland is a new dimension in the agricultural credit program.

Dr Mukherjee in his treatise has made a great attempt in providing comprehensive analysis of the critical dimensions of food security, covering almost all aspects of food security. This is a must read book for scholars, students, development practitioners and policy makers who are interested in exploring the issues of food security further. Finally, I would like to conclude that enhancing food security and reducing undernourishment are crucial for promoting pro-poor economic growth in Asia. I believe, this book by Dr Mukherjee will open a new window for promoting food security agenda in Asia.

Dr Atiur Rahman

Acknowledgements

In writing this book, I have received assistance and help from a large number of institutions and individuals. They are too numerous to be mentioned here, but I express my deepest gratitude towards all of them.

Special mention, however, needs to be made of some individuals and institutions. My deepest debt is to late Dr Neela Mukherjee, who happened to be my best friend for 38 years and my wife, for all her insights and forbearance.

I am grateful to the following experts (in alphabetical order) for their wisdom, which I had the privilege of accessing during our several rounds of discussions: Alekisanita Sisifa, Secretariat of the Pacific Community, Noumea; Anuradha Rajeevan, United Nations Development Programme Office, Regional Centre, Bangkok; Anthea Webb, World Food Programme, Beijing; Arsenio Balisacan, South-East Asian Regional Centre for Graduate Study and Research in Agriculture, Manila; Avinav Kumar, Skillpro Foundation, India; Bakhodur Eshonov, Centre for Economic Research of Uzbekistan, Tashkent; Bina Agarwal, University of Delhi, Delhi; C. Upendranadh, Institute for Human Development, New Delhi; Chaplygina, Russian Institute of Energy and Finance, Moscow; Chee Yoke-Ling, Third World Network, Penang; Douglas Southgate, Ohio State University, Columbus, Ohio; Durga Prasad Paudyal, Centre for Integrated Rural Development for Asia and the Pacific, Dhaka; Elenita C. Dano, Third World Network, Penang; Eric Roeder, United Nations Asia-Pacific Centre for Agricultural Engineering and Machinery, Beijing; Jean d'Cunha, Regional Programme for East and South-East Asia, UNIFEM, Bangkok; Kinlay Dorjee, Food and Agriculture Organization Regional Office for Asia and the Pacific, Bangkok; Katinka Weinberger, Head, Centre for Alleviation of Poverty through Sustainable Agriculture, Bogor, Jakarta; Li Ou, China Agricultural University, Beijing; M. Aslam Khan, University of Peshawar, Peshawar; Nancy Zhang, United Nations

Asia-Pacific Centre for Agricultural Engineering and Machinery, Beijing; Nimal Ranaweera, Independent Consultant, Sri Lanka; Pramod Kumar Aggarwal, Indian Agricultural Research Institute, Pusa, New Delhi; Ran Tao, Chinese Academy of Sciences, Beijing; Sanjay Srivastava, Regional Advisor of Disaster Risk Reduction, United Nations Economic and Social Commission for Asia and the Pacific, Bangkok; Sanjeev Ghotge, World Institute for Sustainable Energy, Pune; Sarah Barber, World Health Organization/China; Shamika Sirimane, Chief, Trade Felicitation Section, United Nations Economic and Social Commission for Asia and the Pacific, Bangkok; Syed Ayub Qutub, Pakistan Institute for Environment-Development Action Research, Islamabad; Shiladitya Chatterjee, Senior Consultant, Asian Development Bank, Manila; Sombath Somphone, Participatory Research and Development Training Centre, Vientiane; Somesh Kumar, Indian Administrative Service, Government of Andhra Pradesh, Hyderabad; Swarna S. Vepa, Madras School of Economics, Chennai; Thelma Paris, International Rice Research Institute, Los Baños, the Philippines; Thiyagarajan Velumail, United Nations Development Programme Regional Centre/Bangkok; Tommaso Cavalli-Sforza, Tony Hazzard and other colleagues in the World Health Organization Western Pacific Regional Office/Manila; and Tulus Tahi Hamonangan Tambunan, University of Trisakti, Jakarta.

I am thankful for technical discussion at various points in time to Abhijit Sen, Member, Planning Commission of India, New Delhi; Alak Narayan Sharma, Director, Institute for Human Development, New Delhi; Amitabh Kundu, Professor, Jawaharlal Nehru University, New Delhi; Andrew Shepherd, Overseas Development Institute, London; Biraj Patnaik, Principal Advisor to the Commissioner for Food Security, Supreme Court of India, New Delhi; Deep Joshi, Ramon Magsaysay Award Winner, India; Ganesh Thapa, Senior Expert, International Fund for Agricultural Development, Rome; Jean Drèze, Senior Professor, G. B. Pant Social Science Institute, Allahabad; Kaushik Basu, Professor, Cornell University, Ithaca; N. C. Saxena, Commissioner for Food Security, Supreme Court of India, New Delhi; P. V. Sateesh, Deccan Development Society, Hyderabad; Raghabendra Jha, Professor, Australian National University; Ravi Srivastava, Professor, Jawaharlal Nehru University, New Delhi; Robert Chambers, Fellow, Institute of Development Studies, Sussex; S. Mahendra Dev,

Indira Gandhi Institute of Development Research, Mumbai; Sumiter Broca, Food and Agriculture Organization Regional Office for Asia and the Pacific, Bangkok; V. S. Vyas, Emeritus Professor, Institute of Development Studies, Jaipur; Vandana Shiva, Director, Research Foundation for Science and Technology, New Delhi; Y. K. Alagh, Sardar Patel Institute of Economics and Social Research, Ahmedabad.

I wish to place on record my appreciation to my colleagues, namely, Zenebe Bashaw Uraguchi, V. Wilde Tom Schaetzel, Suniti Neogy, Sheryl Hendriks, Ruchi Tripathi, Mary Corbett, Lawrence Haddad, Jennie Dey De Pryck, Isatou Jallow, Iris Krebber, Haven Ley, Fiona Quinn, Fatema Rajabali, Eva Majurin, Elaine Mercer, Cheryl Doss, Cathy Farnworth, Catherine Bertini, Caroline Sweetman, Carl Jackson, Bina Agarwal, Andre Croppenstedt, Alyson Brody, Alexandra Spieldoch, Alessandra Galie, Aida Jamangulova, Ashley Aakesson and Morwenna Sullivan, for the insights they provided in the course of an e-discussion conducted by the Institute of Development Studies, Sussex.

I also express my gratitude to the Asian Development Bank, Manila; Asian Institute for Technology, Bangkok; Centre for Poverty Alleviation through Sustainable Agriculture, Bogor, Jakarta; Chatham House (The Royal Institute of International Affairs), London; Food and Agricultural Organization, Regional Office, Bangkok; International Fund for Agricultural Development, Rome; Overseas Development Institute, London; United Nations Asia-Pacific Centre for Agricultural Engineering and Machinery, Beijing; and the United Nations Economic and Social Commission for Asia and the Pacific, Bangkok, for all their help and for permission to use their facilities. I am particularly grateful to Chatham House, London, for permission to use materials from A. Evans' *The Feeding of Nine Billion* (2009) for parts of Chapter 7.

1

Setting the Context and the Problematic

Tell me what you eat and I will tell you who you are.

—Jean Anthelme Brillat-Savarin

This chapter begins by laying down the meaning of the key terms used in the study, such as food security and sustainable agriculture and the relationship between the two. To contextualize the study, a brief account of the state of food security in the Asia-Pacific region is presented, followed by the contours of the current food crisis. The underlying causes are then explored. This is followed by a description of the chapters and how they are linked with one another. A brief account of some of the major international and regional responses to current food security issues is presented. In conclusion, a justification is provided for the study as it stands, if one is needed at all for a region that has been beset with food security issues for decades.

Conceptual Specifications

The concept of food security has evolved over time. The World Food Conference of 1974, held at a time of global food crisis, focused attention on ensuring the availability and price stability of basic foodstuffs at both the international and the national levels. Food security was thus defined from the supply side perspective as the 'availability at all times of adequate world food supplies of basic foodstuffs to sustain a steady expansion of food consumption and to offset fluctuations in production and prices' (United Nations, 1975).

Towards the late 1970s and early 1980s, when the evident success of the Green Revolution in increasing the availability of food did not automatically lead to dramatic reductions in food insecurity (read hunger and malnutrition; see Box 1.1), Amartya Sen came out with one of the most powerful critiques of the food-availability argument through his entitlement and depreciation thesis (Sen, 1981). This led to the demand-side perspective being

---

BOX 1.1   Some Concepts Related to Food Security

Undernourishment describes the status of people whose food intake does not include enough calories (energy) to meet minimum physiological needs.

Malnutrition/undernutrition is defined as a state in which the physical function of an individual is impaired to the point where he or she can no longer maintain natural bodily capacities such as growth, pregnancy, lactation, learning abilities, physical work and resisting and recovering from disease. The term covers a range of problems from being dangerously thin (*underweight*) or too short (see *stunting*) for one's age to being deficient in vitamins and minerals or being too fat (obese).

Stunting reflects shortness-for-age, an indicator of chronic malnutrition and calculated by comparing the height-for-age of a child with a reference population of well-nourished and healthy children.

Wasting reflects a recent and severe process that has led to substantial weight loss, usually associated with starvation and/or disease, calculated by comparing weight-for-height of a child with a reference population of well-nourished and healthy children. It is often used to assess the severity of emergencies because it is strongly related to mortality.

Underweight is measured by comparing the weight-for-age of a child with a reference population of well-nourished and healthy children.

Hunger is a state in which people do not have enough food to provide them the required nutrients (carbohydrates, fats, proteins, vitamins, minerals and water) to have an active and healthy life. Hunger is an outcome of food insecurity. It is the body's way of signalling that it is running short of food and needs to eat something. Hunger can lead to malnutrition.

Source: Author.

---

added, redefining the concept of food security as 'ensuring that all people at all times have both physical and economic access to the basic food that they need' (FAO, 1983). The World Bank defined food security as 'access of all people at all times to enough food for an active, healthy life' (World Bank, 1986), the operative word having changed from 'availability' to 'access'.

In the mid-1990s, two more elements were added. First, it was recognized that it is not only access to food that matters, but also its safety and nutritional content. Second, the social and cultural acceptability of different types of food was also brought into the picture. These additional elements were reflected in the 1996 World Food Summit definition:

> Food security exists when all people, at all times, have physical and economic access to sufficient, safe and nutritious food to meet their dietary needs and food preferences for an active and healthy life. (FAO, 1996)

The Food and Agriculture Organization (FAO, 2001) further refined the definition in *The State of Food Insecurity in the World 2001* by adding that people should have social, as well as physical and economic, access to food. The definition of food security used in the present study, based on Mukherjee (2004), goes a step further by specifying that the food available should be culturally acceptable. In sum, this book understands food security as a state where all the following statements are true:

- Food is systemically available at all times;
- Food that is available is culturally acceptable,[1] where culture is defined broadly to include religious beliefs, customs, usage and practices;
- People have economic access to food;
- People have physical access to food;
- People have social access to food;
- Food that people consume has the requisite nutritional value for a healthy life; and

---

[1] The phrase 'culturally acceptable' food is to be widely interpreted to also mean, amongst other things, food that meets the edicts of religions, such as *halal* food, *kosher* food, vegetarian food and Jain food.

- People have access to potable water, for absorption of food by the body.

Food insecurity exists when one or more of these conditions are unmet. It is important not only to ensure availability of food, but also to guarantee that people have adequate physical, social, cultural and economic access[2] to food to meet their dietary needs and food preferences for an active and healthy life.[3]

Concept and Evolution of Food Security

'Food security' as a concept is younger than 'the right to food'. The right to food had already been recognized in Article 25 of the Universal Declaration of Human Rights (UDHR) in 1948, and is enshrined in Article 11 of the International Covenant on Economic, Social and Cultural Rights (ICESCR) of 1966. The concept of food security was developed in the 1970s. Yet food security, and not the right to food, was the topic of public discourse for a few decades until recently when the focus is again on the individual.

Food security has acquired a number of different meanings over time. Some estimate that approximately 200 definitions and 450 indicators of food security exist (Sage, 2002: 128, 129).[4] In order to better understand the status quo regarding the concept of food

---

[2] Lack of access, of all forms, could be temporary—for example, due to natural disasters, economic collapse or conflict—or permanent, due to persistent poverty or lack of economic development.

[3] At the household level, the food security status of each household lies somewhere along a continuum ranging from high food security to very low food security, and can be divided into four kinds. *High food security*: households have no problems, or anxiety, about consistently accessing adequate food. *Marginal food security*: households have problems or anxiety at times about accessing adequate food, but the quality, variety and quantity of their food intake are not substantially reduced. *Low food security*: households reduce the quality, variety and desirability of their diets, but the quantity of food intake and normal eating patterns are not substantially disrupted. *Very low food security*: at times during the year, the eating patterns of one or more members of the household are disrupted and food intake reduced because the household lacks money and other resources for, and faces conditions not conducive to, accessing food.

[4] See Maxwell (1996: 155, 170) for a list of 32 different definitions of food security from 1975 to 1991.

security, the following section will trace how this increasingly complex concept evolved. The focus will be on public policy documents, as they represent the negotiated consolidation of a diversity of states' views and commitments, and give guidance to the activities of states and international organizations.

Availability and Stability of Supply at the Global and National Levels

As stated at the time of the 1974 World Food Conference, food security was defined as the 'availability at all times of adequate world food supplies of basic foodstuffs . . . to sustain a steady expansion of food consumption . . . and to offset fluctuations in production and prices' (FAO, 1974a).[5] The focus of the debate was on strengthening food production to increase the availability and stability of world food supplies of basic foodstuffs, particularly cereals, to meet increasing demand. Such demands were triggered by population growth and the occurrence of a drought across many major grain-producing countries, which had led to heavy demands on international grain markets (Sage, 2002: 128, 129ff.; Stringer, 2002: 3ff.) and sharp rises in world food prices (Maxwell, 1996). Very low levels of cereal stocks and high cereal prices had evoked a perception that the world was moving towards overall food shortages that required crop protection measures. There was debate as to how the international community could ensure that developing countries had access to adequate flows of staple foods, and new international bodies were proposed: the World Food Council, the FAO Committee on World Food Security and the International Fund for Agricultural Development. The FAO adopted an International Undertaking on World Food Security (FAO, 1974b: 11, in Tomasevski, 1987: 219ff.).

In line with the focus on food shortages, the Universal Declaration on the Eradication of Hunger and Malnutrition adopted at the 1974 World Food Conference proclaimed that 'every man, woman and child has the inalienable right to be free from hunger and malnutrition'.[6] The more comprehensive 'right to adequate food'

---

[5] The Universal Declaration on the Eradication of Hunger and Malnutrition itself contains no definition.

[6] Adopted by the World Food Conference, Rome, 5–16 November 1974, UN Doc. E/CONF.65/20, para. 1.

6 FOOD SECURITY IN ASIA

that had been enshrined only eight years earlier in Article 11 of the ICESCR was not mentioned, and the implications of recognizing a right were not developed in the declaration.[7]

After the recovery from the world food crisis at the beginning of the 1970s and due to difficulties relating to the negotiation of a new international grains agreement, measures at the national level received increased attention. In 1979, the FAO Council adopted a 'Plan of Action on World Food Security' (FAO, 1979: 7ff., in Tomasevski, 1987: 219ff.; see also Sen, 1981). It urged governments to 'take full advantage of the relatively ample world supply situation for cereals in order to build up stocks', to 'adopt and implement national cereal stock policies, and targets or objectives'. It suggested criteria for the management and release of national stocks and encouraged governments to give high priority to the formulation and implementation of national food security programmes.

Access at Household and Individual Levels

In 1981, Amartya Sen's seminal *Poverty and Famines* (Sen, 1981: 8)[8] shattered the theory that food insecurity was mainly a result of lack of availability of foodstuffs, by proving that *individuals'* food security was primarily dependent on their possibilities of *accessing* food, their '[ability] to establish entitlement to enough food' through production-based, labour-based, trade-based, transfer-based or other entitlement relationships (Clay, 1997: 7). Sen showed that individual food security can be severely constrained despite sufficient national supplies, and that some of the worst famines took place due to entitlement shifts with no significant decline in food availability per capita.

In the relatively more stable international food market environment, it became clear that not food availability, but poverty and lack of access (a problem of effective demand), caused food insecurity. The food security agenda was broadened from focusing

---

[7] There is only one further reference to 'the right to life and human dignity as enshrined in the Universal Declaration of Human Rights'.

[8] Maxwell points out that Sen was not the first to discover the importance of access to food. However, it was his study that brought it to the centre of development thinking (Maxwell, 1996: 157).

on instability and acute crisis to the problem of chronic hunger.[9] In 1983, the FAO explicitly expanded its definition of food security to include 'security of access to supplies on the part of all those who need them' (FAO, 1983). People in general, and the household in particular, became of interest. The World Bank used a definition that took people's needs as its starting point. The 1985 World Food Security Compact (FAO, 1985; see Tomasevski, 1987: 224) adopted by the FAO Conference to bring together general principles and suggestions for action by governments, organizations and individuals mentioned food security at the national, household and individual levels. It acknowledged that 'achievement of the "fundamental right of everyone to be free from hunger" depends ultimately on the abolition of poverty' (ibid.: para. 2).

The 1980s also witnessed interest in the complex relationships between chronic, seasonal and temporary food insecurity, peoples' coping strategies, their priorities when making choices as to how to spend overall insufficient resources, food security as part of a wider livelihoods concept and the relationship between household and individual food security.

It became clear that even with household food security, *individual* food security is not guaranteed. Within the household, access to food by individuals is linked to their control over household income and household resources, often to the disadvantage of women and children, particularly girl children (Stringer, 2002: 5). Food-insecure individuals can be found in food-secure households and, likewise, food-secure individuals can be found in some food-insecure households, depending on the—in many cases 'gendered'—power relationships within the 'black box' household (ibid.).

At about the same time as food security research revealed the importance of looking at the individual, the right to food received particular attention for the first time.[10] With the end of the Cold War, the perception of the division of human rights into 'western' civil and political rights, and 'eastern' and 'southern' economic,

---

[9] The definition used by World Bank was 'access by all people at all times to enough food for an active, healthy life' (World Bank, 1986: v).

[10] Cf. the appointment of A. Eide as Special Rapporteur on the Right to Food in 1983, and the publication of the first collection of articles dedicated to the right to food (Alston and Tomasevski, 1984).

social and cultural ones, had also begun to fade away. Asbjørn Eide, in his capacity as Special Rapporteur on the Right to Food of the UN Sub-commission on the Prevention of Discrimination and the Protection of Minorities, submitted a groundbreaking study in 1987 (Eide, 1989). It was the first Sub-commission study to explore the nature and content of a socio-economic right. It analyzed the content of the right and clarified the corresponding national and international obligations of states.

Health and Related Factors

The food security agenda was further broadened by health and nutrition research, which highlighted the fact that reciprocal and synergetic linkages exist between food intake and nutritional well-being (De Rose and Millman, 1998: 8). Disease leads to deterioration in nutritional status at the same time as malnutrition increases susceptibility to disease (ibid.). Gastrointestinal infections particularly can impair the body's ability to absorb both calories and micronutrients (FAO, 2001: 32–33). A life-cycle approach to understanding the long-term and intergenerational consequences of malnutrition was developed (Commission on the Nutrition Challenges of the 21st Century, 2000).[11] It was also shown that food quality and nutritional value mattered since, even if food is consumed in quantities sufficient to meet caloric and protein needs, micronutrient deficits, in particular those of vitamin A, iron and iodine, heavily impact mental functions and vulnerability to disease (ibid.).

Non-food causes of food insecurity were looked into, such as inadequate care—particularly of young children who need not only sufficient, healthy food, but also somebody to feed them. Some strands of the literature have considered food security as only one component of the wider goal, 'adequate nutrition' or 'nutrition security', adequate care and adequate prevention and control of diseases being the other objectives (Eide et al., 1991: 415, 445ff.).

---

[11] See also A. Eide, UN Commission on Human Rights, Sub-commission on the Prevention of Discrimination and Protection of Minorities, Updated Study on the Right to Food, submitted by Mr Asbjørn Eide in accordance with Sub-commission decision 1998/106, E/CN.4/Sub.2/1999/12, paras 19 ff.

The World Food Summit Approach

The definition of food security most widely used and accepted at present is a complex definition adopted at the World Food Summit in 1996: 'Food security, at the individual, household, national, regional and global levels is achieved when all people, at all times, have physical and economic access to sufficient, safe and nutritious food to meet their dietary needs and food preferences for an active and healthy life.' It comprises the elements of earlier definitions, but adds 'safe' and 'nutritious', recognizes dietary needs and the importance of cultural factors, and sees an active and healthy life as the broader goal beyond food security.

According to the FAO, the definition is in keeping with three classic aspects of food security: availability of staple foods, stability of supplies and access for all to these supplies. But it also introduces the idea of adapted food, i.e., of the 'biological utilization' of food, which depends, *inter alia*, on cooking methods, ways of consuming food and the state of a person's health. It also addresses food security as an issue from the individual to the global level.

The notion of food security has evolved, developed and diversified to become more and more encompassing and multilayered. At least three important shifts in thinking about food security can be identified since the World Food Conference of 1974: a shift in the unit level of analysis; a shift from quantity alone to 'quantity and quality'; and a shift from looking at food alone to including the broader context. Attention has widened from the international, to the national, to the household level, and finally to the individual level, once research showed that even behind the veil of household food security, food-insecure individuals could be hidden. While initially the main concern was availability of staple food supplies, gradually the importance of other factors such as food quality, safety and micronutrients was recognized. Finally, non-food factors relevant for food security such as adequate care, health and hygiene practices have received attention (Mechlem, 2004).

The State of Food Security

The Asia and the Pacific region has seen dramatic decrease in the incidence of poverty over the past decades (UNESCAP, 2007). Poverty is certainly a major cause of deprivation leading to food insecurity. Indeed, the interdependence between consumption and income is the most basic interdependence in understanding food deprivation. Food insecurity is 'primarily a problem of general poverty and of deprivation of food entitlements and adequate health and social care' (Sen, 1997). Thus, a dramatic fall in poverty would, *ipso facto*, imply that the food security situation has also improved in the Asia-Pacific region, in the aggregate sense. However, for the Asia and Pacific region as a whole, despite efforts to accelerate economic growth and reduce poverty, the state of progress towards reaching the target of halving the number of people who live in hunger by 2015 (the first of the Millennium Development Goals [MDG]) is setting off alarm bells.

At the end of 2002, there were 583 million people in Asia who did not meet the nutritional requirement and about 545 million people still undernourished, comprising 65 per cent of the world's ill-fed (FAO, 2004). An estimated 119 million hungry people lived in China. India alone is home to 233 million hungry people (UNDP, 2003),[12] way ahead of the number of all the hungry people of Africa taken as a whole, which was 183 million. Currently, more than 60 per cent of the world's undernourished[13] population lives in Asia and the Pacific. South Asia has 300 million undernourished people. East Asia and South-East Asia have 160 and 65 million undernourished people, respectively, as against 242 million undernourished people in Sub-Saharan Africa, the Near East and North Africa taken as a whole (FAO, 2006). Box 1.2 provides summaries of some of these facts.

There are wide differences among sub-regions and countries in their success in reducing poverty and hunger, which are closely associated with their economic performance and investment in social capital. Poverty and hunger are particularly serious in South Asia and in small island states in the Pacific. Parts of Afghanistan, China, Bangladesh, DPR Korea, India, Indonesia,

---

[12] We recognize that there is a difference between hunger and food security, but, given that all hungry people are food-insecure, we take this as a rough approximation. See Box 1.2 for an overview of different related terms.

[13] We recognize that food security and malnutrition are different concepts. See Box 1.2 for definitions of the different concepts related to food security.

BOX 1.2   Magnitude of Food Insecurity in Asia-Pacific

- More than 60 per cent of the world's undernourished population lives in the Asia-Pacific.
- South Asia has 300 million undernourished people.
- East Asia and South-East Asia have 160 and 65 million undernourished people, respectively.
- There are 242 million undernourished people in Sub-Saharan Africa, and in the Near East and North Africa taken as a whole.

Source: Author.

Kazakhstan, Nepal, Pakistan, Tajikistan and Uzbekistan suffer from food insecurity in some forms. And then there are some countries, such as Myanmar and North Korea, where large-scale food insecurity and malnutrition can persist, unless food aid flows to these countries.

If food insecurity has been a problem in the Asia-Pacific region for most of the 20th century, and volatility in food prices have occurred in the past, some of them even worse than the one in 2008–09, why is the concern for food security in the continent so pervasive now? Why are governments, international agencies and civil society organizations currently so concerned about food security? Why have the issues revolving around food security agitated great minds at the current time as never before? The answers are several-fold.

First, there is the need to build the 'resilience' of countries to future shocks and risks that could plunge the Asia-Pacific region into food insecurity of the kind experienced in 2007–08, and even worse. The food security challenge for the Asia-Pacific is not merely about how to attenuate the impact of the spike in prices on the most vulnerable groups; the challenge is how to continue making progress in ensuring food security in a context where the production of food will be increasingly stressed for a range of reasons. That is, the goal is to transform (for want of a better word) food systems for greater resilience and to fully meet these challenges.

Second, the longer-term challenges that plague food production remain as valid today as ever before (see Box 1.3). Land and water constraints remain for the most part unaddressed; investments in rural infrastructure and agricultural research are largely low; costs of agricultural inputs are high relative to farm-gate prices; environmental degradation leading to loss in natural resources available for growing food continues to be worrisome; and the need to adapt to

---

BOX 1.3 The Big Picture

- 900 million people are without electricity.
- 600 million people have no access to safe water.
- 1.8 billion people have no sanitation facilities.
- 40 per cent of urban residents in Asia-Pacific live in slums.
- Population density in Asia-Pacific is one-and-a-half times the world average.
- Arable land per capita in Asia Pacific is 80 per cent of the global average.
- Productive area per capita is only 60 per cent in the Asia-Pacific region.
- Freshwater availability per capita in Asia-Pacific is the lowest in the world.

Source: Author.

---

climate change is more urgent now than ever before. It is, therefore, important to seize the opportunity now, when the crisis in food security is not at its zenith, to reflect on how to avoid subsequent crises by addressing the longer-term challenges. Some of the most important of these challenges are as follows.

First, the population (in the UN Economic and Social Commission for Asia and the Pacific [ESCAP] region) is projected to grow from 4.7 billion in 2005 to 5.1 billion by 2050. To feed a population of 5.1 billion, food production must increase dramatically by 2050. For example, in India, it is estimated that rice yields must increase by at least 45 per cent by 2020 to meet the future demand (Kumar, 1998; Mall and Aggarwal, 2002). Additionally, the entire growth of population in the Asia-Pacific region will take place in developing countries and wholly in urban areas; by 2050 in the ESCAP region, the urban population will swell to 3.41 billion people (up from 1.68 billion in 2007) as rural populations contract to 1.75 billion (from 2.39 billion in 2007). This implies that a smaller work force in the rural areas will have to be far more productive than ever before to produce more food from fewer resources. These estimates imply that there is an urgent need to augment productivity in agriculture, which requires more investment in agricultural research and development (R&D) and infrastructure, more extension services, more and improved machinery and implements among other requirements on a growing list, together with farmers who have higher technical knowledge and better skills, as well as better-functioning supply chains.

Second, because a smaller number of farmers will have to feed a more populous region with fewer resources, one way out could be to expand the area under agriculture. (The area currently in use for agriculture is 0.73 billion hectares for the ESCAP region, compared to 1.5 billion worldwide.) But that seems unlikely, because it will lead to further environmental damage and increased greenhouse gas emission, both of which are not acceptable. An alternative avenue could be to tap into yet-unused yield-enhancing resources, which could double productivity for many crops in many countries, provided farmers have better access to improved inputs, apply scientifically better fertilizers in greater abundance, make use of better seeds, improve their farming and management skills and expand land under irrigation.

Finally, in addition to rising resource scarcity, Asian agriculture will have to cope with the burden of climate change. The Intergovernmental Panel on Climate Change (IPCC) has documented the likely impact of climate change on agriculture in great detail. For instance, yields could decline by 20–40 per cent. In addition, severe weather occurrences, such as droughts and floods, are likely to intensify and cause greater crop and livestock losses. For example, in the year 2010, droughts and severe winter in Mongolia have left the country reeling under pressure. According to the International Federation of Red Cross & Red Crescent Societies, 4.5 million livestock or nearly 10 per cent of Mongolia's animal population have died since December 2009, as temperatures plunged in January 2010 to minus 40 degrees Celsius for three straight weeks (*Bangkok Post*, 2010a). Similarly, millions of people face drinking water shortages in south-western China because of a once-a-century drought that has dried up rivers and threatens vast farmlands. It is reported that the drought has gripped huge areas of Guizhou, Yunnan and Sichuan provinces, the Guangxi region and the mega-city of Chongqing for months, with rainfall 60 per cent below normal since September 2009. Guizhou province has been particularly hard-hit, with 86 out of its 88 cities located within the drought zone and more than 17 million people short of drinking water, so much so that local villagers in some areas were lining up to obtain emergency water supplies distributed by the government. Yunnan was said to be facing the worst drought in 100 years in some areas. In an attempt to ease the situation, the government announced in the second week of March 2010 that it had initiated hundreds of cloud-seeding operations in the region, using rockets fired into the sky or chemicals dropped from aircraft in a bid to induce rainfall, but the efforts seem to have been largely unsuccessful due to a lack of moisture in the skies. Media reports also said that millions of livestock and huge farming areas were short of water. Meteorologists have predicted that the situation could worsen in the coming months, as hot and dry weather was expected to continue and the water demand to rise as farmers would soon turn to their spring planting (*Bangkok Post*, 2010b).

As to the current conundrum of the food crisis, it is true that food prices have started to decline after steep climbs in 2007 and much of 2008, but that is no cause for cheer. The observed drops in food prices reflect the deepening and the persistence of the international economic and financial crisis. This must be dealt with decisively. Moreover, a long-run view of the factors underlying the price hikes in 2007 and 2008 indicates that food prices will continue to be high or at least unlikely to return to the low levels seen in the past. Thus, even assuming for the purpose of argument that the recent hikes in food prices were an overreaction, still the prospect of higher food prices in the long run and the unfolding of a 'global tragedy' in food insecurity are even more relevant today than ever before for most of the countries of the Asia-Pacific region.

Furthermore, a fleeting glance at the most recent production data also reveals that most of the production increase in 2007 and 2008 occurred in developed countries, and hence, that the benefits of higher prices have not accrued to producers in many developing countries, for their supply response was small in 2007 and virtually zero in 2008. Indeed, according to one study, only 9 per cent of price increases actually flowed to farmers in the Asia-Pacific region. The higher prices of key agricultural inputs, such as fertilizers, seeds and energy, made it all the more difficult for many farmers to step up production, especially the small and marginal farmers, who have been confronted with higher input prices without producing a marketable surplus that would earn them higher revenues. This places at risk the possibility of increasing food production consistently over the long term. Moreover, export taxes (such as in Indonesia) and restrictions imposed in some countries (like India, Vietnam and Thailand during the 2008–09 food crisis) meant that high international prices were not always or fully transmitted to domestic markets, burdening even commercial farmers (not to speak of small and marginal farmers) with higher costs and stagnant output prices. The policy response to soaring food prices in developing countries was indeed wide-ranging. There is a need for corrective action.

Even more worrisome is the fact that falling food prices in 2009 had little to do with recovering global supplies, as they have been caught in the downward spiral by slowing demand. This was evidenced by the fact that almost all commodity prices were declining in unison alongside a deteriorating global economic scenario. To the extent that falling food prices reflected an anticipated slowdown in economic growth that constricts demand, lower prices will probably be associated with more, rather than less, poverty and food insecurity in the foreseeable future.

Thus, if a truly long-run view is taken, not just for the next 5 or 10 years but for the next 50 years or so, it is clear that there are serious impending risks to our future capacity to ensure food security for all, most importantly due to changes in demography, the depletion of water resources, climate change and natural disasters.

A Note on Right to Food

There are various approaches to ensuring food security. For instance, Sen advocates the entitlement approach to ensuring food security, or its obverse, eliminating deprivation (Sen, 1981). One of the most debated approaches to food security is the right to food approach,[14] which has been advocated in recent years by a large number of civil society organizations. The notions of food security and the right to food are both parts of the current discourse. The global discourse is articulated in terms of both food security and right to food.

As stated earlier, 'food security' as a concept is younger than 'the right to food' concept. Even at the cost of being repetitive, it needs to be underscored that both the notions of food security and right to food are parts of the current discourse. The discourse

---

[14] See Mukherjee (2004) for a discussion of the various approaches, especially chapters 1 and 3.

globally is both in terms of food security and right to food. To cite some examples, the United Nations Secretary-General appointed a High-Level Task Force on Food Security in 2008 following the food crisis triggered by food price hikes. The FAO's *State of Food Insecurity in the World* is one of the most important publications on the subject. The MDG 1 enjoins reducing the number of 'hungry people' by half and ensuring right to food of half of the hungry people.

The right to food is important as a matter of discourse, but the present book is designed to deal with food security rather than the right to food. A book on right to food would have to deal with a different set of issues like governance, judicial activism, governance structures and so on, which this book does not aim to do.

In fact, the right to food security is a long way away from being accepted by states in Asia-Pacific. Certainly in major countries like China, right to food is nowhere in the horizon. Even in countries like India where the rights approach may be debated, its implementation and legislation are far away. Moreover, most countries in Asia will not have the wherewithal to implement a right to food law. It cannot work in the poorer countries, like Afghanistan, Pakistan, Lao PDR, Bhutan, Nepal and Myanmar. Introducing concepts and discussing them is very good intellectual exercise, but it will not add value in terms of finding solutions to the problem of food security in the short and medium run. This recalls the cliché: 'In the long run we are all dead.' This book, therefore, attempts to deal with political and financial issues that are 'doable' in the contemporary environment in Asia, eschewing discussion of issues which are ideally correct but whose time has not yet arrived, practically speaking.

In any case, the concepts of food security and the right to food relate closely to each other (see Box 1.4). Evidently, it is in the circumstances of food security that the right to food is most likely to be realized. Over time, the concept of food security has become more and more similar to that of the right to food, by paying more attention to food security at the individual level. Moreover, a right to food approach does not claim to achieve food security faster than other approaches, or to diminish the importance of experiences gained with food security policies. Any attempt to realize the right to food needs to be integrated with existing experiences with food security policies, programmes and strategies.

**BOX 1.4   Right to Food and Food Security:
A Quick Comparison**

Objective

Overall Objective

Food security aims at achieving food security; right to food aims at fully realizing the right to food. Both concepts cover food availability, accessibility, safety, cultural acceptability and conditions in other fields. *De facto*, both also have in common the fact that the attention paid to food availability largely outweighs food safety concerns, although the safety of food is an element in both concepts.

Justification of the Objective

The aim of food security can be based on a number of grounds. These range from moral grounds to more economic approaches which emphasize the idea that the overall costs of hunger are prohibitive to society as a whole. Human rights are based exclusively on the idea of human dignity. They require *a priori* recognition of the universal, interdependent and inalienable character of all human rights.

Nature of the Objective

Food security up to now has remained a policy concept. Hence, striving for food security means striving towards enactment of policies and their implementation that ensure food security. In international law, the concept is used in non-binding instruments such as those adopted at the World Food Summit and the World Food Summit Five Years Later, and has not been given a normative content. Being a policy-oriented rather than a legal concept, food security is subject to easy redefinition, and means implementing an implicit obligation of the state. Conversely, aiming for the realization of the right to food means complying with a legal obligation.

Method

Unit of Analysis of the Objective

The concept of food security applies to various levels: global, regional, national, local, household or individual. Looking at each level can be of

(BOX 1.4 *Continued*)

(BOX 1.4 *Continued*)

use, depending on the type of information and measure sought. If food security exists at the individual level, all larger units are food-secure as well. The right to food as a human right applies to the individual and, in specific cases, also to the group level. There is no right to food of a nation-state that can be fulfilled or violated, nor of a household. The main 'unit' of relevance is the individual.

Permissible Methods

Since food security as such is not a subject of special international law, states have wide discretion to choose their own ways of managing food production, availability and supply, regulating markets, dealing with access to natural resources, etc. Legal restrictions from various fields of law, such as provisions on access to natural resources or on trade, do, however, apply. Also, human rights law constrains choices. Realization of the right to food requires looking at outcomes and means. While it leaves to states the liberty to design their own food security strategies so as to suit best the national context, it also sets minimum requirements that must be met to make a right to food policy out of a food security policy.

In conclusion, it can be said that the concepts of food security and the right to food relate closely to each other. The right to food approach adds new dimensions and poses some methodological limits to them. To an extent, it reduces arbitrariness.

Source: Kerstin Mechlem (2004) and various other publications.

Organization of the Chapters

Building on the foregoing remarks, Chapter 2 discusses the major causes of food insecurity in the Asia-Pacific region. If we look at the causes that have resulted in food insecurity, it would be apparent, as Sen argues, that not only availability of food but a range of causes from the economic to the social and the climatic affect food security. Thus, the causes that lead to food insecurity make for a complex constellation of forces. The problem is compounded further by the fact that the causes of food security can lie at various levels, starting from basic causes to the more proximate ones. It is important to distinguish between these for a correct set of public action and community responses; the 'one glove fits all' approach may not yield optimum results, especially in such a diverse set of countries as existing in Asia.

Among the groups that are food-insecure, there are some that are more food-insecure than others. Women are among the most food-insecure people, and the largest as a group. Gender discrimination and its proximate causes have attracted considerable attention in recent times, thanks to a robust women's movement in the region. Chapter 3 discusses the eight kinds of food insecurities faced by women, which emerge from gender-based discrimination, as articulated by Sen (2001). This brings home the important fact that people who are food-insecure do not constitute a homogeneous group, and hence public action for removing food insecurity needs to take a differentiated approach.

Given the complexity of the issues involved in solving food insecurity in Asia, the governments of different countries have attempted to introduce various policy measures from time to time. The policies have also evolved over time in different countries. Given that millions of people are still hungry in different parts of Asia, especially in South Asia, it is obvious that there is a disjuncture between the policies adopted and the causes that initiate food insecurity in the first place, and then entrench and perpetuate it. Chapter 4 describes a suite of major policies pursued in some of the major countries of Asia. This list is non-exhaustive, because an analysis of all policies is beyond the scope of the present work; but the suite is broad enough to draw meaningful conclusions in order to derive policy outcomes that may work at the ground level and be doable without too much exertion in resource mobilization.

While the governments in different countries have implemented a range of policies to meet the challenge of food security, some meeting with success, communities have also taken measures to deal with food insecurity at the household level. Because much food insecurity often occurs due to idiosyncratic risks, community-based responses have played a significant role in tackling food insecurity in the various countries of Asia-Pacific. Chapter 5 documents some of the major community-based responses known in Asia to meet the challenges of food insecurity. These community-based responses to food insecurity have important policy implications and indicate possible public action in this regard.

The response to food insecurity has to be differentiated. Certain classes of people who need the protection of the state, like the aged, infirm, disabled, women, small farmers, landless workers and informal-sector workers, who enjoy few or no social security benefits, need to be treated differently. The role of the state in dealing with these special classes of people is of paramount importance. Chapter 6 examines what the state could do for such special classes of people.

Chapter 7, the final chapter, outlines an agenda for action to ensure food security in Asia. Given the heterogeneity and diversity of countries in Asia *inter se* and among regions within countries, the agenda needs to be adapted to the special conditions of each country.

References

Alston, P., and K. Tomasevski (eds). 1984. *The Right to Food*. Dordrecht: Martinus Nijhoff.

Bangkok Post. 2010a. 'Mongolia: Aid Call for Herders amid Harsh Winter'. 30 March.

———. 2010b. 'China Drought Affects Millions'. 17 March. http://www.bangkokpost.com/breakingnews/171896/china-drought-leaves-millions-short-of-water (accessed on 30 March 2010).

Clay, E. 1997. 'Food Security: A Status Report of the Literature'. Overseas Development Institute (ODI) Research Report. London: ODI.

Commission on the Nutrition Challenges of the 21st Century. 2000. 'Ending Malnutrition by 2020: An Agenda for Change in the Millennium'. Final Report to the UN Standing Committee on Nutrition. Available at http://www.unsystem.org/scn/Publications/UN_Report.PDF (accessed on 10 October 2003).

De Rose, L. F., and S. R. Millman. 1998. 'Introduction', in L. F. De Rose, E. Messer and S. R. Millman, *Who's Hungry? And How Do We Know It? Food Shortage, Poverty and Deprivation*. Tokyo: UN University Press.

Eide, A. 1989. *Right to Adequate Food as a Human Right*. New York: United Nations, Human Rights Study Series No. 1, Sales No. E.89.XIV.2.

Eide, A., A. Oshaug and W. Barth Eide. 1991. 'Food Security and the Right to Food in International Law and Development'. *Transnational Law and Contemporary Problems*, vol. 1, no. 2, pp. 415–67.

FAO (Food and Agriculture Organization). 1974a. 'Report of the World Food Conference', 5–16 November. Rome: FAO.

———. 1974b. 'Council Resolution I/64'. 'Report of the Council of FAO, Sixty-Fourth Session', 18–29 November. Rome: FAO.

———. 1979. 'Council Resolution I/75'. 'Report of the Council of FAO, Seventy-Fifth Session', 11–22 June. Rome: FAO.

———. 1983. 'Resolution 2/83 on World Food Security'. Report of the Conference of FAO, Twenty-Second Session, 5–23 November. Rome: FAO.

FAO (Food and Agriculture Organization). 1985. 'The World Food Security Compact'. Report of the Conference of FAO, Twenty-Third Session, 9–28 November. Rome: FAO.
———. 1996. 'Rome Declaration on World Food Security and World Food Summit Plan of Action'. World Food Summit, 13–17 November. Rome.
———. 2001. *The State of Food Insecurity in the World 2001*. Rome: FAO.
———. 2004. *The State of Food Insecurity in the World 2004*. Rome: FAO.
———. 2006. *The State of Food Insecurity in the World 2006*. Rome: FAO.
Kumar, Praduman. 1998. 'Food Demand and Supply Projections for India', Agricultural Economics Policy. New Delhi: Indian Agricultural Research Institute.
Mall, R. K., and P. K. Aggarwal. 2002. 'Climate Change and Rice Yields in Diverse Agro-environments of India, I. Evaluation of Impact Assessment Models'. *Climatic Change*, vol. 52, no. 3, pp. 315–30.
Maxwell, S. 1996. 'Food Security: A Post-modern Perspective'. *Food Policy*, vol. 21, no. 2, May 1996, pp. 155–70.
Mechlem, K. 2004. 'Food Security and the Right to Food in the Discourse of the United Nations', *European Law Journal*, vol. 10, no. 5, September 2004, pp. 631–648.
Mukherjee, Amitava. 2004. *Hunger: Theory, Perspectives and Reality*. Hants and London: Ashgate Publishing and University of London.
Sage, C. 2002. 'Food Security', in E. Page and M. Redcliff (eds), *Human Security and the Environment: International Comparisons*. Cheltenham: Edward Elgar.
Sen, Amartya. 1981. *Poverty and Famines: An Essay in Entitlement and Deprivation*. Oxford: Clarendon Press.
———. 1997. 'Hunger in the Contemporary World', Suntory and Toyota International Centres for Economics and Related Disciplines, London School of Economics.
———. 2001. *Development as Freedom*. Oxford, New York etc.: Oxford University Press.
Stringer, R. 2002. 'Food Security in Developing Countries'. Policy Discussion Paper No. 0011. Adelaide: Centre for International Economic Studies.
Tomasevski, K. (ed.). 1987. *The Right to Food: Guide through Applicable International Law*. Dordrecht: Martinus Nijhoff.
World Bank. 1986. *Poverty and Hunger: Issues and Options for Food Security in Developing Countries*. Washington, D.C.: World Bank.
United Nations. 1975. 'Report of the World Food Conference', Rome, 5–16 November 1974. New York.
United Nations Development Programme. 2003. 'Millennium Development Goals: A Compact Among Nations to End Human Poverty', Human Development Report 2003. New York: UNDP.
UNESCAP. 2007. *Economic and Social Survey for Asia and the Pacific*. Bangkok: UNESCAP.

2

Food Insecurity and Its Causes

And he gave it for his opinion, that whoever could make two ears of corn,
or two blades of grass, to grow upon a spot of ground where only one grew
before, would deserve better of mankind, and do more essential service to
his country, than the whole race of politicians put together.

—Jonathan Swift, *Gulliver's Travels*

Food insecurity and poverty are intricately intertwined, and the alleviation of both is strongly correlated. Nowhere are these linkages more real than in Asia and the Pacific. Over recent decades, millions of people have come out of poverty in Asia, with China leading the charge. In East Asia and the Pacific, for example, the incidence of extreme poverty ($1.25/day), fell from 79 per cent of population in 1981 to 18 per cent in 2005 (Chen and Ravallion, 2008). Major strides have been made towards improving food security of millions of people in Asia. For example in 1969–71, 762 million people in Asia were undernourished, it declinedto an estimated 722 million in 1979–81, to 582 million in 1990–92, and to 535 million in 1995–97, despite significant population growth during the same period. Estimates for 2003–05 indicate a slight rise, to 542 million and in 2011, it stands at 578 million (thanks to the price volatility and fuel crisis of recent years), which means more than 60 per cent of the world's undernourished population lives in Asia Pacific. South Asia alone is the home to 300 million under-nourished people, while East Asia and South-East Asia have 160 and 65 million undernourished people.

Notwithstanding the correlation between economic growth and reductions in *aggregate* poverty and hunger, progress toward food security in Asia has not been uniform, both internationally and intra-nationally. Take the case of three of Asia's largest and poorest nations—India, Indonesia and Pakistan—which recorded increases in undernourishment between the late 1990s and the early 2000s, and these increases exceeded the declines registered in other countries of the continent, such as in Bangladesh, China and Vietnam. More worrisome is the fact that even countries which experienced rapid economic expansion and declines in the overall prevalence of food insecurity, sub-populations (such as women and children) in those countries chronically or periodically suffer from food insecurity, because these sub-populations live in socio-economic environments and have characteristics which prevent them from reaping the fruits of economic expansion. Constellations of forces that combine to produce food insecurity in such sub-populations include poor land quality, exposure to environmental risks, isolation from labour and other markets, and of course, gender inequality. Overall, various processes—including population growth, the development of new agricultural technologies, environmental degradation, etc., policies—like inequity in ownership and inheritance of land, and institutions—exacerbate or cause food insecurity.

Causes of Food Insecurity[1]

Food insecurity can occur because land, labour and complementary productive resources at the disposal of a household, region or nation are not adequate to produce all the food it needs.[2] One possible reason might be land scarcity. However, such food insecurity can also be a consequence of technologically challenged farmers, lack or paucity of non-land inputs, or adverse climatic conditions, such as droughts, hailstorms and floods. By definition, food insecurity exists for urban communities and even for entire nations

---

[1] I am thankful to Douglas Southgate and Ian Coxhead for the permission to use materials from their paper entitled "Food Security and its Determinants in Asia Pacific" prepared for the 65th Commission Session of the United Nations Economic and Social Commission for Asia and the Pacific, held in Bangkok, 2009 for the sections on 'Population and Growth in Demand for Food' and 'Agricultural Intensification'.

[2] Experts call this 'structural food insecurity'.

(like Singapore) which suffer from shortage of arable land. But whether or not the inhabitants of these areas actually experience food insecurity also hinges on market- and policy-related factors. Indeed, the incidence of food insecurity is inconsequential if the non-agricultural output of urban areas can be exchanged for adequate amounts of food, or what Sen called 'exchange'- and 'trade-based entitlements'.[3] Food insecurity can also be a problem for rural people living on marginal areas or micro-environments, such as drought-prone areas in western China and north-eastern Thailand; high-altitude areas in Nepal and India, and in the mountains that divide China from its southern neighbours; and coastal regions, like those in the central Philippines and central Vietnam, that are highly susceptible to typhoons. The broad processes that have added to or diminished food insecurity in Asia and the Pacific are discussed first. These include population and demand growth, the intensification of agricultural production and introduction of new technologies, and more recent stresses associated with environmental degradation. We shall then turn to other factors, including social and cultural issues.

Population and Growth in Demand for Food

Current problems of food security in Asia must be viewed in the context of changes in demand for food which in turn is predicated on population growth and economic expansion in the continent. In 1950, Asia's population was 1.34 billion, equal to 54 per cent of the global total. During the next quarter of a century, human numbers in the continent went up by 2.1 per cent per annum, reaching 2.26 billion in 1975. During the next 25 years, annual growth was slower, averaging 1.7 per cent. But because there was less demographic expansion in Europe, North America and other affluent continents, Asia's share of the global population was larger at the turn of the 21st century than it had been 50 years earlier: 3.46 billion out of 6.06 billion, or 57 per cent (UNDP, 2002: 162–65) of the world's population live in Asia.

However, in Asia as in other parts of the world, deceleration in population growth has been quite marked, due primarily to

---

[3] See Appendix 2.1 for a definition of these concepts.

declines in human fertility rates. In China, for example, the total fertility rate (TFR) fell from 2.5 births per woman in 1980 to 1.8 births in 2006 (World Bank, 2008b). If the latter rate, which is lower than the replacement level of 2.1 births per woman, is sustained for a few more decades, China's population will contract. Comparable declines have occurred throughout Asia, falling to or below the replacement level in nearly every country where per capita income equals or exceeds $3,000 (PPP),[4] and registering sizable reductions elsewhere.[5] After being equal to 5.0 births per woman in 1980, India's TFR has now fallen to 2.5 in the first decade of the 21st century (World Bank, 2008b).

Natural increase (the difference between birth and death rates) will continue in Asia for a few more decades, although the phantom of unbridled growth in human numbers has receded in the world's largest and most populous continent. In various countries, including but not limited to China, growth in food demand is now driven both by demographic expansion and by the dietary diversification and improvement resulting from better standards of living.

In this connection it needs to be underscored that the impact of increased earnings on food security is not inconsequential in the Asian context. Notwithstanding intra-regional variations and the impacts of periodic slow-downs, the general trend in Asia is one of increased household earnings which affect what people eat in multiple ways. Whereas penurious households use most of their available resources to satisfy caloric requirements—usually by consuming the cheapest available carbohydrates—diets quickly diversify as poverty decreases. More fruits and vegetables are eaten, for example, and animal protein intake increases. The most important consequence of improved living standards in Asia and the Pacific is increased consumption of meat, dairy goods and eggs which in turn causes additional corn and other grains to be fed to cattle, poultry and other livestock.

---

[4] Examples include Indonesia, Sri Lanka and Thailand, with TFRs in 2006 of 2.2, 1.9 and 1.8 births per woman, respectively (World Bank, 2008b).

[5] Between 1980 and 2006, the number of births per woman fell from 7.0 to 3.9 in Pakistan and from 6.1 to 2.9 in Bangladesh (ibid.).

Agricultural Intensification

As reported earlier, aggregate food insecurity in Asia has abated in recent decades despite increases in human numbers and food consumption. Interestingly, increases in population and food demand, which have been outstripped by growth in the supply of food, did not result in an unrestrained expansion of agricultural land use, the main reason being the agricultural technological developments taking place in Asia since the middle 1960s.

Since agriculture and civilization have had such a long history in China, India, Java, Philippines and other countries in Asia, population density is rather high in many parts of the continent. In addition, opportunities to increase crop and livestock production by expanding the area under agriculture are limited, as supply of arable land is by and large severely restricted. Even where opportunities for expanding the area under agriculture exist (on some Indonesian islands for example) agricultural extensification (the expansion of area under cultivation) has adverse environmental impacts, such as loss in biodiversity that occur as farmers impinge on micro-environments, forests and other species-rich habitats. Because of these adverse environmental impacts and the direct expenses farmers must incur to clear away trees and other vegetation, as also existence of laws that prohibit incursions into forest and range lands, major increase in agricultural land use is not anticipated in the region (Fischer and Heilig, 1997) in the future.

Higher population densities have induced the kind of agricultural intensification which is characteristic of environments, marked by shortage of land for agricultural use. According to the hypothesis of induced innovation, agricultural development can come about by raising the productivity of either labour or land or some combination of the two, depending on the scarcity of the factor of production. In places, where a combination of land scarcity and labour abundance exists, as is the case today in much of Asia, adoption of measures to raise land productivity, including increasing use of fertilizers, irrigation water and biological improvement is typically seen (Hayami and Ruttan, 1985).

The Green Revolution, in many parts of Asia since the mid 1960s, is a classic illustartion of an increase in land productivity induced by the relative scarcity of natural resource for agricultural production.

This increase in land productivity was made possible by research and testing in seed varieties carried out over many years which led to the discovery of new varieties of rice and wheat that produced more grain than traditional strains when fertilizer and irrigation water were applied to farm fields (Dalrymple, 1985).

It is significant that during the Green Revolution the small farmers adopted the new technology about as readily as larger farmers wherever the production environment was favourable to such adaption (David and Otsuka 1993). More importantly, the new technologies greatly increased labour requirements per hectare and per year due to multiple cropping requiring more labour that was made possible by the new technology. As a result, more employment was created for the poorest of the rural poor, who were generally landless and earned their primary incomes through employment on others' farms. The greatest impact of the Green Revolution was to raise cereal yields—not just in Asia but throughout the world. With yields and output going up, food prices fell, including for people who otherwise would have starved (Dalrymple 1985; Southgate, Graham and Tweeten 2007, pp. 110–11).

Fertilizer use and use of irrigation water have continued to increase in Asia since the Green Revolution and so have agricultural yields. There are only a few countries (such as Malaysia and Sri Lanka, where agricultural yields had reached high levels three decades ago) where crop production per hectare did not rise by at least 50 per cent between 1980 and 2000. During the last two decades of the 20th century, China, India and Vietnam registered yield gains of 60 per cent, 81 per cent and 114 per cent, respectively.

Thanks mainly to yield, agricultural output increased faster than population growth during the last three decades of the 20th century, as shown in Table 2.1, though there are some exceptions to this trend. In addition, fruit and vegetable production went up rapidly, in response to higher demands for these products consequent upon the dietary diversification resulting from improved living standards. Of course production of these items grew so rapidly, more because additional land was brought under the cultivation of these commodities than because of yield increases, except for output of rice, the staple crop throughout Asia, which grew faster than population mainly because of yield increases, both during and since the Green Revolution.

TABLE 2.1   Rural Population Density, Fertilizer Use, Irrigation and Cereal Yields in Selected Asian Nations, 1980 and 2000

Country (ranked by average income)	Rural population density in 2001 (persons/km² of arable land)	Fertilizer use (kg/ha) 1980	Fertilizer use (kg/ha) 2000	Irrigation (% of arable land) 1980	Irrigation (% of arable land) 2000	Cereal yield (kg/ha) 1980	Cereal yield (kg/ha) 2000
Malaysia	554	427	670	6.7	4.8	2,828	3,132
Thailand	326	18	112	16.4	27.1	1,911	2,654
Philippines	564	64	134	12.8	14.6	1,611	2,692
China	561	149	256	45.1	36.3	3,027	4,845
Sri Lanka	1,607	180	277	28.3	33.6	2,462	3,520
Indonesia	591	65	124	16.2	14.4	2,837	4,141
India	460	35	107	22.8	32.2	1,324	2,390
Vietnam	923	30	341	25.6	37.6	2,049	4,375
Pakistan	438	53	136	72.7	81.6	1,608	2,266
Mongolia	87	8	3	6.7	4.8	573	751
Bangladesh	1,228	46	166	17.1	49.6	1,938	3,312
Laos	495	4	11	13.1	18.2	1,402	3,140
Cambodia	274	5	0	5.8	7.1	1,615	2,178
Nepal	668	10	26	22.5	36.2	2,521	3,453

Source: World Bank (2008b).

Gains made in land productivity in Asia during the Green Revolution were instrumental in reducing food insecurity in the continent. But the future remains uncertain because future growth in yields faces two easily identifiable challenges. One, the environmental damage, including land degradation that the Green Revolution ushered in does not seem to be abating. And two, the diminishing rate of technological innovation, resulting in part from reduced budgets for agricultural research and development, show no sign of reversal as of now. We shall have occasions to revert to these later in greater detail.

Degradation of Environmental Resources

Land

The availability of natural resources is critical in determining the extent of current and potential food security. Vast areas of croplands, grasslands, woodlands and forests in Asia and the Pacific have been critically affected by various forms of land degradation. Land degradation is a complex process, which takes different forms and has different levels of intensity, influenced mainly by topography, soil characteristics, climatic conditions, vegetative cover and human activities (ESCAP, 2005).

The full impact of land degradation has been more severe in dryland ecosystems, where it has caused desertification. For example, in South and South-East Asia, around 74 per cent of agricultural lands were severely affected by wind and water erosion as well as by chemical and physical deterioration (Woods et al., 2000). Central Asia is most seriously affected by desertification and erosion. In Kazakhstan alone, around 66 per cent of the total land area is desertified (ESCAP, 2005: Chapter 6).

The adverse impact of these phenomena on the productivity of agricultural lands has affected the food security of about a billion people in the Asia-Pacific region. In China, soil erosion is a threat to food supply and increases the risk of floods, according to a report published on 20 November 2008 by the Ministry of Water Resources. China has more than 3.5 million sq km of eroded land, of which 1.6 million sq km is due to water and 1.9 million due to wind (see Figure 2.1). During 2000, economic losses of a total of

---

BOX 2.1   Some Quotable Quotes

1. H.E. Mr Chen Lei, Minister of Water Resources, Government of the P. R. China, quoted by Xinhua News Agency: 'In recent years, China has been losing 15,000 sq km of land per year to erosion.'
2. Sun Honglie, a member of the Chinese Academy of Sciences, was quoted by Xinhua as saying that 'agricultural and forestry exploitation, and highway, railway and urban construction projects are the major causes of land erosion, accounting for 78 per cent of the total'.
3. H.E. Mr E. Jingping, Vice-minister of Water Resources, Government of the P. R. China, said: 'China has a more severe soil erosion problem than India, Japan, the United States, Australia and many other countries.'

Source: Collected and compiled by the author.

FIGURE 2.1   Disappearing Land in China

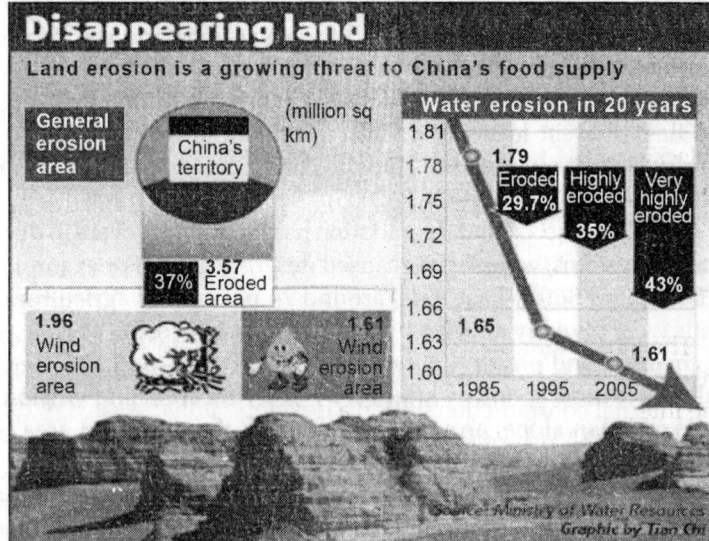

Source: Xie Yu (2008).

200 billion Yuan ($29 billion), 2.25 per cent of the country's GDP, were attributed to land erosion (Xie Yu, 2008).

The problem of erosion is worst in hillside areas. China has about 200,000 sq km of hillside land, which is 17.5 per cent of its total arable land area (Xie Yu, 2008).[6] As a consequence, crop output could fall in north-eastern China by as much as 40 per cent over the next 50 years, if erosion continues at its current rate. In southwest China, if land degradation continues at the same rate, over the next 35 years about 100 million people will be at risk of losing their land, and hence, of vulnerability to food insecurity.

BOX 2.2 Desertification

Iran is losing its battle with the desert. Mohammad Jarian, who heads Iran's Anti-Desertification Organization, reported in 2002 that sand storms had buried 124 villages in the south-eastern province of Sistan-Baluchistan, forcing their abandonment. Drifting sands had covered grazing areas, starving livestock and depriving villagers of their livelihood. Neighbouring Afghanistan is faced with a similar situation. The Registan Desert is migrating westward, encroaching on agricultural areas. A United Nations Environment Programme (UNEP) team reports that 'up to 100 villages have been submerged by windblown dust and sand'. In the country's north-west, sand dunes are moving onto agricultural land in the upper reaches of the Amu Darya basin, their path cleared by the loss of stabilizing vegetation from firewood gathering and overgrazing. The UNEP team observed sand dunes 15 metres high blocking roads, forcing residents to establish new routes. Dust storms carry away topsoil, a resource that will take centuries to replace through natural processes. Overgrazing is the principal culprit.

Source: Brown (2008: 65–66).

The adverse impact of natural resources degradation on food insecurity, especially in rural areas of Asia, can often be traced to the internal contradiction within a development approach, which promotes the use of productive resources but does not make adequate provision to arrest the process of 'loss' through degradation of resources. For example, as cited earlier, a report of the Ministry of Water Resources, Government of P. R. China, says that in 2000, economic losses of a total of 200 billion Yuan ($29 billion), 2.25 per cent of the country's GDP, were attributed to land

---

[6] On the Loess Plateau, for example, for each kilogram of crops produced, between 40 and 60 kg of earth is lost.

erosion. In contrast, the level of investment in stopping it is too low, with just 1.63 billion Yuan, 0.012 per cent of GDP, spent in 2004 (Xie Yu, 2008).

Most developing countries in the Asia-Pacific region have adopted a strategy of agricultural intensification with significant negative environmental trade-offs, which affect the integrity of natural ecosystems and their future potential. For example, the regional production and use of mineral fertilizers as a proportion of global production has been increasing, and is dominated by north-east Asia and South Asia. Countries such as India, Lao People's Democratic Republic, Myanmar, the Philippines, Sri Lanka, Thailand and Vietnam intensified their use of mineral fertilizers by as much as 90 per cent over the period 1992 to 2002 (ESCAP, 2005). Misuse and excessive use of mineral fertilizes is responsible for land degradation, soil nutrient imbalances, eutrophication and algal blooms in freshwater systems and coastal waters. The misuse of pesticides and herbicides has not only impacted insect diversity and contaminated water supplies but has threatened the health of farmers. Organochlorines have not only killed the targeted insect pests but also their natural predators (ADB, 2000), causing harm to food production.

Cropping patterns and agricultural practices have also had a substantial impact on the environmental resources necessary for food production. For example, cassava, a crop that is well adapted to marginal lands and is produced extensively in countries like Indonesia, can nevertheless rapidly deplete the soil nutrients. In order to meet their basic food needs, smallholders and the rural poor cultivate annual food crops, such as maize, intensively on steep slopes. This has accelerated soil erosion as well, except in cases where it is combined with the planting of perennial crops and other conservation measures. Also, the excessive focus on rice production in the lowlands neglects upland poor in the agro-ecologically marginal areas of Java. The population process leading to increased marginalization also has a profound impact on the environment, as weaker members of the community are pushed into ecologically fragile areas. In China, about three-quarters of the country's poorest people live in areas affected by land erosion (Xie Yu, 2008). Similar has been the case in the mountain ecologies, the Philippines and Thailand.

The case of degradation of land due to shrimp cultivation also warrants mentioning. Shrimp farming is primarily an export industry carried out in developing countries, including countries in Asia, where shrimp farm operations are poisoning local environments. Shrimp farm operations need large volumes of brackish water to fill pond enclosures, which has in the past limited the cultivation of tiger shrimp (a marine species) to a relatively narrow band of coastal land. Farmers have discovered, however, that it is both feasible and profitable to grow tiger shrimp in areas far removed from the coast by bringing hyper-saline water inland and mixing it with freshwater drawn from irrigation canals or natural streams. Small-scale tiger shrimp farms are now common in traditional rice-growing areas such as Chachoengsao, Prachinburi, Suphanburi and Nakhon Pathom in Thailand and coastal Tamil Nadu in India.[7]

Once shrimp ponds become operational, a host of ecological impacts is generated with devastating consequences for the affected environment and human communities in the area. Pollution from shrimp farming has severe side-effects for local people who inhabit and use the surrounding environment to maintain their food and water supplies. Shortages of freshwater have resulted in many shrimp-producing areas such as southern Thailand and Tamil Nadu.

To maintain the overcrowded shrimp population in intensive production systems, and to attain higher production efficiency, copious amounts of artificial feed, pesticides, chemical additives and antibiotics must be continuously added. These compounds, together with excrement from the shrimp, make the wastewater from the ponds poisonous. The polluted wastewater is generally pumped back into the surrounding environment in order to save costs, poisoning coastal waterways and the sea, fresh groundwater supplies, native flora and fauna and adjacent communities. In addition, shrimp pond effluents are often high in organic matter, with a resulting high biological oxygen demand that can cause oxygen depletion in receiving waters. The combination of surplus

---

[7] The development of low-salinity culture techniques, the higher economic profits vis-à-vis rice cultivation, government policies that promote shrimp farming as a means of rural economic development and the belief that inland freshwater areas are free from virulent shrimp pathogens have contributed to the proliferation of shrimp cultivation.

organic matter and increased salinity from pond effluents can cause severe problems, especially for fish populations and other sea life that inhabit the receiving waterways. Saltwater in the ponds also seeps into the local groundwater, and the increased salinity damages drinking water supplies and surrounding agriculture land, making alternative cropping (such as rice) nearly impossible.

The lifespan of an intensive shrimp farm is between five and 10 years (many are forced to shut down within two to three years after choking on their own self-generated pollution). Once the farm is abandoned, it is expensive and difficult, if not impossible, to rehabilitate the land for any other purpose (e.g., farming, or replenishment of destroyed mangrove forests). This, in itself, is an immense problem. For example, in Thailand, less than 5 per cent of the initial farms set up in the Gulf of Thailand (Thailand's first shrimp-growing region) remain in operation today. The bottom soil of an abandoned shrimp pond that has been used for intensive culture is usually too saline for agriculture or other uses, so that the destruction of land by shrimp farming may, for practical purposes, be irreversible. Shrimp cultivation also has impacts on mangrove forests, to which we shall return later.

Many countries in the region have attempted to increase their arable and permanent croplands, pushing the balance of ecosystems to the limit (ESCAP, 2005). Constraints in arable land are compounded by soil and slope constraints. Steep slopes (more than 8 per cent slope incline) and poor soil condition characterize many of these agricultural lands. In addition, the fertility of many of these areas has declined significantly after years of overuse and misuse of fertilizers and intensive irrigation. These conditions are particularly critical for small-scale and marginalized farmers, many of whom are poor, and who are dependent on the natural fertility of the soil. With little fertile lowland to cultivate, many poor farmers have moved towards the uplands, shifting pressure.

Poor farmers have often been pushed into increasingly arid, infertile and erosion-prone areas. Such farmers have adopted several coping mechanisms, one of which has been reducing the length of fallow periods, and ploughing upland previously reserved for grazing. Such expedients have led to declining yields, accelerated degradation of the environment and acute shortages of animal traction because agriculture expands at the expense of grazing. The strategy of shortening fallow periods has been reported for a long time in Bhutan, Indonesia, Laos, Nepal and Papua New Guinea. This practice does not allow the land to regain its nutrients and hence, has caused a decline in food output in the longer run. Extensive livestock keeping has also put rangeland under increasing pressure in the Asia-Pacific region. Animal survival rates being low during a drought, families tend to keep large herds for food security and for rebuilding stock. Overstocking has been aggravated by 'open access' to resources, including common property resources (CPRs), and by a lack of technology to improve the carrying capacity of arid and semi-arid lands. Changes in the system of tenure, population pressure and encroachment by powerful factions in society have led to a breakdown of the communal systems of range management, without an appropriate substitute. This has not only made the upkeep of herds more difficult, with adverse impact on food production, but has also impinged on availability of cultivated food from these areas.

Water

Water is a limiting factor for agricultural production in Asia and the Pacific, with drought conditions and lowered aquifer levels depressing agricultural productivity across every sub-region. Drought conditions diminished food security and affected more than 600 million people across the region between 1995 and 2004. In 2005, Afghanistan was in the sixth year of its worst drought in 30 years; in 2004, the drought had reduced cereal production by an estimated 25 per cent. Australian farmers, seeking to make a living on the driest inhabited continent, were in the grip of its worst drought in 20 years; high rural suicide rates were linked to this drought, a situation which was replicated in India.

The Asia-Pacific region suffers from a double jeopardy of dwindling water resources and deteriorating water quality: many of the countries with the least available water per person also have some of the worst water quality. The water resources of Azerbaijan, China, India, the Islamic Republic of Iran, Pakistan, Thailand, Turkey and Uzbekistan are among those under the most pressure in the region. And these are the countries where the water quality is also poor. For example, in China, some 52 urban river stretches may be so contaminated that they cannot be used for irrigation (Warford, 2004). A 2001 survey of water quality in Islamabad and Rawalpindi,

Pakistan, showed 94 per cent of samples unsuitable for drinking due to bacteriological contamination, 34 per cent affected by faecal contamination and 12.8 per cent of samples unsuitable for drinking due to high nitrate levels (Pakistan Water Gateway, 2001).

While water quality and patterns of resource exploitation have been reducing the region's ability to meet water needs in several countries, the economically accessible freshwater endowment has also been decreasing as river systems, freshwater lakes, floodplains, wetlands, forests and other vegetative cover in river basins and aquifers come under threat from development. High losses of watershed forest cover are increasing run-off rates, reducing aquifer recharge and increasing the variability of water flow.

The increasing pressure on natural water infrastructure threatens critical ecosystem goods and services. River systems and other inland water bodies are important as freshwater fisheries, often providing the primary source of protein for rural communities. The lower Mekong river basin produces two million metric tons of fish and other species annually for human consumption. Two-thirds of this amount comes from natural wetlands. Wetlands provide groundwater recharge, waste treatment and detoxification services, and potentially reduce nitrate concentrations by more than 80 per cent. The declining ecological integrity of freshwater systems is signalled by the decline of freshwater biodiversity. As shown by the Living Planet index, freshwater vertebrate species have declined most rapidly, and most consistently, compared to other species groups (Millennium Ecosystem Assessment, 2005).

Poor surface water quality and localized, periodic or seasonal surface water scarcity mean that groundwater is increasingly being tapped. The exploitation of groundwater resources is leading to a rapid lowering of water tables across China, the Philippines, India, Pakistan and the Islamic Republic of Iran, and to the growing exploitation of deeper aquifers (ESCAP, 2005). Sinking groundwater tables have resulted in diminished grain harvests in India and China. Groundwater depletion does not affect only agricultural harvest; poor communities that depend on shallow drinking-water wells, and urban centres that depend on groundwater, also pay the price of overly paid extraction and access poor-quality water leading to water-borne diseases that impair food security. In Jakarta (Indonesia) and Dhaka (Bangladesh), a large proportion of water is supplied from aquifers (Worldwatch Institute, 2001); Quetta (Pakistan) may run out of water by 2018, based on the rate at which its water table is falling (Brown, 2003). The situation with regard to potable water is also not good. In the Asia-Pacific, in 2002, an estimated 665 million people (almost one in five people) were without access to improved water, and some 1.9 billion (almost one in two people) were without access to improved sanitation (WHO and UNICEF, 2000). Between 1990 and 2002, the number of people without access to sanitation increased in some countries, such as Indonesia, the Islamic Republic of Iran, Nepal, Papua New Guinea, Turkey and Uzbekistan. During the same period, infrastructure development to provide safe drinking water did not keep pace with population increases in Bangladesh, Papua New Guinea, the Philippines, Uzbekistan and Vietnam (ESCAP, 2005). Lack of access to potable water in several countries of the region causes water-borne diseases, which is in turn germane to food insecurity, as said earlier.

Forests

Deforestation in almost all the countries in the region has been a common phenomenon. Though the losses of natural forest in countries like Cambodia, Papua New Guinea, the Russian Federation and Vietnam were relatively modest in the period from 1990 to 2000, yet FAO's *Global Forest Resources Assessment* (FAO, 2005a) indicates that substantial losses occurred in those countries between 2000 and 2005. Natural forest losses in Sri Lanka and Indonesia continue to occur at higher rates. As fossil fuel prices are high for the common people, there has been mounting pressure on natural forests as people turn away from increasingly costly fossil fuels to wood from forests which partakes of the nature of a free good. Indeed, the demand for fuel-wood has been as much as three times more than what was grown. Use of fuel-wood is not the only factor to be blamed for forest cover loss. The FAO *State of the Forests 2005* report makes the link between deforestation and the illegal trade in timber. The losses of natural forest in countries such as Myanmar, Sri Lanka and Nepal seem to support this conclusion. While official Russian Federation estimates put illegal felling at no more than 5 per cent of overall production, estimates as high as 20 per cent have been made (United Nations ECE and FAO, 2005). Estimates of illegal production of both hardwood and softwood in China are as high as 30 per cent of production in Indonesia, and 5 per cent of production in Malaysia (ESCAP, 2005). Deforestation has been one of the major causes of desertification in many countries.

Food insecurity, malnutrition and undernutrition are closely linked to the degradation of the environment from another perspective. As natural resources (like forests and water bodies) are depleted, the suffering of the poor, who depend on these resources for their food, fibre and fodder, is accentuated. For example, the depletion of forest cover in the quest for cultivable land and the consequent loss of what is called 'uncultivated food' has had an adverse impact on availability of food in South Asia (Mazhar et al., 2008). And fuel-wood foraged and gathered by people dwindled significantly, in turn threatening agricultural sustainability, adversely affecting food production.

Mangrove forests are unique ecosystem features of tropical and subtropical coastlines, and fulfil critical functions in both conservation and providing livelihoods for communities. The Asia-Pacific region accounts for about 50 per cent of the total mangrove forest area in the world (FAO, 2003), with South-East Asia accounting for about 78 per cent of the mangrove forests in Asia and the Pacific. Mangrove forests are under severe strain. For example, in Lampung, South Sumatra (Indonesia), at least 85 per cent of approximately 12,000 hectares of mangrove forests was reported to have been severely damaged by locals who used the wood to build traditional hatcheries. Meanwhile, in Tangerang, Banten province, mangrove forests are being destroyed by domestic and industrial waste. In Jakarta, the remaining mangroves in Muara Angke are under serious threat as domestic and industrial waste discharged into the capital's rivers finds its way to these forests. The destruction of mangrove forests also means that vast areas are losing a natural barrier to coastal erosion by the sea.

The conversion of mangrove ecosystems for aquaculture cultivation in countries like Bangladesh, India, Indonesia and Thailand has had the most serious effects, since this activity not only induces loss of vegetation but also leads to the deterioration of water quality and the loss of biodiversity, and contributes to the decline of fish stocks (Millennium Ecosystem Assessment, 2005). It thus affects the food security of a large number of people. Indonesia's mangrove forests, accounting for 18 to 23 per cent of mangroves worldwide and covering as many as 4.25 million hectares (supposed to be the world's largest mangrove forest area), have been cut in half by decades of export-oriented shrimp cultivation.[8] The Indonesian government has allocated up to 1 million hectares of land, mostly in mangrove forests, for the shrimp hatchery industry. Similarly in Trang province of southern Thailand, in the fishing communities located along the Andaman Sea, the environment has been ruined beyond remediation through the cutting of the mangroves to make way for shrimp farms, which makes it impossible for the fisher folk to earn their livelihood and to support their families from their polluted seas and degraded lands.

Natural Disasters and Related Processes

The Asia-Pacific region is among the most disaster-prone regions in the world and is subject to hydro-meteorological (floods, cyclones and droughts), geological (earthquakes, landslides and volcanoes) and other disasters, such as epidemics, insect infestations, hot and cold waves and forest fires. In 2004, Asia remained the most frequently hit continent (International Federation of Red Cross and Red Crescent Societies, 2005). The United Nations Environment Programme estimates that 80 per cent of all natural disasters worldwide occur within Asia and the Pacific (UNDP, 2004). 'Figures for cost of damage are notoriously unreliable' (International Federation of Red Cross and Red Crescent Societies, 2005: 197). However, some estimates are available. For example, in the year 2004, 245 of the 641 natural hazards events recorded globally occurred in Asia and the Pacific, accounting for US$73 billion of total global economic losses valued at US$145 billion (Munich Re Group, 2005).

Natural disasters cause immense suffering and have a devastating long-term impact on food production (See Box 2.3). Erratic weather events and storms such as cyclones can unleash flooding, lash coastal areas, produce heavy rain that leads to landslides, and wipe out coastal fishing infrastructure devastating all who are dependent on the sea for their livelihoods. In the aftermath of

---

[8] Not all mangrove areas can be converted into shrimp farms, but many businessmen often go ahead and clear the forests anyway out of sheer ignorance.

BOX 2.3   Impact of the Asian Tsunami

The 26 December 2004 Asian tsunami tragedy demonstrates the destructive capacity of natural disasters and their impact upon food security. According to early impact assessments, agricultural and fishery losses due to the tsunami were severe. For example, in the state of Tamil Nadu in India, 59,000 fishing vessels were destroyed, affecting the livelihood of nearly 700,000 fishers as well as their families. In Aceh province of Indonesia, income losses in agriculture and fisheries made up more than one-third of total losses due to the disaster, while in the worst-affected districts of Sri Lanka, 80 per cent of fishing vessels were destroyed or seriously damaged.

Source: Field notes of the Author.

---

disasters, repairing damaged infrastructure, compensating for personal loss, and re-habilitating the landscape consume resources that could otherwise be devoted to improving nutrition levels and food security. Natural disasters also have a cumulative impact, causing successive loss of resilience in both the environment and the society (FAO, 2005c).

Droughts have been observed frequently in Asia. The impact of droughts differs widely between developed and developing countries because of the influence of such factors as water supply and water use efficiency. The majority of the estimated 500 million rural poor in the Asia-Pacific region are subsistence farmers occupying mainly rain-fed land (ESCAP, 1995). They face dire consequences from droughts. The drought-prone countries in the region are Afghanistan, Iran, Myanmar, Pakistan, Nepal, India, Sri Lanka and parts of Bangladesh. In India, about 33 per cent of the arable land (about 14 per cent of the total land area of the country) is considered to be drought-prone, and a further 35 per cent can also be affected if rainfall is exceptionally low for extended periods (ibid.). The six years of continuous drought in Australia from 2004 onwards, which led to a food crisis in many countries dependent on imports of food from Australia, is another case in point. Droughts make for food insecurity both through loss in food availability and through volatility of food prices, which makes economic access to whatever food is available even more difficult, something witnessed in most parts of Asia-Pacific during the 2007–08 food price crises.

The Asia-Pacific region alone has recorded 70 per cent of the world's earthquakes measuring 7 or more on the Richter scale, at an average rate of 15 events per year (ESCAP, 1995). The countries of the region badly affected by earthquakes include Japan, the Philippines, India, Nepal, Afghanistan, the Islamic Republic of Iran and the Pacific Islands. About 80 per cent of China's territorial area, 60 per cent of its large cities and 70 per cent of its urban areas with populations over one million, are located in seismic zones (ibid.). The devastating earthquake in China in 2008 which claimed over 68,000 lives is still fresh in public memory.

Volcanic eruptions occur in the Philippines, Indonesia, Japan, Papua New Guinea, New Zealand, Solomon Islands, Tonga and Vanuatu amongst which those most affected include Indonesia (129 active volcanoes), Japan (77 active volcanoes) and the Philippines (21 active volcanoes) (ESCAP, 1995). The eruptions destroyed surrounding forests, caused massive siltation of rivers and coastal areas and deposited volcanic ash in areas adjoining the sources of erution and even across continents (UNEP, 2001).

BOX. 2.4   Glacial Melt: Turning Off the Tap to Asia's Water Tank while Inundating Valuable Agricultural Lands along the Coasts

The Qinghai-Tibetan (Qingzang) plateau is a vast, elevated plateau in Central Asia covering most of the Tibet Autonomous Region and Qinghai province in the People's Republic of China and Ladakh in Kashmir. It occupies an area of around 1,000 by 2,500 kilometres, and has an average elevation of over 4,500 metres. Sometimes called 'the roof of the world', and Asia's 'water tank', it is the highest and largest plateau on earth, with an area of 2.5 million square kilometres.

Currently the plateau is undergoing rapid transformation due to climate change, rising temperatures, deficiency and changes in precipitation, and loss of biodiversity that will have devastating and catastrophic impacts on agriculture, food security, livelihoods and population migration around region. Glaciers and wetlands on the plateau serve as headwaters for several important rivers in Asia and play pivotal roles in agricultural productivity, water donation and storage. Currently, many rivers are being altered at an unprecedented rate. For example, the headwaters of the Yangtze and Yellow rivers are

(BOX 2.4 *Continued*)

(BOX 2.4 *Continued*)

located in the Qinghai-Tibetan plateau. Being the two longest rivers in China, they account for 27 per cent of the country's total land area and 36 per cent of its run-off , and annually provide 25 per cent and 49 per cent of the run-off for the rivers, respectively. While some studies show increased run-off in the short term, within the next few decades the 'water tank' will be vastly reduced once glaciers have greatly declined and wetlands exposed to forces such as desertification.

Glaciologists report that the Gangotri glacier, which supplies 70 per cent of the ice melt that feeds the Ganges river during the dry season, could disappear entirely in a matter of decades. What could threaten world food security more than the melting of the glaciers that feed the major rivers of Asia during the dry season, the rivers that irrigate the region's rice and wheat fields? In a region with half the world's people, this potential loss of water during the dry season could lead not just to hunger but to starvation on an unimaginable scale. Asian food security would take a second hit because its rice-growing river deltas and floodplains would be under water. The World Bank tells us that a sea-level rise of only 1 metre would inundate half of the rice-land in Bangladesh. While a 1-metre rise in sea level will not happen overnight, what is worrisome is that if ice melting continues at today's rates, at some point such a rise in sea level will no longer be preventable. The melting that would cause this is not just what may happen if the earth's temperature rises further; this is something that is starting to happen right now with the current temperature. As summer neared its end in 2007, reports from Greenland indicated that the flow of glaciers into the sea had accelerated beyond anything glaciologists had thought possible, which will lead to the flooded coasts noted earlier.

Sources: Shuqing An et al. (2006); Brown (2008).

Violence and Militarism

Violence and militarism are two elements that cause food insecurity in the Asia-Pacific region. Violence is of two broad forms: one, harmful actions of institutions or individuals against persons or property, which are easily visible; two, destructive actions which do not necessarily involve a direct relationship between the victim and the institution or person responsible for the harm, actions which are largely invisible. Discriminations based on social norms, race, gender, sex, age, class, caste, colour, language and religion are forms of institutional violence. In this regard, militarism is one the most extreme forms of institutional expression of violence. Our recent history is replete with instances of conflicts and war. For decades, the Cold War drained resources towards war-preparedness and building numerous weapons of mass destruction to kill people in even greater numbers. Small wonder that military expenditure increased throughout the Cold War period in Asia in sympathy with global trends. Even today, after the end of the Cold War, all over the region and within each country, war-preparedness continues. Resources expended on war-preparedness are obviously diverted from alternative uses such as in producing food, providing basic education and primary health care, and providing for R&D and extension services to farmers to improve food production, food distribution and people's access to food.

Diversion of resources to war and war-preparedness and civil strife has serious consequences for food security and hunger (Koomson, 2000). Koomson notes four consequences of war and civil strife. First, countries at war invariably hack down social-sector spending, which includes both human and economic resources. Such curtailment has an adverse impact in the long run economic access of the people to food. Second, civil strife affects the timing of increase in the intensity of war and, therefore, food production. The intensity of war often increases in the periods before the rainy season, which disrupts agricultural activities that must essentially be completed before the onset of rains. Such disruption dislocates food production. Third, war and civil strife take a heavy toll on infrastructure, including roads, hospitals, schools, water sources, drinking-water supply, food storage facilities and sewer systems. These are contributory factors in worsening the food security situation (see Box 2.5).

Let us take the case of East Asia to illustrate these points. According to one study by Feffer (2008), between 1999 and 2006, the Republic of Korea raised its defence spending by over 70 per cent, and the government in Seoul planned to increase this figure by 7 to 8 per cent every year for the next dozen years. Military spending by China and Russia increased even more over the same period. North Korea attempted to keep pace, increasing military spending by 25 per cent (in local currency) between 2004 and 2007. This arms race in East Asia has had specific, regional implications. With

> ### BOX 2.5 Violence and Militarism Exacerbated Food Insecurity in Afghanistan
>
> The war in Afghanistan exacerbated food insecurity, with an unprecedented increase in armed attacks on humanitarian aid convoys carrying out food distribution activities. At least 26 attacks on World Food Programme (WFP) food aid trucks had been recorded by November 2008, mainly in the insecure south and south-west. Food to feed tens of thousands of hungry people was looted and/or wasted in the attacks, WFP said. The humanitarian food aid was in response to the country's inability to engage fully in food-based agricultural activities due to war and the on-going threats of violence.
>
> More than 1.6 million children under 5 and hundreds of thousands of vulnerable women were exposed to acute malnutrition, and some could have died in the winter of 2008 due to food insecurity and lack of medical care. 'Around 1.6 million children under 5 and 625,000 child-bearing-age women are at risk of dying this winter due to malnutrition,' the Ministry of Public Health (MoPH) said in a statement (in English) on 25 November. The government said that the food crisis had been exacerbated as well by drought, high food prices and loss of livestock across the country. 'We fear that a humanitarian crisis will be imminent and villagers in those districts might lose a big number of their livestock in the coming winter,' the statement said. Food insecurity is also making vulnerable people, mostly children and pregnant women, more prone to diseases, the MoPH said.
>
> *Source:* IRIN (2008).

Russia and China moving closer together in the Shanghai Cooperation Organization, and the United States and Japan strengthening their bilateral alliance, a new Cold War divide is emerging in the region. This has prompted more resources to be devoted to war-preparedness.

This is no mere regional issue. In East Asia, the largest militaries in the world—the United States, China, Russia and Japan—all face one another. The countries participating in the Six-Party Talks that tackle nuclear proliferation are responsible for 65 per cent of global military spending. These developments in East Asia mirror a global trend: world military expenditures increased by 45 per cent over the last decade.

The same is true of South Asia. South Asia is among the poorest and yet the most militarized regions in the world. The arms purchased by India and Pakistan (the two countries account for 93 per cent of total military expenditure in South Asia) were and are enormous. As one commentator remarked: 'India ranked by the World Bank at 142 in terms of per capita income, ranks first in the world in terms of arms imports. Pakistan is not far behind, being ranked 119 in terms of per capita income, and tenth in the world in terms of arms imports' (Hussain, 2000). South Asia was home to 40 per cent of the world's poor and yet had an annual military expenditure of US$14 billion in 2000.

The military expenditures in South Asia have been huge (Skons et al., 2003). In a situation where 53 per cent of the children in South Asia are malnourished and 36 per cent of the population deprived of safe drinking water, the trade-offs between military expenditures and the provision of basic services are worth considering. For example, a modern submarine with associated support systems costs US$300 million, which would be enough to provide safe drinking water to 60 million people in the countries of South Asia.

The economic burden of nuclear deterrence is equally high in the region. It is difficult to calculate the costs or benefits of the Indian and Pakistani nuclear programmes. One expert argues that, based on likely labour, facility and material costs, India and Pakistan would have each allocated more than $1 billion to design and manufacture a small number of nuclear-capable missiles (Prithvi and Agni for India, Ghauri and Shaheen for Pakistan) (Lavoy, 2003). Additionally, each of these neighbours would have spent five times that amount for the production of fissile materials and the manufacture of a few nuclear weapons, noting that these are at best only partial estimates of the costs involved in nuclear and missile programmes in India and Pakistan (ibid.). Indian defence expert K. Subrahmanyam is quoted to have revealed that, way back in 1985, the Indian military attempted to calculate the expenditure required for a 'balanced deterrent programme'. It was estimated that a force of warheads 'in low three digit figures' with aircraft and missile delivery systems would cost 70 billion rupees (180 billion in 1999 rupees, or $5 billion). Indian analysts calculated that New Delhi must spend $1 billion a year for the next 10 years to field a nuclear deterrent force like the one contemplated as far back as 1985 (ibid.). The cost for Pakistan to assemble a similar deterrent arsenal would have been about the same.

According to one Indian estimate, a single Agni missile costs as much as the annual operation of 13,000 health care centres. More than 3,000 public housing units could be built for the price of one nuclear warhead. The expenditures required to develop India's 'minimum' deterrent could meet 25 per cent of the yearly costs of sending every Indian child to school. Nearly all Pakistani children could be educated and fed for the cost of the nuclear and missile arsenal that is being created for their 'protection'. Thus, India and Pakistan might be able to finance their deterrent programmes, but there are huge opportunity costs. Although they have relatively modern industrial sectors with expertise in nuclear energy, missile development and armaments production (and space, satellite communications and software design for India), India and Pakistan are afflicted by some of the world's worst poverty. Widespread unemployment, outdated infrastructure, rising food prices and low living standards beset each society.

Wars and civil wars destroy investments, real and potential, which could help fight food insecurity conditions, as well as perpetuating food insecurities in the future as well. While wars generate large numbers of refugees in other countries, civil wars produce large numbers of internally displaced persons and refugees in the homeland. The internal refugees are hard to identify and very often feel insecure, because they have neither land to grow food on nor money to finance farming operations. Additionally, they do not have either wages or employment to buy food.

There is, however, a much more fearsome dimension of the relationship between food security and war; that is, fighting attacks food security ruthlessly. Stable environments are also essential for helping households escape hunger. Research has shown that many of the large outliers with comparatively high global hunger index (GHI) scores are countries that have experienced long-lasting wars in the past 15 years. And even this is an incomplete picture, since those countries most affected by conflict, such as Afghanistan, are those without hunger estimates. Conflict impacts hunger both during the conflict and after it has ceased, as evidenced by the experience of some of the countries considered here that have recently experienced conflict (such as Sri Lanka). Hunger is also sometimes used as a weapon when combatants cut off food supplies with the aim of starving opposing populations into submission.

The most direct impact of conflict on well-being is the loss of human life. In addition to the immediate distress this causes, loss of life has a long-term impact on a household's welfare. The loss of able-bodied members limits the household's earning ability and deprives it of its economic access to food. In north-eastern Sri Lanka (where civil conflict raged for nearly two decades), one of every 12 households reported that a family member had been killed as a result of the conflict and, by implication, that their economic ability to access food was to that extent, diminished or even extinguished. In some cases, the poorest and most vulnerable are more likely to become combatants and risk loss of life, with consequent impact on economic access to food.

When people are compelled to leave their homes as a result of conflict, they are cut off from their usual sources of income and food and become very vulnerable to food insecurity. In refugee camps, they are most often subject to overcrowding, poor sanitary conditions and inadequate food supplies. Outbreaks of micronutrient deficiency diseases have been frequently reported in refugee camps. The living conditions in the camps facilitate the spread of infectious diseases, which jeopardize food security through poor food absorption. The disruption of markets, roads, crops, livestock and land that warfare brings, also has an immediate and long-term impact on the incomes of those in the affected areas and, consequently, on their economic access to food.

The destruction of livestock and the loss of life during strife have long-term economic impacts, jeopardizing food security. In Sri Lanka, there has been some compensation for asset loss resulting from the civil war, but those who have not benefited from this are the poorest households. Provision of basic services is difficult during and after conflict, when institutions are absent, many service providers are missing and security cannot be guaranteed. Persistent hunger becomes more likely when basic services are absent. Schools are destroyed during conflict and teachers are killed, compromising the education of a whole generation, especially in long-lasting civil wars, which in turn affects economic access to food in the long run.

Health care services are also jeopardized in times of conflict. Deliberate destruction of health care facilities has been reported during conflicts. More generally, health care systems suffer from a lack of public funding, lack of medical supplies and personnel losses during times of conflict. These factors affect food security adversely. The impact of conflict on hunger, in turn, makes conflict more likely. Regression estimates suggest that halving the income of a country doubles the risk of civil war. This and the fact that conflict is also likely to recur—half of all civil wars are post-conflict relapses—generate a 'conflict trap' in which countries embark on a downward spiral of increasing impoverishment, hunger and violence. Achieving peace, an equitable society and economic growth are clearly important elements of a hunger-reduction strategy.

Fighting leads to breakdown of transportation networks, disruption of fuel and food supplies, rise in food prices, fall in investments, rising unemployment and decline in both nominal and real incomes. On a long-term basis, if warfare continues, food production per capita starts to fall. Frances Stewart demonstrates that in almost all the countries she studied, food production dropped by more than 15 per cent on account of war (Stewart, 1993).

Violence and militarism, thus, almost inevitably lead to food insecurity and, where food insecurity already exists, exacerbate it beyond recognition.

Racism and Ethnocentrism

Racism and ethnocentrism contribute significantly to the food insecurity of millions of people in the Asia-Pacific region today. Ethnocentrism describes the belief that one's own patterns of behaviour are preferable to those of all other cultures. Because people are taught the values of the culture in which they grow up, they tend to view their own patterns of behaviour as being right, normal and best. Other cultures, as a corollary, are viewed as wrong, or irrational and misguided. Ethnocentrism and racism are the foundation of many famines, wars and ethnic conflicts, communal violence, tribalism and colonialism. Racism and ethnocentrism are associated with power, and both are apparent and easily detectable when one group dominates another. Some of the most serious consequences of racism and ethnocentrism are primarily economic. In the Asia-Pacific region, under the guise of development, many ethnic groups have been impoverished and suffer food insecurity.

Competition for scarce resources is inherent in the modern development of the Asia-Pacific region as it moves toward the position of an economic power-house of the world. This process has led to the diminution and maybe even extinction of ethnic groups, the exodus of marginalized people (such as indigenous peoples, tribal people or nomads) or the destruction of local ecosystems upon which these groups depend. A dominant group often 'develops' the area for its own purposes, making it uninhabitable for the local populations, which often happen to be the marginalized groups. The events in West Bengal, India, in 2008, where the state government acquired the agricultural land of farmers for a major automobile company to build a car factory, provide a recent case in point. In an ethnically divided society, the government often intervenes in ethnic struggle by distributing services and resources; but these interventions are either not always enough or do not meet the needs of marginalized groups, as the events in West Bengal testify.

Countries in Asia such as Sri Lanka and East Timor, which suffer from societal stratification in terms of racism and ethnocentrism, are also countries that face considerable poverty and food insecurity emanating from different forms of racism and ethnocentrism. For example, the northern provinces of Sri Lanka and some of the eastern states of India, namely, Bihar, Uttar Pradesh and Orissa, as well as Rajasthan in the west, which suffer from ethnocentric discrimination, also face severe problems of food insecurity. Similarly, a section of the population in Myanmar suffers from food insecurity because of ethnocentrism. On the other hand, societies which live in harmony, while divided along ethnic lines, achieve considerable development and succeed in having food security. Vietnam, Indonesia (until lately) and Maldives are examples of this.

When disaster strikes, racism and ethnocentrism play a vital role in perpetuating food insecurity. Indeed, disaster can reinforce social discrimination in a hierarchical society, where opportunities and resources are not fairly distributed between and among various social and ethnic groups. For example, following the South Asia earthquake in October 2005, discrimination was noted in affected areas of the North West Frontier Province, Pakistan, and in Pakistan-administered Kashmir. A research report by an international non-governmental organization (NGO) and a national NGO noted that in the village of Charan Gada, Muzzaffarabad district, the Sawati castes were discriminated against by the Sayeds, who received the bulk of the food relief (International Federation of Red Cross and Red Crescent Societies, 2007: 30–31).

Age and Vulnerability

Two age groups suffer more hunger than others, more so if they are poor. These two groups are children, and among them, the younger ones, particularly girl children, and the old who are long past their more productive years.

In the late 1990s in India, about 37 per cent of children of age 4 or below suffered from second-degree malnourishment or 'stunting', which is higher in Karnataka, Madhya Pradesh, Gujarat, Andhra Pradesh, Bihar, Uttar Pradesh, the north-eastern states, Haryana and West Bengal than in Kerala and Tamil Nadu. About 29 per cent of the children in the age group 5–12 are stunted. Stunting is found almost across the board and has little to do with parental incomes, and the landholding pattern of the households and social groups to which the children belong has no bearing either on the wasting of children (Shariff, 1999). This is the prevalent situation in India, despite the fact that the country has been running the Integrated Child Development Services (ICDS) since 1975, in 3,702 community development blocks, which is 70 per cent of the country (VHAI, 1997). The ICDS provides a package consisting of supplementary nutrition to children, pregnant mothers (during the last trimester of pregnancy) and lactating mothers, immunization, health check-ups, referral services, non-formal education and nutrition and health education.

Children's needs and voices are not always heard in the countries of the Asia and the Pacific, despite the UN Convention on the Rights of the Child to which most of these countries are signatories (Goonesekere, 1999). Within children as a class adversely affected by food insecurity, street children, the children of single parents and child workers are especially vulnerable groups. And the girl child is even worse off. Research has shown that boys receive better treatment than girls: they get more medical attention and more food. Food insecurity in children has struck silently, because only up to 2 per cent of children exhibit visible signs of malnutrition, though in reality, millions under the age of 5 suffer chronic malnourishment. This makes them vulnerable to illness and to physical as well as mental underdevelopment.

Even in democratic countries, despite a free press, malnourished children largely go unnoticed. Neither the print nor the electronic media pick them up very often, barring honourable exceptions.

It is often not realized that hunger among children leading to child malnutrition is one of the world's greatest problems in the 21st century, with one in every three children in developing countries stunted physically and underdeveloped mentally because of malnutrition.

While there is a prevalence of malnourished children in poor households that face food shortages before and during harvests and during natural calamities (earthquakes, floods, droughts, epidemics, etc.), civil strife and war, many malnourished children are found in homes that apparently do not suffer from food shortages, particularly the girl child. The principal cause of this is illness, especially that caused by water-borne diseases, notably diarrhoea, which afflicts poor households. This is also true of poor households throughout the Asia-Pacific region. And it is more acute for girl children because they have lower access to health care. These diseases in turn emerge on account of the fact that poor communities do not have potable water even in 'normal times', and the condition of sanitation is abysmal. In times of stress, the situation is even worse.

Globally, 1.3 billion people are deprived of access to safe drinking water.[9] The situation in individual countries gives an even more difficult picture. For instance, in Delhi, thousands of people who live along and around the river Yamuna consume the highly polluted water from the Yamuna, which is dangerous. All indications are that consumption of poor-quality water will continue with all the concomitant consequences. The situation is pretty much the same for sanitation as well. Worldwide, 2.6 billion people are without access to sanitation facilities.[10] In Mumbai, half of the total population is without sanitation facilities.

Poor sanitation and poor quality of drinking water are the principal causes of water-borne diseases that affect children. Street children who are found at railway stations, bus terminuses and busy street crossings in cities and towns, children abandoned by

---

[9] Earth: A Graphic Look at the State of the World, 'Human Conditions'. http://www.theglobaleducationproject.org/earth/human-conditions. php#5 (accessed on 18 January 2012).

[10] World Water Council, 'Water Supply and Sanitation'. http://www. worldwatercouncil.org/index.php?id=23 (accessed on 18 January 2012).

their parents, orphaned, kidnapped and made to beg, suffer additionally from more diseases. Respiratory infections, tuberculosis and leprosy diseases are common. Children cared for by a single parent are often involuntarily neglected for a wide variety of factors: lack of resources (including time), energy and conflicting pulls. Consequently, their food needs as well as educational needs and their health care requirements suffer. This, again, not only leads to a fall in food intake but is also germane to lower food absorption, illness and disease.

Children who toil as child labourers are, ironically, a food-insecure class even though they waste their childhood probably in order that others may eat. There is no need to debate about the state of child labour. Most of these children are not only denied their basic right to education, but also work under stress in agriculture, hazardous industries, firecracker-manufacturing units, mines, quarries and so on, and are very often, the victims of occupational diseases. They suffer from repetitive impoverishment (UNICEF, 1995). And when disease strikes and nutrition suffers, the worst consequences of diseases are magnified.

Then, there are the aged and elderly people. They suffer food insecurity in silence. Young adults generally have priority over the old. The classic logic for this has been that the bread winner must be fed better. But in reality, it is a power relationship. Thus, it has been rightly observed: 'Intra-household resource allocation shows bias not only by gender, but also by age and by sibling hierarchy. The house hold power relations determine claims to consumption' (UNDP, 1999). There is a need to disaggregate, and to look at the disaggregated data to locate the causes of women's hunger, irrespective of their poverty status. Women who are past their most productive years, face a bleak future (Rajan et al., 1999). The growth of the region's elderly and aged population, that is, people aged 60 years and above, has been identified by the WHO as one of the major challenges of the 21st century. 'The persistence of poverty combined with ageing in countries still tackling basic problems of development has no precedent in the history of mankind' (WHO, quoted in Ridge, 1999).

It is estimated that the proportion of persons aged 60 years and older in the world will double, between 2000 and 2050, from 10 to 21 per cent (i.e., from 600 million to 2 billion in absolute numbers).

Among the world's population aged 60 years and above, 52 per cent lived in Asia and the Pacific in 2002, and this is projected to increase to 59 per cent in 2025, making the Asia-Pacific, the most rapidly ageing region of the world (ESCAP, 2003: 1).

The progress in medical science and health delivery systems (however much criticized) has resulted in the higher possibilities of people living longer. In demographic terms, this has resulted in more persons living not only to adulthood, but surviving into old age. For example, in India, life expectancy at birth is now 63.7 years (UNDP, 2007), from only 53 years in 1978. Life expectancy at birth of people in China has increased to 72.5 years. For East Asia and the Pacific, it is 71.7 years, while it is 63.8 years in South Asia.

Two types of ageing are discernible in the Asia-Pacific region: biological ageing and social ageing. While biological ageing has to do with the number of years that a person lives, social ageing is dependent upon the role that society reserves for people as they progress in years. These two types of ageing react with one another to create a reinforcing spiral of problems for the elderly and force them into food insecurity.

The state of food security experienced by the elderly depends, on the one hand, on cultural beliefs, practices, customs, age, degree of social integration, extent of physical well-being, economic situation and mobility. On the other hand, public policies with regard to health and social security greatly determine the food intake of senior citizens. Thus, the triumvirate of family, society and the government have their respective roles in determining the status of food security of the aged and elderly. With increased life expectancy in Asia, the food security of the elderly will assume even greater importance in the years ahead.

The breaking down of the joint family system and the decline of the extended family culture have led to the emergence of the nucleus family, comprising the husband, wife and their children. Old parents and, at times, grandparents, have all fallen out of the picture. With the emergence of 'nuclear' families, the elderly remain relatively uncared for, and hence suffer from food insecurity among other deprivations. Unless the couple is a working couple with children to be cared for, parents may not be wanted in the household, or get care and be fed well.

Lifestyles have also changed. Newer opportunities and a newer range of careers have opened up for those with access to education, often taking them to distant places in search of a job. For the majority who are not educated, especially for those in rural areas, their traditional family vocation/profession/land can no longer support them (there is already widespread disguised unemployment and open unemployment). They have to move out of their homes, villages and towns to eke out a living in distant places. The massive surge in rural–urban migration and the growth of urban slums all over the continent provide some pointers to this phenomenon. Thus, in either case, people have moved out to distant places and cities in search of meaningful employment. People who leave their villages in search of a better life in the cities leave the elderly behind, often to fend for themselves. Additionally, for many, years of seasonal migration has not led to long-term increase in assets or reduction in ill-being (Black et al., 2005). Moreover, the households suffer from a loss of social position and status where the emigrant is absent for a long time and fails to take part in important events, leading to increased marginality from credit and social networks having strong bearing on food security of the elderly.

Even otherwise, with increases in industrial employment and growing population pressure, male and able-bodied people often move out of agriculture and come to urban centres for work, returning only during the cultivation and harvesting seasons. The millions of people from rural China and India, for instance, who come to work in urban areas, return to their villages during the cultivation season. The elderly in the villages not only fend for themselves as mentioned earlier, but also look after their property, home and hearth, without adequate command over resources, leaving them little choice but to suffer in silence.

Social and professional life has changed dramatically in a competitive environment in the Asia-Pacific region. Most families suffer from a three-tier generational trap, with deep chasms between each tier. The elderly belong to one generation, the adults and able-bodied members belong to the third generation. The range and intensity of interests of the third generation are so varied and so fundamentally different from the interests of the first and the second generations that there is no common ground. This has led to a general isolationism among the elderly, even where they live with younger people. The everyday family dinners where the whole family would meet and talk about their lives' joys and sorrows are almost gone for the elderly.

Inappropriate policies and inadequate implementation of some policies and institutions also cause food insecurity for senior citizens. In developed countries, social security benefits are the mainstay of survival for the elderly, though there are also private pensions. However, in the developing countries of the Asia-Pacific region, these systems do not exist or do not work. For instance, in India, as much as 87 per cent of the labour force is in the unorganized sectors, with no access to any form of social protection (Government of India, 2006). A committee set up by the Government of India to look at the social security of the elderly came to the conclusion that the present pension provisions cover less than 11 per cent of India's workers. Thus, the pension scheme and the system of paying provident fund (in the organized sector) in India cover only a very small segment of Indian society. Even the implementation of existing policies leaves much to be desired. Elderly people find it difficult to derive benefits from the policies designed for the old.

War, civil strife and communal disturbances cause a range of special problems for the aged and the elderly, just as they do for other vulnerable groups. These groups often flee to refugee camps and to displaced persons' shelters/homes, or remain in their homes, in the belief that they will be spared by the intruders, without enough provision for food or health care facilities. When food arrives in the affected areas, their physical condition and social standing prevent them from competing for food, leaving them hungry. In the unfortunate event of the able-bodied members of the family being killed in such disturbances, the elderly are left to fend for themselves, pushing them into states of poverty and food insecurity, because they can play an economically productive role only at the margin. The experiences of parents of soldiers killed during war are a case in point.

The old and the aged are also susceptible to certain diseases that hit them the most and knock them out of the work force. Alzheimer's disease is one example. Old-age blindness (millions suffer from cataracts), broken limbs and failing general health are some crippling diseases the old suffer in our region. The crippling effect of non-communicable diseases such as cardiac diseases is far-reaching. This is particularly so where the state of health care facilities for the non-rich is far from perfect (VHAI, 1997). Diseases have a three-fold effect on food security for the elderly. First, diseases reduce the earning capacity of older people, and hence they suffer even more food insecurity. Second, the expenditure on treatment takes a heavy toll on the slender financial capacity of the old, and is often met by cutting down on the consumption of food. Third, diseased older people are additionally vulnerable to criminal attacks, robbing them of whatever assets they may have, thereby reducing their earning capacity and hence their capacity to have access to food. In fact, disease alone is one of the major causes of isolation of older persons.

Older people are often bypassed in aid distribution during disasters, first, because they have lower lung power, and second, because they are physically weaker than others in the struggle to get their hands on aid. For example, after the Indian Ocean tsunami of December 2004, 9,000 older people were missed out in the rush of assistance (International Federation of Red Cross and Red Crescent Societies, 2007: 66). This adversely affected their access to food. In Sri Lanka, older people received no monetary compensation to help them restart their livelihoods after the tsunami if they were living with adult children. The older people thus lost a vital opportunity to restore their purchasing power to access food. 'The assumption by aid organizations that all relief materials (including food) will be shared equally within multi-generational families often leaves older people without the support they need' (ibid.: 67). Old age also prevents the elderly from getting adequate medical attention during disasters. For example,

> [among] those affected by the 2005 South Asia earthquake, mental concerns were more prevalent among people aged 60 and over, including 'increased isolation, feelings of being a burden more than an asset, inter-generational conflict, and the reality of major losses that will not be able to be restored in their lifetime', according to the author of a psychosocial needs assessment conducted in September 2006. (Ibid.: 75)

Seasonality

Seasonality is a hugely important factor causing food insecurity. It is common knowledge that, in large parts of Asia-Pacific, most farmers are small and marginal farmers, and most of the rural population depends for food on these small and marginal farmers. The food produced by small and marginal farmers can support the food needs of their families and rural people for periods ranging from three to nine months, usually after the harvests. Thus, people suffer food insecurity for three to nine months during a year, depending on the place and environment in which they are situated. During these seasons of food insecurity, people take recourse to various measures.

A longitudinal participatory study of food insecurity and community coping mechanisms was conducted in village Rakshit Chak (inhabited by landless families of the Lodha tribe), Midnapore district, West Bengal, over the years 1993, 1995 and 1998, using food calendars with women. The study demonstrates that seasonality affects their food security seriously due to non-availability of food. It shows that the food security situation would have been worse but for the common property resources and the rights of the poor to access them, which act as an insurance against fall in food availability in certain times of the year (Mukherjee and Mukherjee, 2008). The study of the food calendar (Table 2.2) reveals that there are four phases of food insecurity in the village under study. Phase one covers the period from mid-November to mid-February, when people eat the most. Food insecurity sets in during phase two, spread over two months from mid-February to mid-April, when people eat less. Mid-April to mid-September marks phase three, when food consumption is very low. Phase four spreads over the period from mid-September to mid-November, when a typical diet of poor households consists mostly of rice.

It can be seen from tables 2.2 and 2.3 that there are significant seasonal variations in the availability of foods like rice, potato and pulses, which are from the 'primary system of food' produced using technology in the economic sense. For example, the consumption of rice, the staple food, reaches its peak between mid-January and mid-February, after which it declines during mid-March to mid-April, remaining stationary through mid-September to mid-October. The consumption of rice remains low throughout the summer months. Secondary food, which comes primarily from CPRs, provides valuable additions to the food basket of the poor in the form of fibre and animal protein, adding nutrition to the diet of food-insecure villagers. In the lean period (mid-March to mid-September), when food from the primary system is low, the villagers access CPRs to

TABLE 2.2 Seasonal Food Calendar of Village Krishna Rakshit Chak, Midnapore, 1993

Month	Rice	Potatoes	Pulses	Vegetables	Fruits	Food from water sources	Others from the wild
1	2	3	4	5	6	7	8
Magh (mid-January to mid-February)	**************	**************	***	Cabbage	–	–	Wild borums and wild rabbits
Phalgun (mid-February to mid-March)	**********	***********	**	Spinach	–	–	Neem leaves
Chaitra (mid-March to mid-April)	****	****	**2	Pumpkin	–	Fish and wild water plants	–
Baisakh (mid-April to mid-May)	****	***	*	Pui leaves and herbs	Mango and jackfruit	Fish, snails and wild water plants	–
Jyastha (mid-May to mid-June)	****	****	**	Lota, leaves and herbs	Mango and jackfruit	Fish, wild water plants	–
Asardh (mid-June to mid-July)	****	***	***	Jhinge (similar to sukini), green papaya	–	–	–
Srabon (mid-July to mid-August)	****	****	*****	Green papaya	–	–	–
Bhadra (mid-August to mid-September)	****	***	***	Green banana	–	Fish and snails	–
Ashwin (mid-September to mid-October)	****	***	**	–	–	–	–
Kartick (mid-October to mid-November)	**	***	**	Radish, leaves	–	–	–
Ahgrayan (mid-November to mid-December)	****	****	******	Tomatoes	–	–	–

Source: Mukherjee and Mukherjee (2008).

Note: The stars in the cells represent the number of stones that the villagers placed in the relative cell, as ordinal ranking of the quantity of the item consumed. This participatory methodology is called scoring and ranking. For more details of the methodology used, see Neela Mukherjee (1993). *Participatory Rural Appraisal*, New Delhi, Concept Publishing Company for the details of the methodology used.

TABLE 2.3 Seasonal Food Calendar of Village Krishna Rakshit Chak, Midnapore, 1995

Month	Rice	Potatoes	Pulses	Vegetables	Fruits	Fish	Snails	Others from the wild $
1	2	3	4	5	6	7	8	9
Magh (mid-January to mid-February)	*****************	***************	***	–	–	–	–	–
Phalgun (mid-February to mid-March)	**********	************	**	–	–	–	–	Neem leaves
Chaitra (mid-March- to mid-April)	*****	***	–	–	–	–	–	–
Baisakh@ (mid-April to mid-May)	****	***	–	–	–	–	–	–
Jyastha (mid-May to mid-June)	****	***	–	–	–	–	–	–
Asardh (mid-June to mid-July)	****	**	–	–	–	–	–	–
Srabon (mid-July to mid-August)	****	***	–	–	–	–	–	–
Bhadra (mid-August to mid-September)	**	–	–	–	–	***********	–	–

Source: Mukherjee and Mukherjee (2008).

Note: The stars in the cells represent the number of stones that the villagers placed in the relative cell, as ordinal ranking of the quantity of the item consumed. This participatory methodology is called scoring and ranking. For more details of the methodology used, see Neela Mukherjee (1993). *Participatory Rural Appraisal*, New Delhi, Concept Publishing Company for the details of the methodology used.

cope with food shortages. Access to CPRs like water bodies can make a significant difference to the quality and quantity of food available. For example, in the village, a huge tank called Rajbandh was not auctioned in 1998 by the local body due to a dispute, which allowed local villagers access to the tank from which they could collect a number of food items. While the auctioning of CPRs like tanks provides resources to local bodies, it also means that insecure households are deprived of food in lean periods.

Incidentally, the months when food availability is low are often also the months when the energy needs of people are high and their purchasing power to access food (economic access) is low. For example, in the case just cited, food availability is lowest during the summer for villagers in Krishna Rakshit Chak, and these are the months when employment opportunities are also low, especially in rural areas, as there are no agricultural operations.[11] Villagers thus suffer a double jeopardy of seasonality, with lower food availability and lower economic access to food.

Seasonality has been a cause of food insecurity in many other countries in Asia as well. For example, in the Democratic People's Republic of Korea, household access to food exhibits seasonal variation depending on the production cycle, resulting in a higher level of access during the harvest period and lower access during the lean (or hungry) season (FAO and WFP, 2008). Food access is most difficult during the months from May to July, consistent with the seasonal calendar of food production which shows that the main harvest of maize and rice takes place between September and October each year. Vegetable production for the winter, *kimchi* also takes place around this period, resulting in relative improvements in the amount of food during the winter period. The second harvest (mainly of potatoes) occurs during spring and contributes to a modest improvement in food access during this period. A seasonal access to food can be seen clearly, with a gradual decline from peak availability in November to a trough between May and July, and a gradual recovery thereafter till the spring harvest season of potatoes, wheat and barley. The food availability curve is roughly a U-shaped curve with an elongated bottom.

---

[11] The National Rural Employment Guarantee Scheme (NREGS) of the Government of India, which guarantees employment for 100 days in a year for poor people, has mitigated the situation to a good extent.

Shocks and Risks as Causes of Hunger

Among the causes that lead to hunger, risks and shocks faced by poor people are the foremost. Risks, in the context of hunger, denote an uncertain outcome of events that can damage the well-being of people by pushing them into hunger or deepening hunger, while shock means an adverse realization of a stochastic variable. Risk is thus an *ex ante* concept, and shock is an *ex post* one.

Knowing that a crisis may descend any time, and not knowing whether one will be able to cope with it, are all part of poor people's landscape of life. Poor people are exposed to a wide array of risks and shocks. A forbidding amount of literature has been generated on the various aspects of risks and shocks in developing countries, including their sources and kinds, their effects on poor households, and individual, household and community responses to managing these events.

There are various ways in which risks and shocks can be classified. For example, they can be classified as political, environmental, economic and social risks (Holmes and Jones, 2009). Risks and shocks can be classified as macro, meso and micro risks. They can also be categorized broadly into idiosyncratic and covariate risks and shocks (Bhattamishra and Barrett, 2008). All these classifications have their own advantages and disadvantages. Moreover, there can also be various combinations and permutations of such risks and shocks. For example, economic risks and shocks can be either idiosyncratic or covariate. Similarly, idiosyncratic and covariate risks and shocks can be at the macro, meso or micro level. Table 2.4 presents the various combinations of risks and shocks. The issue of hunger and risks in terms of idiosyncratic and covariate risks will be discussed here, as it helps both to keep the discussion uncomplicated and to formulate policy options without much complexity.

Now, what do we mean by idiosyncratic and covariate risks and shocks? Idiosyncratic risks and shocks are those risks and shocks where the experience of one household in this regard is typically unrelated to the experience of neighbouring households. Covariate risks and shocks are those risks and shocks which many households in the same locality suffer. For ease of understanding, we shall discuss the link between hunger and shocks only, given their *ex post* character.

TABLE 2.4 Main Sources of Idiosyncratic and Covariate Risk

Type of risk	Idiosyncratic		Covariate
	Risks affecting an individual or a household (micro)	Risks affecting groups of households or communities (meso)	Risks affecting regions or nations (macro)
Natural	Rainfall, landslide, volcanic eruption	Earthquake, flood, drought, high winds	
Health	Illness, injury, disability, old age, death	Epidemic	
Social	Crime, domestic violence	Terrorism, gang activity	Civil strife, war, social upheaval
Economic	Job loss, wage cut	Unemployment, resettlement, harvest failure	Changes in food prices, growth collapse, hyperinflation, fuel, financial or currency crisis, technology shock, terms of trade shock, transition costs of economic reforms, etc.
Political	Riots	Political default on social programmes, coup d'état	
Environmental		Pollution, deforestation, nuclear disaster	

Source: World Bank (2001).

Idiosyncratic shocks commonly arise due to chronic or sudden illness (usually not infectious diseases), crop yield shocks associated with microclimatic variations, localized wildlife damage, loss of draught power due to death or maiming of animals used in cultivation/transportation, pest infestation and one-off events such as property loss due to fire, theft or vandalism, sudden death or loss of limbs of the bread-winner, or even a temporary job loss, where employment is largely in the informal sector, and domestic violence. On the other hand, *covariate shocks* occur due to natural disasters, like the Indian Ocean tsunami that led to 400,901 people being displaced in Indonesia alone (OCHA, 2005), clearly swelling the ranks of the hungry; due to war (like the war in Afghanistan); civil strife, like the one in Sri Lanka which has created 300,000 internally displaced persons who face hunger of different degrees; food price instability such as that seen in 2008 that added nearly 41 million people to the number of those who were hungry (FAO, 2008); and financial crises such as the one currently gripping the world, which threw 24.8 million workers out of jobs, jeopardizing their economic access to food.[12] Covariate shocks are those to which everyone in a community is vulnerable.

Generally, idiosyncratic risks are covered by insurance programmes such as life insurance, health insurance and draft animal insurance, but they are very expensive for those who are most likely to suffer from such risks, namely, the poor. Insurance against covariate risks is not generally available, barring a few publicly funded schemes like crop insurance in India. This is because the risks are too large; and where such schemes are available they are prohibitively expensive.

It may be underscored that though the distinction between idiosyncratic and covariate shocks is critical for policy making, traces of idiosyncratic shocks can be found even in largely covariate shocks, because there are discernible differences between households in their exposure and capacity to respond to shocks. Additionally, the extent to which a risk is covariate or idiosyncratic is a function of the underlying causes. For example, job loss due to heavy rain can be an idiosyncratic risk for a casual mason, or job loss can be

---

[12] Especially when less than 20 per cent of the population is covered by unemployment benefits.

covariate risk for a whole community of casual workers if it is the outcome of macroeconomic turbulence such as that as seen in 2008. Similarly, the risk of infection by HIV/AIDS can be an idiosyncratic risk germane to causing hunger, or it can be a covariate risk if it becomes a pandemic, contributing to hunger in a community or even a nation like in some African countries. However, the bottom line is that a growing body of empirical evidence suggests that idiosyncratic risks predominate over covariate risks in Asia, like in Africa (Deaton, 1997), and hence idiosyncratic shocks are the dominant cause of hunger in this region. The main sources of idiosyncratic and covariate risks are shown in Table 2.4.

There are basically five main clusters of idiosyncratic risks: (a) sickness, injury and disablement; (b) natural disasters; (c) ageing; (d) crime and domestic violence; and (e) unemployment and other labour market risks. We shall discuss them briefly here.

Sickness, Injury and Disablement

Poor people often live and work in environments that expose them to greater risk of illness or injury, and they have less access to health care (Prasad et al., 1999). Their health risks are strongly connected to the availability of food, which is affected by almost all the risks the poor face (natural disasters, wars, harvest failures and food price volatility). Communicable diseases like SARS, bird flu and the H1N1 virus that have afflicted Asia since 2003 are concentrated among the poor. Respiratory infections are considered by some experts as the leading cause of illness and even death (Gwatkin et al., 2000). In India, for instance, it has been found that the poor are 4.5 times as likely to contract tuberculosis as the rich, and twice as likely to lose a child before the age of 2 (World Bank, 1998).

Illness and injury in the household have both direct costs (for promotive health, preventive health care and curative health care) and opportunity costs like lost income or schooling while ill. Direct costs eat into the economic power of the poor to access food, while opportunity costs affect both the short-term loss of economic access to food for the poor and long-term loss in economic access in term of higher capability poverty. The timing, duration and frequency of illness also affect its impact on the food security of the people. Poor households can compensate for an illness during the slack agricultural season, but illness during the peak season leads to a heavy loss of income, especially on small farms, leading to diminution of economic access to food, usually necessitating taking out loans at usurious rates of interest, making further inroads into the economic access to food, something that triggered the global phenomenon called the 'Grameen Bank' of Bangladesh.

Natural Disasters

Natural disasters cause hunger to millions of people. Take just two examples. The destructive capacity of natural disasters and their impact upon food security has already been discussed. The destruction wrought in Asia by the December 2004 tsunami has been described in Box 2.3.

More recently, on 3 May 2008, Cyclone Nargis devastated Myanmar's low-lying Irrawaddy delta region, leaving more than one million people homeless. An estimated 80,000 people died in the delta's Labutta district alone. Myanmar had been expected to export 600,000 tons of rice in 2008, including exports to Sri Lanka and Bangladesh, but it could not live up to these expectations after the cyclone flooded 5,000 sq km (1,930 square miles) of farmland in the country's main rice-growing area, exacerbating a food crisis even outside of Myanmar. Additionally, an 80 per cent loss recorded in the December 2008 harvest (CARE, 2009) meant that food distribution had to continue in Myanmar itself to prevent hunger and starvation.

Ageing

Many risks are associated with ageing: illness, social isolation, inability to continue working, inability to cook, uncertainty about transfers (usually from sons and daughters) being able to provide an adequate means of livelihood and so on. The incidence of poverty among the elderly varies significantly. In many countries in Central Asia, like Georgia and Armenia, the incidence of poverty is above average among the elderly, particularly among people 75 and older. There is also a gender dimension to the problem of poverty associated with old age. Women, because of their higher life expectancy at birth, constitute the majority of the elderly, and they tend to be more prone to hunger in old age than men (World Bank, 2000). With better health care facilities and rising income in many

Asian countries, the number of elderly people in the developing countries in Asia and the Pacific will increase significantly in the foreseeable future, adding to the number of the food-insecure.

Crime and Domestic Violence

Crime and domestic violence reduce earnings, take away income and destroy assets, which entrenches hunger. Unlike the rich, the poor have few, if any, means of protecting their life, property, assets and income against crime, all of which help fight hunger. Crime also hurts the children of poor people: children exposed to violence may perform worse in school, thereby reducing their income-earning capabilities, which may, in the long run, be germane to lower purchasing power and hence lower economic access to food. A study of urban communities in the Philippines, among other countries, showed that difficult economic conditions destroy social capital as involvement in community organizations declines, informal ties among residents weaken and gang violence, vandalism and crime increase (Moser, 1998). Violence and crime may thus deprive poor people of two of their best means of reducing vulnerability to hunger—human and social capital, which act as 'informal' insurance against transitory hunger. Rich and poor women alike are victims of domestic violence, but the incidence is often higher in poor households.

Unemployment and Other Labour Market Risks

Labour market risks include unemployment, falling wages and taking up hazardous and low-quality jobs in the informal sector as a result of macroeconomic crises or policy reforms. The first workers to be laid off during cutbacks in public-sector jobs are usually those with low skills, who are then exposed to hunger, among other things, and are forced to migrate to urban areas, adding to the already staggering numbers of urban poor. This is a pattern observed in many countries during the structural adjustment reforms of the 1980s and early 1990s. Economic crises such as the East Asian crisis in 1999 and the crises of 2007–08 also had significant effects on labour markets, with real wages and non-agricultural employment falling in all affected countries.

Fluctuations in demand for labour often affect women and young workers disproportionately. These groups are therefore at greater risk of facing hunger as a result. Most public-sector retrenchment programmes have affected women's employment more than men's (World Bank, 2000). Women are more likely than men to work for small firms, which tend to be more elastic to demand fluctuations and pay lower wages, both of which have a strong bearing on women's purchasing power and economic access to food. As incomes fall, poor households try to respond by cutting consumption of food, pulling out children from school and pruning health care, especially for women and girl children.

Liberalization of markets often boosts the price of staples, which could benefit small farmers if they are net sellers of food. But in many cases they are net buyers of food. Thus, the urban poor, rural landless, poor people, artisans, herdsmen, fisher folk and farmers are confronted with seasonality, selling food immediately after the harvest when food is plentiful and prices are low and buying food when it is scarce and prices are high. All these categories of poor are exposed to different degrees of hunger. Traders in food can step in and equalize prices over the year through arbitrage, but rural areas lack transport facilities and related infrastructure to facilitate the process, apart from the question of incentives.

For the rural poor, crop diversification and income diversification into non-farm activities can help reduce food price and harvest risks, and eliminate hunger due to successive harvest failures from various natural causes. For example, because of insufficient monsoons in India in 2009, indebtedness increased in 80 per cent of households in 160 districts declared drought-affected. According to the best available count, as many as 246 districts spread across 10 states, or 40 per cent of the total number of districts, were declared drought-hit. The situation is tailor-made to stoke inflationary fires, deepen rural distress and cause hunger, especially when, according to the Government of India's *Economic Survey*, the per capita consumption expenditure of 71.9 per cent of the rural population was less than ₹20 per day in 2004–05 (Gangadharan, 2009).

There are informal and formal mechanisms that are in place to deal with these shocks. The various ways that the state (read government), individuals, households and communities deal with the different kinds of shocks are tabulated in Table 2.5.

TABLE 2.5 Mechanisms for Managing Shocks

	Informal mechanisms		Formal mechanisms	
Objective	Individual and household	Group-based	Market-based	Government-provided
Reducing the impact of shocks	□ Preventive health practices □ Migration □ More secure income sources	□ Collective action for infrastructure, dykes, terraces □ Common property resource management		□ Sound macroeconomic policy □ Environmental policy □ Education and training policy □ Public health policy □ Infrastructure (dams, roads) □ Active labour market policies
Mitigating the effects of shocks Diversification Insurance	□ Crop and plot diversification □ Income source diversification □ Investment in physical and human capital □ Marriage and extended family □ Sharecropper tenancy □ Buffer stocks	□ Occupational associations □ Rotating savings and credit associations □ Investment in social capital (networks, associations, rituals, reciprocal gift giving)	□ Savings accounts in financial institutions □ Microfinance □ Old-age annuities □ Accident, disability and other insurance	□ Agricultural extension □ Liberalized trade □ Protection of property rights □ Pension systems □ Mandated insurance for unemployment, illness, disability and other risks
Coping with shocks[a]	□ Sale of assets □ Loans from moneylenders □ Child labour □ Reduced food consumption □ Seasonal or temporary migration	□ Transfers from networks of mutual support	□ Sale of financial assets □ Loans from financial institutions	□ Social assistance □ Workfare □ Subsidies □ Social funds □ Cash transfers

Source: Adapted from Holzmann and Jorgensen (2000) and World Bank (2001).

Notes: The white area shows household and community responses through informal mechanisms to improve risk mitigation and coping. The dark shaded area shows publicly provided mechanisms for insuring against risk and coping with shocks—the social safety net.

[a] Publicly provided coping mechanisms can also serve risk-mitigating purposes if they are in place on a permanent basis.

Paucity of Investment in Agricultural Research and Development (R&D)

As discussed earlier a key cause of improved food security over the past generation in Asia and the Pacific was the steady flow of productivity-enhancing agricultural innovations, embodied in the Green Revolution. Of late, this has been on the down swing, due to paucity of public investment in agricultural research, in as much as following the food and fuel crisis on 2008, at the global summit on food security convened in Rome in June 2008, UN Secretary-General Ban Ki-Moon argued for continued supportto agricultural research and development, and maintained that the current level of funding in this area is deficient.

Indeed, public expenditure on R&D in agriculture for Asia appears to have peaked during the 1980s (Pardey and Beintema 2001). During the last decade of the 20th century, while government funding in the developed world for agricultural R&D remained almost unchanged, it actually declined in the developing world, presumably because, *inter alai*, governments regarded technological investment in agriculture a low priority so long as food prices were low. Support for the activities of the international agricultural research centers also diminished. For example, the International Rice Research Institute, Manila, which was the source of much of the technological breakthroughs that ushered in the Green Revolution in Asia during the 1960s and 1970s, experienced a fall in its budget in inflation-adjusted terms from $US55m per year in 1992 to under $US30m in 2004 (Otsuka, 2005).

In the developed countries like the United States where intellectual property rights regimes are robust, private agribusiness firms (like Monsanto and Cargill) spend large sums on agricultural R & D, especially in biotechnology. Such investments from the private sector, though propelled by profit motive, substituted to an extent the falling public sector investment towards improved crop and livestock production. In this connection, the governments of three developing nations—Brazil, China, and India—also provided substantial support for agricultural research and development, including biotechnology but elsewhere in the developing world, this support has dwindled to very low levels (Pardey and Beintema, 2001). Small wonder the Secretary General Ban has called for a renewed commitment by member countries to agricultural research and development noting that the 'overall price tag for national governments and international donors could exceed $15 to 20 billion annually, over a number of years' (Ban 2008).[13]

Market-related Causes of Food Insecurity

Given the nature of natural resources endowment, there will always be countries—and subpopulations within countries—which cannot be self-sufficient in food. Such countries and sub-population within countries nevertheless will have to legally access food from other countries and places having surplus food, and food must be made available in the market. This is possible only through purchase of food. Thus, markets and food prices have a strong bearing on this issue. We must, therefore, now turn to market-related causes of food insecurity in the Asia, which could be terms of trade (the relative prices at which exchange takes place), market failures, public action in the form of trade and pricing policies and the involvement of state agencies (like Parastatal) in production, storage and distribution of food.

Markets and Food Insecurity

Even in the countries of Asia which export food, households that are net sellers of food comprise a very small fraction of the total population. Vietnam, for example, is the world's second largest exporter of rice. Yet even in the country's 'rice basket' areas, less than half the population sells more rice than it buys: 47 per cent in the Mekong delta and 45 per cent in the Red River delta (Glewwe and Linh, 2008).

For those countries which, or those individuals who, do not produce enough food to supply their own food requirements, food security depends greatly on the terms of trade at which they exchange their production (non-food output) or their assets (say, labour) for food and other basic needs. If the terms of trade they

---

[13] Future improvements in agricultural technology depend on factors other than financial support. For example, conservation of biodiversity, which is particularly threatened in insular environments (e.g., remote islands in the Pacific Ocean), is also needed.

face are unfavourable, or move in an unfavourable direction, individuals or communities can exist in, or fall into, a state of food insecurity. Conversely, an improvement in the terms of trade, for example, due to falling food prices or rising wages, can be a source of improved food security among those who need to buy food to meet their demand.

The physical conditions of agriculture, including infrastructure and technology, typically change quite slowly over a period of time. Moreover, many food-insecure people face such adverse production environment (such as dependence on rain-fed food production such as in large parts in India) that even technological and infrastructural improvements may not carry them very far towards self-sufficiency in food. Under such circumstances, markets (and the institutions and policies that influence their operation) play a major role in determining who is food-insecure and for how long. Importantly, markets transmit the effects of macroeconomic growth (or of the failure to grow) to food-deficit areas and communities. The problem of food insecurity, like that of poverty, is frequently traceable to macroeconomic conditions and market failures rather than to chronic structural deficits. Sen (1981) notes that the Bengal famine of the early 1940s was caused, *inter alia*, by flawed distribution, not underproduction, because starving people who produced food themselves died in front of granaries full of food grains. Similarly Ravallion (1987) found that the 1974 famine in Bangladesh was not because of shortage in food production but due to problems of access to food. China's famine, during the Great Leap Forward of 1958 to 1961, saw that 20 million or more people died because of starvation or hunger-related causes while the country continued to sell grain on the international market (Short, 1999: 486–505).

Wages, Labour and Migration

In the long run, the security of food supplies at every level of the economy can be assured only by guaranteeing that the purchasing power of the poor is adequate to cover food costs, and that markets and market-related infrastructure are in place to meet their demands in a timely fashion. As labour is typically the most important economic asset of the poor, it is not surprising that labour markets are of singular importance as factors determining the escape from poverty.

In some ways, economic development can raise the productivity of rural labour *in situ*. The Green Revolution was an example of technological innovations with complementary investment (irrigation, etc.) that raised the productivity of on-farm labour. But more and more, growth occurring *outside* of agriculture is of the greatest importance for the welfare of the poor. Modern economic growth in many parts of Asia is increasingly characterized by urban and industrial job expansion. This induces internal migration, especially the movement of labour out of regions that tend to be structurally deficient or are routinely vulnerable to supply shocks. Myanmar, north-eastern Thailand, central and western China, northern and north-central Vietnam, the central Philippines, and rural Java and large parts of Sumatra in Indonesia, are all experiencing out-migration, primarily by workers who are rural, poor, landless or land-deprived.

The Asian and Pacific region has witnessed a dramatic increase in international migration in the past few decades, with 58 million migrants in 2005 (United Nations, 2006). There is a high level of migration from China, the Philippines, many South Asian and Pacific Island countries and some central Asian countries. Increased disparities between countries with regard to income and opportunities and demographic imbalances among countries encourage people to move so as to improve their lives. In many cases, the financial remittances sent by migrants enable families to purchase more and better-quality food. Such remittances are a very important contributor to poverty alleviation and overall socio-economic development. In 2007, over $121 billion worth of remittances were received in the Asia-Pacific region, more than double the figure in 2000.[14] Migration also occurs within countries, the trend generally being from rural to urban areas, again by those seeking greater opportunities, including the landless poor. The implications for food security are significant, with more people becoming dependent on market transactions for the purchase of food, and thus susceptible to price fluctuations.

---

[14] Using data from the World Bank. www.worldbank.org/prospects/migrationandremittances, accessed 6 November 2008.

Although remittances can make a positive contribution, various negative consequences of migration remain. One of the main negative fallouts relates to the loss of human capital through out-migration, or the 'brain drain'. While less developed countries tend to lose resourceful workers to more developed ones, the same happens with rural areas losing out relative to urban ones. In some countries, due to conflict or the out-migration of men, women are heading rural households and producing food. However, they tend to be disadvantaged in terms of access to credit, training or agricultural inputs. A shortage of human capital in key sectors, including education, health and agriculture, may act to reduce productivity and economic growth and hinder the provision of social services in areas of out-migration. Another negative consequence of migration relates to the treatment of migrants. Migrant workers frequently constitute vulnerable populations in the host country or in the urban areas they move to, because they are often not entitled to government assistance, may be very dependent on their employer for food (such is the case for domestic workers), and may lack the social networks which they had in their communities of origin. As a consequence, they face heightened risks of food insecurity, while there is also the potential for conflict with other groups when food and other commodity prices rise and put pressure on the ability to maintain certain standards of living, including nutritional intake.

It should be noted, however, that migration (or internal displacement) is also often a response to livelihood or food insecurity, landlessness, loss of land tenure or crises. Conflicts, such as those in Afghanistan and Timor-Leste, have displaced millions of people and contributed to greater food insecurity. On the other hand, migration is often seasonal and is used as a coping mechanism to improve food security, such as in the case of some Central Asian countries as well as other countries in Asia and the Pacific. Income earned through labour migration can be an important means of acquiring land, agricultural machinery and food, while the benefits to food security vary from place to place. For example, in the case of Tajikistan, one study found that 'labour migration has allowed a significant part of the Tajik population to escape from starvation stemming from economic stagnation and massive unemployment' (Olimova, 2005). Mongolia provides another example of migration related to food security. Summer droughts, extremely harsh winters, overgrazing and locusts as well as unaffordable, privatized veterinary services have at times caused the death of livestock and pushed herders off the land and into urban poverty, even though undernourishment is more severe here than in rural areas. A similar situation exists in many parts of South Asia, where high levels of urban poverty and related food insecurity often prevail.

As a group, countries in Asia and the Pacific have experienced high rates of urbanization and are expected to continue to become increasingly urbanized in the years to come. For instance, from having an urban population of 42 per cent in 2008, the region is projected to have 51 per cent of people living in urban settings by 2025 (ESCAP, 2008a). The expansion of urban areas often involves encroachment on arable land that could otherwise be used for food production, and hence threatens food security. In addition, within urban areas, there is increasing pressure on public services, including those related to hygiene. Millions of people live in slums where sanitation and water are inadequate, increasing the burden of disease and exacerbating malnutrition. Furthermore, in Asia and the Pacific, demand for food is especially high in poorly planned and managed cities, which pushes up the price of food, given limited supply, and makes the poor and other vulnerable groups especially susceptible to food insecurity due to lack of access to food. Growing urban populations spur changing food habits, frequently increasing dependence on imported food at the expense of locally produced food.[15] Due to urban economies being highly monetized, income security is especially crucial for people living in cities, and is one of the prerequisites of improving food security.

In the Pacific, with its high levels of internal migration and poverty in towns and cities, poverty generally does not imply the kind of hunger or destitution experienced in certain other parts of Asia. It means, rather, a continuous struggle to meet essential living expenses in urban areas in particular, especially those expenses that require cash payments. Families constantly have to make choices between the competing demands for expenditure on food and other

---

[15] Comprehensive Framework for Action, page 24. This was developed by the UN Secretary-General's high-level task force on global food security crisis, established in 2008. http://www.un.org/en/issues/food/task-force/pdf/framework.pdf (accessed on 2 March 2012).

basic needs, given the limited availability of cash income. Often trade-offs are made, for instance between food or school fees, and there are struggles to purchase adequate and suitably nutritious food. As seen in other parts of the region (see also Chapter 5), households may borrow regularly from 'loan sharks' who charge them very high rates of interest for small unsecured loans to meet basic family commitments and community obligations. As a consequence, many families are frequently, and some constantly, in debt (Abbott and Pollard, 2004). Moreover, in a more monetized economy they are at greater risk of food insecurity. Improving income opportunities and social protection, including access to affordable credit, is hence fundamental to enhancing food security for the most vulnerable.

Migrants' principal motives may be negative (distress, displacement) or positive (opportunity). But since each of these motives results in the same action, it is very difficult to distinguish between the two. Often the best an analyst can do is to make inferences by examining outcomes after the fact—in particular, by assessing whether migrants are better off after the move than before. One proximate indicator is the existence and growth of employment opportunities in the destination. High or rising underemployment and unemployment, falling earnings, widespread labour exploitation and other obvious signs of hardship among migrants are clear evidence of distress-related motives for migration, and poverty and food insecurity in urban slums are no less severe for being in an urban setting.

For example, the slums and squatter communities that pervaded Manila in the 1980s and 1990s reflected both persistent anti-agriculture policies and sluggish growth for the economy as a whole, resulting in widespread economic and social dislocation and insecurity. In Asia's most deprived populations, especially in Myanmar (until recently) and North Korea, cross-border migration is a gamble of utter desperation by those for whom staying home almost certainly means severe undernutrition. About 1.5 million from Myanmar workers in Thailand have crossed the border to work in jobs that are 'dirty, difficult, and dangerous' (not to mention degrading), and that pay only two-thirds of the wages offered to Thai workers (Kulkolkarn et al., 2007). In Vietnam, some of the migration in recent years has been distress-driven. This is especially true of women from rural areas whose circumstances have changed due to the death of a spouse or divorce, and who must relocate to find work to support dependent children, aged parents or invalids (GSO, 2005; Kabeer and Tran, 2000). Similar experiences can be observed in many other Asia-Pacific economies.

The departure of out-migrants has mixed effects at home. Certainly there is a loss of labour, often the most productive labour. But to offset this, there is both a reduction in mouths to feed at home, which improves food security, and the flow of remittances (when it occurs), which provides an additional source of purchasing power and frequently helps insulate against unanticipated rural income shocks due to crop failures or agricultural price fluctuations (issues explored further in Chapter 5 on community responses). Migration also creates dynamic positive effects, as knowledge about other labour market opportunities flows back to the source population and enters the decision making of future potential migrants (Phan and Coxhead, 2008).

Opportunity-driven migration is clearly a more positive experience. Booms in the growth of export-oriented, labour-intensive manufacturing jobs in China, Thailand, Vietnam and Bangladesh have created direct benefits for relatively poor populations located far away in those same countries—either by raising the wage offered for their labour or by increasing effective employment rates, measured in terms of hours worked per month (Manning and Bhatnagar, 2004; Phan and Coxhead, 2008), both of which enhance the economic access of the migrants and/or their families to food. Under these circumstances, migration and economic growth are complementary, and both work to raise purchasing power for food and other goods and services and reduce income volatility for many poor people, smoothing out the phases of food insecurity. Migrants in foreign-investment-based factories in the Pearl River delta, in Ho Chi Minh City, in Dhaka and in many other locations throughout Asia accumulate savings and send home remittances that help to increase access to food security for a much wider community.

The flip side of migration has been explored elsewhere in this study as well (see Chapter 5 on community-based responses and Chapter 6 on social access and social frameworks for food security in the Asia-Pacific region). Suffice it to say for the present that even in optimistic situations, those at the bottom of the pyramid are prevented from taking part in the general improvement of economic access to food, often because their level of deprivation is such that they cannot afford to take the risky and costly step of sending family members out to participate in other labour markets. This is very often because they lack access to capital markets, as borrowed capital is often needed for the 'investment' of migrating (Phan and Coxhead, 2008; Jalan and Ravallion, 2002). In addition, government policies sometimes restrict internal migration, especially rural-to-urban migration. This is famously the case in China, where the *hukou* residence certificate, which has the effect of discouraging internal labour movement, has been at its most effective in limiting the mobility of the poorest households. By segmenting the labour market and preventing some would-be migrants from taking advantage of opportunities for urban-based employment, restrictions such as these act like a tax on rural earnings (Zhai and Hertel, 2004).

Though data on migration and remittances in Asia are incomplete, it is safe to conclude that there is no one, generally applicable conclusion about the links between labour incomes and food insecurity. Distress-driven migration is *symptomatic* of poverty, deprivation and food insecurity, whereas opportunity-driven migration almost certainly contributes to the alleviation of poverty and food security.

Exploitative Intermediation

The poor are exploited by various kinds of middlemen. Landowners exploit sharecroppers and tenants, robbing them of their produce; moneylenders exploit debtors; and traders exploit small-scale producers robbing them of purchasing power that gives them economic access to food (IFAD, 1991). Farmers borrow from moneylenders at 25 per cent per 100 days (ECLOF International, 2008). All these factors combine to keep millions of people in the grip of food insecurity in the Asia-Pacific region. But, in some cases, cooperatives, banks and government agencies, whose task is to protect the poor, may themselves practise forms of exploitation. There are instances where heavy levies imposed by cooperatives or government agencies have damaged the food security of farmers.

The impact of exploitative intermediation on farmers is best illustrated by the cases of farmers' suicides in India. Due to a variety of factors, farmers' indebtedness has assumed serious proportions in recent years. Of the 89.33 million farmer households in India estimated in 2003, 43.42 million or 48.6 per cent were indebted (Government of India, 2005), implying that more than half of the farming households did not borrow, probably because they were financially excluded (Government of India, 2007). These excluded farmers would have borrowed at rates of interest often reaching 100 per cent per annum. In many cases, this indebtedness of farmers and their failure to discharge their liabilities became one of the important causes of farmers' suicides in India.

It is well known that poor people, especially farmers, are adversely affected by seasonality of various kinds, including seasonality in food, employment and income (Chapter 6 provides a fuller discussion of this issue). The impact of seasonality often translates into food deprivation, as shown in some examples cited in Chapter 5. Famines in India and China during the last half-century and the 1974 Bangladesh famine have been closely related to seasonal food shortages as well as to lack of employment and income. Such seasonal food shortages are often aggravated by exploitative intermediation, because poor farmers (small and marginal farmers in particular) are forced to sell their harvest immediately after the harvest when the prices are at their lowest, often to middlemen or to resourcelenders, to meet pressing needs and often to repay loans taken for the very same agricultural operation. Subsequently, the same farmers have to buy some of the agricultural produce they grew at a much higher price. Thus, availability of food and money are compromised, and the smallholders' long-term food security and indeed well-being are put at serious risk. Seasonal food shortages also create indebtedness, as discussed in greater detail in Chapter 5. To feed themselves, many of the poor have to borrow money at exorbitant rates of interest, sometimes exceeding 100 per cent.[16] They then have to devote much of their energies and resources to debt servicing, further reducing their chances of achieving food security. One of the causes of bonded labour especially in parts of South Asia is the inability of the poor to repay their debts. The labourers who borrow from moneylenders are compelled to

---

[16] This is the phenomenon which gave birth to the Grameen Bank in Bangladesh and to SEWA, Ahmedabad and Working Women's Forum, Chennai, India.

work to 'repay' their debts to their creditors. 'Conditions of work are made so that their "debt" is increased through the application of extortionate rates of interest; and wages which are not high enough to allow the worker to repay the advance except through additional work' (Asia Foundation, 2007).

Corruption is an excruciating form of exploitative intermediation in Asia:

> Across the region, corruption comes in many forms, but the commonest overall distinction is between 'grand' and 'petty'. Grand corruption typically involves relatively large bribes from contractors or other corporations, generally associated with high level politicians or officials. Petty corruption involves smaller amounts but more frequent transactions—lower-level public officials demanding 'speed money' to issue licenses, for example, or to allow full access to schools, hospitals or public utilities. It tends to affect the daily lives of a very large number of people, and more so the poor. (UNDP, 2008)

Corruption undermines efforts as it hits the poor by diverting elsewhere the goods and resources targeted to them. The proverbial diversion of food from the public distribution system in India denies poor people access to food. The poor also lose economic access to food in terms of the bribes they pay to access basic services like primary and maternity health care, basic education and water supply. The amounts of money that the poor have to dole out to obtain access to basic services, to which they are otherwise entitled, imposes an 'informal tax' on the people.[17] Indeed, some Asian nations (such as Afghanistan, Cambodia, Laos, Kyrgyzstan, Myanmar, Tajikistan, Turkmenistan and Uzbekistan) have been rated as among the world's more corrupt nations, according to Transparency International's Corruption Perceptions Index 2008. And some countries whose ratings have significantly worsened since 2006 include Bhutan, Laos, Papua New Guinea and Thailand. Heading the list of those perceived to be the most corrupt in the region are Bangladesh, Indonesia, Vietnam, Pakistan, the Philippines and India.

---

[17] The UN General Assembly approved—unanimously—the Convention Against Corruption in 2003, but fighting corruption will be a long haul.

Large and inefficient bureaucracies are also indicative of an exploitative relationship. Most countries of the Asia-Pacific region have large bureaucracies which are not always efficient. Two examples would illustrate the point. Take the case of India:

> According to the Congress-led government's own estimate, most development spending fails to reach its intended recipients. Instead it is sponged up, or siphoned off, by a vast, tumorous bureaucracy. That is why, despite India's commitment to universal health care, water and education, only five countries have a lower portion of health spending in the public sector; over half of urban children are educated privately; and nearly all investment in irrigation is private. (*Economist*, 2008)

BOX 2.6   Jailed in Marja, Afghanistan

A sharecropper in Marja district in Helmand province, Afghanistan, had reportedly taken an advance payment of US$1,600 on the condition that he would repay the loan through ten kilograms of opium at harvest time in 2003. However, his crop was destroyed in the eradication campaign during the 2002–03 growing season and he failed to repay the loan. In response, his landlord took back the land, blaming the sharecropper for failing to bribe the eradication team a sum of US$200 necessary for them to spare the crop. The sharecropper maintained that he did not have that kind of money to pay the bribe, and because his crop had been destroyed he could not repay his outstanding debts either.

In the 2003–04 growing season, the sharecropper obtained five *jeribs* of land from a different landlord for cultivation. However, his creditor from the previous growing season had him imprisoned for twenty-three days for defaulting in repaying his loans. The sharecropper's mother and the current landlord appealed to the district administrator for his release, insisting that the women of the family would help him in the field so that he could repay his old debts. The sharecropper was released but was ashamed. He stated that 'no wife or mother works on the land in this district but mine are working with me. My nine-year-old daughter and my two younger children are also working with me. They cannot go to school as they help me on the land—this is the curse of debt.' He was cultivating all five *jeribs* of land with opium poppy in the 2003–04 season.

Source: Mansfield (2008).

As for China, about 90 per cent of Chinese are sick of bureaucratic muddle, an online survey for a leading newspaper revealed as recently as in 2008. The *China Youth Daily* released a report of its online poll on the efficiency of government officials, saying that 90.3 per cent of those surveyed were unsatisfied. In a letter published in *Nanfang Weekly* on 24 August 2004, the party general secretary of Qipan township, Hubei, described to the state council, China's cabinet, the heavy burdens shouldered by farmers in Qipan—a township with a population of 40,000—on account of exactions by the bureaucracy. In addition to the basic 200 Yuan ($41) per hectare land fee which every farmer must pay, each household paid a variety of fees. For a farm household, that can add up to several thousand Yuan a year—an astronomical sum for farmers whose annual net income might not exceed 1,000 Yuan a year. The result was that 80 per cent of the farmers were in debt, as reported the in *South China Morning Post* (*Strait Times*, 2004).

These exploitative intermediations cause food insecurity for the people, especially the poor.

Food Prices

The other side of the purchasing power ratio defined at the beginning of this section is the price of food sought by poor households. Whereas higher wages and better labour productivity improve real income for food-deficit households, increases in food prices reduce it. From the Green Revolution until the early years of the 21st century, the real price of rice (as measured by the deflated Bangkok export price) declined substantially. Between 1975 and 2002, it fell from about US$800/ton (in 2002 prices) to about $200/ ton. This represents, on average, a very substantial improvement in purchasing power for populations that spend between one-third and two-thirds of total earnings on basic food items. During the same period, undernourishment (as a percentage of the population) fell in East Asia from 45 per cent to 12 per cent; in South-East Asia from 39 per cent to 12 per cent; and in South Asia from 37 per cent to 21 per cent.[18]

---

[18] Data for earlier years are available at http://faostat.fao.org/site/562/default.aspx (accessed on 2 March 2012).

Soon after the turn of the 21st century, however, the long decline in real food prices reversed. Having fallen to $200/ton in 2002, the price of rice in world markets began to climb, reaching nearly $700/ton during the first half of 2008. This and related food price trends, which have been dramatic enough to grab newspaper headlines and spur talk of a global 'food crisis', also exceed the gradual increases that many economists forecast could happen, depending on future changes in demand and supply.

An important trigger for higher food prices has been the increase in oil prices. The very straightforward reason for this is that the latter increase is a direct consequence of monetary devaluation in the United States. International petroleum prices are always expressed in dollars and, as the US currency has lost value, exporting countries have demanded more dollars for every barrel they supply. Expensive energy has affected the food economy in various ways, generally driving up prices of edible products. Certainly, production costs are sensitive to energy prices where agriculture is mechanized. But these prices matter even in settings where tractors and other machinery are rarely used. Chemical synthesis of nitrogen fertilizer, which has been a critical ingredient for yield growth, requires a lot of energy. So does the transportation of inputs and output. Even where crop production is unmechanized, therefore, the cost of food rises and falls as the scarcity of energy varies.

Another linkage between energy and agriculture has to do with the search for alternative energy sources, which gains strength when conventional fuels are costly. Some of these alternative sources are agricultural, including the conversion of commodities such as sugar and corn into alcohol (or ethanol) as well as the production of biodiesel from palm oil. The European Union has set a goal that biofuels should comprise at least 5.75 per cent of all transport fuel by 2010 (Commission of the European Communities, 2006). In the United States, the conversion of corn into ethanol is encouraged by import restrictions and subsidies, which cost the US treasury $7 billion per annum (Doornbosch and Steenblik, 2007). The effects of rising food and vegetable oil prices, and the consequent conversion of agricultural lands, on the world's poor have attracted substantial attention, with a leading UN official describing the diversion of land to oil palm as a 'crime against humanity' (Ferrett, 2007).

Observing that 'a moratorium on grain-based biofuels would quickly unlock these commodities for use as food', the director-general of the International Food Policy Research Institute contends that 'this measure might bring corn prices down globally by about 20 percent'. Joachim von Braun also says that wheat prices would fall by 10 per cent if biofuel development ceased (von Braun, 2008). However, biofuel development is by no means the only cause of higher food prices. Export restrictions have also had an impact. According to the World Bank (2008a), more than 30 nations, including several with the potential to be major suppliers in international commodity markets, adopted export restrictions in late 2007 and early 2008. As a result of these restrictions, prices climbed even higher, to the detriment of food consumers everywhere.

The recent global rise in food prices is the joint outcome of yield slow-downs, diversion of land and cereals to non-staple food uses, export restrictions and other interventions. Although hard data are not yet available, there is a wealth of informal evidence that rising food prices have pushed large numbers of Asia's poor and near-poor into higher risk of food insecurity, offsetting the positive effects of rising employment and labour incomes even in the fastest-expanding economies. For Asia's food-insecure populations, a 'perfect storm' of purchasing power failure looms, if the current global slow-down destroys jobs just as other factors retarding food supply growth keep food prices at or near their current high levels.

The Role of Speculation[19]

While the sections above have developed possible reasons why food prices have soared up in recent years, many observers have pointed out that the speed of the price increases seemed faster than could be explained by them. These observers pointed out to speculation as an additional explanatory factor, though speculative activity cannot be entirely disassociated from the other factors mentioned above. The dramatic drop in food grain stocks in recent years, the

[19] I thank Alberto E. Isgut for his inputs in developing this section. This section is also based on Noemi Pace et al. (2008).

world over, has inevitably reduced the capacity of governments to play a significant role in stabilizing market prices of food grains. It is in periods of low reserves, when speculative attacks become more likely. Because building food reserves necessarily take time, speculators to bet that prices will continue going up at least in the short run. However, experts have different opinions about the relationship between food commodity derivatives markets and price increase and volatility. At the one end some experts argue that speculations by large or small traders do not cause sharp changes in prices or their volatility and, on the contrary, may provide a useful price discovery role (see Chatrath and Song, 1999; Garcia, Leuthold and Zapata, 1986; Haigh, Hranaiova and Overdahl, 2007 and Irwin and Yoshimaru, 1999). At the other end, others maintain that volume of trading in commodities futures affects commodity prices. For example, for Raizada and Sahi (2006), higher volumes in futures markets had significant causal impact on inflation and for Chan, Fung and Leung (2004) the volume of commodities derivatives trading has a positive effect on volatility in the Chinese futures exchanges. Analysis of Yang, Balyeat and Leatham (2005) examined the lead-lag relationship between futures trading activity and cash price volatility for some agricultural commodities and concluded that the causality running from unexpected futures trading volume to cash price volatility of such commodities is typically positive. Thus an increase in cash price volatility may be seen as being caused by an increase in unexpected trading volume.

For sure, commodities have become more and more popular for financial investors in since the crisis of 2007–08. For example, the stock of futures commodity contracts increased by $70 billion more since at the end of 2007 to $400 billion at the end of March 2008. This was twice as much commodity contracts as at the end of 2005 (International Food Research Institute, 2008).

Significantly enough it has relatively recently been seen that a new financial investors, consisting of index funds, hedge funds and exchange traded funds (ETFs), have resorted to heavy investments in futures commodity contracts (often trying to replicate the performance of major commodity-price indexes, such as the Standard & Poor's or Goldman Sachs Commodity Index) as a way to diversify wider portfolios of stocks and bonds, and believed that commodity markets were experiencing 'super cycle'—a long-term trend that would drive prices higher for years to come (International Food Research Institute, 2008. See also Iain Macwhirter, 2008).

Speculation in futures markets per se is acceptable as speculators play a useful role in providing liquidity and facilitating the process of price discovery in the markets where they operate. The problem is that the new breed of futures markets speculators representing index funds, hedge funds, and ETFs, tend to move in and out of the market, not in response to supply and demand developments, according to pre-specified investment targets and have tended to bet only that commodity prices would rise (International Food Research Institute, op cit.)

In a testimony before the Committee on Homeland Security and Governmental Affairs United States Senate, it has been testified that a particular category of investors (comprising corporate and government pension funds, sovereign wealth funds, university endowments and other institutional investors) are contributing to food and energy price inflation. (The Testimony by Michael Masters is available at: http://hsgac.senate.gov/public/_files/052008Masters.pdf)

Collectively, this class of investors now generally accounts for a larger share of outstanding commodities futures contracts than any other market player. A very recent OECD study also states that the sharp increase of prices may be caused by an increasingly large long position (buying contract) placed by institutional investors.

This analysis raises some important issues. If speculation, rather than supply/demand factors, is a major cause, then prices would be expected to fall significantly over the following months.

The key question is: with this substantial influx of new speculators in commodities futures markets, who are betting that commodity prices will increase in future, shouldn't they in fact push those prices up? The answer to this question has been heatedly debated during and since 2008. To better understand the gist of it, consider the following old Wall Street story:

> On a slow afternoon, trader A decided to open a market for a can of sardines. Bidding started at $1. B bought it for $2 and sold it to C for $3. D and E decided to get into the act, with the result that E became the owner for $5. E decided to open the can and discovered

the sardines had gone bad. He went back to A to get his money back, protesting that the sardines were rotten. A smiled broadly, and said, 'You don't understand. Those were trading sardines, not eating sardines. (Smith, 2008)

The argument is that speculation is unimportant because the futures speculators will never take delivery of the actual food; but this is precisely the problem. Purchases of agricultural commodities futures contracts have traditionally been a way by which a limited number of traders stabilized future commodity prices and enabled farmers to finance investments in future crop production. Speculative purchases have no other purpose than to make money for the speculators, who hold their contracts to drive up current prices with the intention, not of selling the commodities in the real market, but of unloading their holdings onto an artificially inflated market, at the expense of the ultimate consumer.

The point often made is that commodities futures markets trade paper, not physical goods. Among other economists, Paul Krugman (2008) argued that for futures market speculation to influence commodities cash prices, future prices should induce hoarding by sellers. In other words, unless future prices affect demand and supply of physical goods, cash prices should not be influenced. But with food inventories at historical lows, he sees no evidence of hoarding in the data. However, hoarding behaviour may not be readily reflected in the inventory data. For instance, in situations where the price of a commodity increases sharply and abruptly, as was the case of rice in the first four months of 2008, households have an incentive to stock up as much rice as possible to protect themselves from future expected increases. At that time, illegal hoarding activities by large warehouses were detected in countries such as India and the Philippines (International Food Research Institute, 2008). Moreover, in the case of oil, an expectation of fast-increasing prices creates an incentive to hold oil underground, so no increase in inventories should take place (Davidson, 2008).

The other argument against the speculation hypothesis is based on the observation that index fund investors are not taking delivery and owning stocks of commodities (Irwin, 2008). As explained in Box 2.7, speculators' participation in futures market consists in

---

**BOX 2.7   Futures Commodity Contracts,
Hedgers and Speculators**

Futures commodity contracts have been traded at the Chicago Board
of Trade since 1865, providing an important risk management tool for
participants in commodity markets. 'A futures contract is a standard-
ized contract, traded on a futures exchange, to buy or sell a certain
underlying instrument at a certain date in the future, at a specified
price. The future date is called the delivery date or final settlement date.
The pre-set price is called the futures price. The price of the underlying
asset on the delivery date is called the settlement price.'

A major purpose of these contracts is to protect buyers of commodities from future price increases or to protect sellers from future
price drops. For instance, a candy maker might buy sugar and cocoa
futures contracts to lock in a price for some portion of its requirements
for these ingredients. If the market price rises, the buyer is insured
from the financial loss of paying a higher market price at the delivery
date. If, instead, the market price drops, the buyer loses money on its
futures contracts, but this loss can be thought of as the cost of insurance
(Millman, 2008).

Market participants such as the candy maker in the example above,
or a farmer selling wheat ahead of the harvest, are called hedgers—their
goal is to hedge out the risk of price changes. While, in principle, all
participants in the futures markets can be hedgers, it would be difficult for the market to function that way, as there might not be enough
participants to match demand and supply for futures.

Another type of market participants called speculators play the very
useful role of providing liquidity to these markets. Speculators are
defined as market participants 'who seek to make a profit by predicting
market moves and buying a commodity "on paper" for which they have
no practical use' (Millman, 2008). These market participants are willing
to take on risks, becoming the counterparties to futures contracts with
hedgers or other speculators. In doing so, they provide liquidity to the
market, making it work more efficiently. They also facilitate the process
of price discovery, by which all available information that can affect
future demand or supply of a commodity is factored in to determine
its future price (ibid.).

Source: NationMaster.com, Encyclopedia. http://www.nationmaster.com/
encyclopedia/Futures-contract (accessed on 23 October 2008).

BOX 2.8   Crop Worth ₹55,600 Crores (556 Billion) Wasted Every Year

NEW DELHI: For a country reeling under an unprecedented food crisis, here is an alarming figure: India loses ₹55,600 crores worth of crops each year after harvesting. This has been admitted by the government before the Parliamentary Standing Committee on Agriculture. Food-grain worth ₹16,500 crores, roughly 10 per cent of the food-grain produced, is lost after harvesting every year. Considering that India has turned into a net importer of pulses as its crop productivity declines, it is shocking that ₹2,000 crores or 15 per cent of the pulse crop is destroyed each year. And, while the government, fearing that staple crop production is failing, pushes for a horticultural mission to compensate for the loss in agricultural growth, 30 per cent of all fruits produced, roughly worth ₹13,600 crores, go to waste. The losses in vegetable crops are as bad: 30 per cent of the vegetables the country produces, worth ₹14,100 crores, are lost due to mismanagement after they are harvested. Livestock and fisheries produce worth ₹8,400 crores too are lost annually.

The Parliamentary Standing Committee, reacting to the information, has reported, 'The wastage of agricultural produce is massive, processing levels are very low, around 2 per cent for fruits and vegetables, 26 per cent for marine, 6 per cent for poultry and 20 per cent for buffalo meat as against 60–70 per cent in developed countries.' While a certain percentage of post-harvest loss is considered natural, the high rates prevailing in India, studies by FAO and others show, are close to the losses recorded even in African nations. Bad harvest timing, inefficient machinery, lack of storage facilities, contamination, inordinate exposure to heat, cold and lack of moisture are the major reasons experts cite for the losses which can be prevented to some extent.

Source: Times of India, 17 April 2008.

---

buying future contracts with the expectation of selling them at a higher price. But for this to happen, there must be other market participants willing to buy the futures at a higher price, which will only occur if cash and futures prices do not deviate too much from each other around the time of delivery. In normal times, arbitrageurs ensure that the two prices converge, although future prices are usually a little higher than cash prices to cover costs of delivering the physical commodity. However, researchers have found evidence of an abnormal relationship between futures and cash prices around the time of delivery over 2006–08 (ibid.). During that period, the future price was occasionally substantially higher than the cash price at the time of delivery. A large gap between futures and cash prices is usually interpreted as a signal for commodity inventory holders to increase their inventories—that is, to hoard (ibid.). Thus, the possibility of futures speculation influencing cash prices is real.

While there is no hard proof of excessive speculation in futures markets, a major concern is that futures markets have not been working as efficiently as they should in facilitating risk management and the process of price discovery. More economic research is needed to fully understand the impact on prices resulting from the large increase in food commodity speculation in recent years. But there is compelling evidence that increased speculation is causing adverse impacts on global food prices, which makes a case for the commodities futures market to be regulated more effectively.

Transportation and Storage

The inadequacy and low quality of post-harvest systems have the potential to decrease the availability of food-grains, and at a far higher environmental cost. Though lack of water, energy, labour and land are the key factors that limit food production around the Asia-Pacific region, lack of efforts to conserve whatever food is already produced, and to bring more of it to market, is an important cause of food insecurity.

Many Central Asian countries have grain storage facilities dating back to the 1930s in design and in construction, and they are inadequate to meet the challenges of the current time. The state of most vegetable and cold stores are—if available—even worse. According to one estimate carried out 10 years ago, losses of cereal grains between farm and table in Ukraine was between 25 per cent and 60 per cent (Williams, 2009). From more recent visits to the region, Williams has 'no reason to believe that these loss rates have improved significantly' (ibid.). At current levels of grain production at 60 million tons in Ukraine and the Russian Federation, the loss from storage of food-grains would be some 15 to 25 million tons annually. Losses in fruit and vegetable crops due to poor or paucity of storage facilities are estimated to be much higher, but, because most of these are grown by small-scale producers, the losses remain largely unmeasured.

In South Asia the situation is as bad. Currently, India produces 178 million tons of rice and wheat (Dev, 2009) and has 50 million tons of food reserves, but admits to losing some 21 million tons of wheat (which is almost equal to the entire production of wheat in Australia) and 40 million tons of fruits and vegetables annually due to inadequate storage and distribution systems. That is, the equivalent of almost 42 per cent of wheat buffer stock is wasted due to faulty storage and transportation. A lack of controlled atmosphere and refrigeration equipment leads to spoilage losses of up to 33 per cent of perishable food (ESCAP, 2008b). In India, various research studies by the Economic Times Intelligence Group and the Investment Information and Credit Rating Agency reveal that large quantities of grain are wasted due to improper handling and storage, pest infestation, poor logistics, inadequate storage and a lack of transport infrastructure. Pakistan and Bangladesh have old storage infrastructure where rats generally have field days. Pakistan has been debating the improvement of its national storage system for many years, but has only just started to implement the programme. In Afghanistan, because the infrastructure is in such poor condition, farmers are forced to sell their crops straight off the fields into a saturated market at whatever price they can get. The off-farm price for first-grade onions was $7–$9 per ton in 2008, which is ridiculously low. By providing groups of small farmers with storage facilities, the same onions could be sold three months later for $375 per ton, helping alleviate rural poverty.

In South-East Asia, an estimated 37 per cent of rice is lost between field and table. In China the figure is up to 45 per cent, and in Vietnam it can reach 80 per cent, which adds up to 150 million tons of rice each year, representing a massive wastage of resources. Inadequate logistics systems not only increase costs but also impact the availability of food to consumers. According to a report by the Pacific Economic Cooperation Council, China's cold storage capacity is estimated to cover only 20–30 per cent of demand (ESCAP, 2008b).

The purchasing policies of major supermarkets for fresh produce encourage waste. They refuse to sign supply contracts with farmers, preferring instead to operate under 'supply agreements' where the benefits are heavily stacked in favour of the supermarkets. Penalties are imposed on farmers for any failure to deliver agreed quantities of fresh fruits and vegetables during the year, which forces farmers to grow much more crop than they need as a form of insurance against poor weather and other factors that may reduce the yield. As a result, some 30 per cent of the UK vegetable crop is never harvested. Even worse, 30 per cent of what is harvested never reaches the supermarkets due to trimming, quality selection, etc., and of the quantity that does reach the supermarket, between 30 per cent and 50 per cent is thrown away by the final purchaser. So increase in production cannot take us very far when, under current commercial practices, 71 per cent of the crops produced are simply thrown away. Even if production were to be doubled, the actual increase in useful food supply would be around 30 per cent.

Development Policy Interventions

Though Chapter 4 will deal with policies that affect food security, for completeness we have to discuss, albeit briefly, agricultural and trade policies here. Economic growth has indirect yet powerful influences on poverty. At the aggregate level, economic access to food never improves, and often deteriorates, in the absence of sustained economic growth. Despite several decades of economic growth in most of Asia, some nations have not put in place the macroeconomic and other policies required for sustained economic expansion and enhancement of economic access to food.

Agricultural Policies

Admittedly, agricultural policy measures have a strong bearing on food insecurity. In contemporary Asia, the two countries with the greatest risk of starvation, North Korea and Myanmar, are both characterized by state control of markets. Though some of these interventions are intended, paradoxically, to ensure food security for part of the population—the urban part—by holding down food prices or otherwise suppressing markets for edible goods, the results have been almost the reverse: food insecurity. Myanmar, for example, formerly one of the world's largest rice exporters, is now one of the poorest countries in Asia and indeed in the world, with high rankings on virtually all correlates of food insecurity. North Korea, which professes a state ideology of 'self-reliance' (*juche*), has paradoxically gained the distinction of becoming the world's largest recipient of rice donations.

Inflation—a regressive tax—has a disproportionate impact on the purchasing power of the poor and their economic access to food. Inflation in the developing countries of Asia and the Pacific rose from a low of 3 per cent a year in the 1960s to more than 10 per cent in the 1970s and 12 per cent in the 1980s and 1990s. Rates have since come down to about 6 per cent, but they remain high in some countries like Pakistan, Sri Lanka, Lao People's Democratic Republic, Samoa and Tonga. Research has found that low inflation has a positive impact on poverty reduction (Chen and Ravallion, 2008). The inflation–poverty nexus signals macroeconomic policy implications for food security, particularly in South Asia and Central Asia, which have higher inflation than East, North-East and South-East Asia. The agricultural sector with low yields bears the brunt of inflation-induced investment cuts.

High interest rates prevailing in many countries in Asia over the last decades has reduced the borrowing capacity of farmers, especially small-scale farmers, curtailing investment and farm cultivation, ultimately showing up as reduced food production. The real lending rate climbed to a historical high exceeding 8 per cent in the mid-1990s and remains high even now. North and Central Asia, the Pacific Islands and South Asia have had higher real interest rates than East, north-east and South-East Asia.

Other elements of macroeconomic policies, such as exchange rates and debt, have also affected the poor and the food-insecure. Indeed, agricultural taxation comes mainly through exchange rate policies and industrial protection, but, because food-insecure people depend heavily on agriculture, they lose the most.

Direct and indirect taxations of agriculture were common in many countries from the 1960s to the 1980s, reaching 40 per cent in some countries and slowing both agricultural growth and overall growth. In Asia, as in other regions, indirect taxation on agriculture, such as through price intervention, is more than twice direct taxation. Access to finance by rural farmers has also been curtailed, particularly since the structural adjustment programmes of the 1980s and the phasing out of subsidized credit schemes. With the changes in monetary policy, the agricultural refinance schemes operated by central banks have ceased. Price stability has become the main objective. Many central banks now set rediscount rates and avoid directly supporting specific sectors. Commercial bank lending for agriculture is naturally limited due to the low returns and lack of collateral. Many countries—such as Indonesia, Malaysia, Thailand and Vietnam—tend to use voluntary savings for financing agriculture. Thailand and Vietnam have had issued bonds for agricultural finance, while Pakistan has used equity issues in recent years, but the amounts remain small (FAO, 1999).

As stated earlier, limited spending on agricultural R&D and extension has constrained productivity growth and hence food availability. Agricultural R&D is one of the main sources of productivity growth. While expenditure on R&D in the Asia-Pacific region has increased gradually, in some countries it has either declined or remained stagnant. In Thailand for example, it has remained more or less stagnant, at 0.4–0.5 per cent since the 1970s, with a small recent improvement. India recently increased its R&D expenditure from 0.18 per cent of agricultural value added to 0.34 per cent.

Export Policies

Trade policies have an important bearing on eliminating food insecurity, especially for countries who seek self-reliance in food. When global food prices start to skyrocket, exporting countries resort to raising trade barriers. For example, in 2008, in the face of rising food prices, governments in a number of countries in Asia, including Bangladesh, Cambodia, China, India, Indonesia, Kazakhstan, Pakistan, Thailand and Vietnam, attempted to hold prices down and maintain domestic supplies by restricting or banning exports or imposing quotas or export taxes and other such measures.

Countries like Thailand, with a larger share of the global export market in rice, even suggested forming a cartel for rice producers, in line with OPEC, with Thailand, Vietnam, India, Pakistan and Cambodia as members. The proposal did not go through, however, as this could harm poor, rice-importing countries. Thus, the impact of export policies on food availability and hence hunger can work in either direction.

The impact of these measures on international prices needs a special mention. Take the case of rice. Only 7 to 9 per cent of the rice produced is traded internationally. So the impact of interventions is felt more quickly on the price of rice than on commodities in 'thicker' markets, such as those for wheat or corn (Slayton and Timmer, 2008). In 2007–08, as export restrictions were imposed by countries like India and Vietnam, food prices rose from $357 per ton in November 2007 to $481 per ton in February 2008, exacerbating the forces that perpetuate hunger. Then massive purchases by a major importer, the Philippines, led to further rise in food prices during the period under reference, to $1,015 per ton in April 2008. After reaching the peak in April 2008, food prices fell, partly because Japan released rice stocks in May 2008 (ESCAP, 2009), and India and Thailand announced that record rice harvests were expected around June 2008, brightening the prospects of increasing food supply and lowering food prices, ultimately lowering the prospect of more hunger. However, though restricting exports in several cases helped domestic consumers, it harmed domestic production as well as consumers in other countries, thus undermining regional and global efforts at reducing hunger. In some cases, restricting exports has even encouraged the smuggling of food, with implications for national food security.

Import Polices

Trade measures can also be taken by countries through import policies. For example, when food prices have been low, Philippines and Indonesia have at times attempted (though perhaps not with much success) to combine measures to boost local production with restrictions on cereal imports (FAO, 2005b). More recently, China has taken actions with seemingly similar objectives. Chinese national authorities, equating food security and self-sufficiency, decreed the preservation of 120 million hectares of farmland (He, 2007) to be their top priority. In fact, keeping food prices high may actually undermine efforts to eliminate food insecurity in the long run, because such measures distort producer incentives and undermine the operation of local markets. Therefore, these measures may have lowered food availability and hence contributed to increasing food insecurity. Import restrictions have also proved expensive for food-insecure people because they are denied the advantage of higher economic access to food through cheaper imports. Similarly, when international food prices are high, food-importing countries reduce tariffs. For example, in response to the price surge of 2007 and 2008, a number of countries (including Azerbaijan, Bangladesh, India, Indonesia, the Philippines, the Republic of Korea, Solomon Islands and Timor-Leste) opted to reduce import tariffs and other import surcharges, along with domestic taxes, on selected food crops including rice, maize, wheat and soybean. However, the net impact of these measures on food prices and food availability has usually been marginal, because by and large food tariffs were relatively low, and they have generally been coming down further over the past years—now averaging around 10 per cent. Most low-income net food importers in Asia and the Pacific, such as Kyrgyzstan, Tajikistan and Nepal, already had low applied tariffs on basic food. Therefore, even when they elected to remove tariffs on foodstuffs altogether, it did not make much difference to retail prices of food in their respective domestic markets, and hence on the economic access to food of food-insecure people.

While food crises such as the price surge of 2007 and 2008 may occasionally justify unusual measures, the record of large food importers in Asia indicates that striving for self-sufficiency may actually undermine efforts to achieve food security in the long run.

At the national level, policies on food transfers have also affected food security. According to FAO, from 2000 to 2004, Indonesia received an average of 131,000 tons of rice food aid per year,[20] the Philippines 67,000 tons, Cambodia 26,000 tons and North Korea a staggering 496,000 tons; the latter was equivalent to 71 per cent

---

[20] This figure does not include aid flows after the Indian Ocean tsunami of 26 December 2004.

and 47 per cent of the Asian and world totals, respectively (International Rice Research Institute, 2008).[21] Food aid is typically applied to break-in existing food shortages and *ex post* food insecurity. However, long-term development assistance and development policy include many other forms of transfer that are intended to tackle *ex ante* food insecurity by raising the purchasing power and hence economic access of the people. The Mahatma Gandhi National Rural Employment Guarantee Scheme in India and the food-for-work programmes in Bangladesh, Cambodia and Laos are cases in point. Moreover, many transfers to food-insecure households do not take the form of food and do not necessarily emanate from government or international aid agencies. During the 1980s' Philippine economic crisis, the proportion of households whose primary income source was transfers or other unearned income rose from 6 per cent (1971, the closest comparable figure) to 18.3 per cent in 1985, which was 'the year of shared poverty in post-war Philippine history' (Rao, 1988).

Incidence of Food Insecurity by Gender[22] and within Households

Insecurity caused by intra-familial distribution of power and gender discrimination are among the most important causes and determinants of food insecurity in the Asia-Pacific region. And these causes have been identified as being germane to greater food insecurity across sub-regions in Asia. For example, in East Asia and the Pacific, where two-fifths of the population were chronically or often hungry as recently as the early 1970s, the prevalence of food insecurity has fallen to approximately 10 per cent, while the prevalence is much higher in South Asia. This difference has to do in part with the direct and indirect impacts of gender-related discrimination. 'The extremely low status of women relative to men

---

[21] Authors' calculations from FAO data compiled by the International Rice Research Institute (http:irri.org/science/ricestat/data/may2008/WRS2008-Table15.pdf, accessed 4 November 2008).

[22] See Chapter 3 on gender and food security for more details.

in South Asia compared with that in Sub-Saharan Africa is thought to compromise women's own health, the subsequent birth weight of their children, and the quality of care their children receive' (Smith et al., 2003). Indeed, the common practice of a household's women eating after the men often leaves women the least to eat and only the leftovers, causing them to be food-insecure. This practice explains, in large part, why five out of every six Indian women are anaemic—as opposed to two out of every five women south of the Sahara—despite average earnings, educational attainment, access to clean water and other indicators of well-being being better in India.

These and related issues have attracted considerable debate in recent times, but a huge amount of work remains to be done. The issues, being very important, are the subject-matter of a separate chapter (Chapter 3 of this book), and are not pursued here any further.

Concluding Remarks

Food insecurity changes over time because of various processes, some in the agricultural sector and others having to do with the entire economy or population or with the natural environment. Growth in food demand in Asia and the Pacific is no longer driven solely by demographic expansion. Instead, this growth is very significantly a consequence of improved living standards, which have caused per capita consumption of livestock products and other edible goods to increase.

Since the 1960s, food supplies in Asia and the Pacific have grown faster than demand. Increases in crop and livestock output have resulted mainly from agricultural intensification during and since the Green Revolution, though the pace of yield growth has been slackening, partly because of environmental constraints on agricultural intensification and partly because of lagging support for agricultural research and development.

Degradation of lands, water pollution and loss in forest cover have taken a heavy toll on food security in the Asia-Pacific region. Natural disasters and the related loss in food availability and the obstacles these events cause in access to food are other major factors.

The role of exploitative intermediation in causing food insecurity is considerable in several countries of the region.

Any increase in the effective purchasing power of impoverished populations, whether resulting from an increase in earnings or transfers or from a decline in food prices, diminishes the incidence of food insecurity. Similar results are obtained by augmenting the effective exchange entitlements of those same populations, for example, through the provision of health and education services. When economic opportunities draw migrants out from impoverished areas, overall food insecurity improves. However, no such outcome is possible if food insecurity and desperation cause people to relocate to places where their economic prospects stand little chance of improvement.

Food insecurity was once endemic in Asia and the Pacific, but is now confined largely to specific subpopulations, as they cannot produce enough food to meet their consumption needs, and they cannot acquire enough food through alternative channels. In some cases, identification of vulnerable populations requires intra-household analysis, focusing for instance on the food insecurities faced by women, which is attempted here in chapter. In large measure, further progress toward food security in Asia and the Pacific requires initiatives that target the needs of these populations.

Appendix 2.1: A Brief Note on the Entitlement and Deprivation Thesis

Food insecurity has to be seen as the characteristic of the person not having enough to eat (Sen, 1981). Except in cases where an individual starves deliberately, to understand food security, one has to go into the structure of ownership of food. *Ownership relations* are one kind of entitlement relationship (ibid.), which, as accepted in a private ownership market economy, typically include the following entitlements, among others:

1. *Trade-based entitlement*, where an individual is entitled to own goods and services which s/he obtains by trading something owned by her/him with a willing party or parties.
2. *Production-based entitlement*, where an individual is entitled to own goods and services that he/she obtains by arranging their production, using resources owned by her/him or by hiring resources from willing parties meeting the agreed conditions of trade.

Production-based entitlement may also be construed as exchange with nature when considered in the context of food security.

3. *Own labour entitlement*, where an individual is entitled to her/his own labour power and thus to the trade-based and production-based entitlements related to the individual's labour power.

4. *Inheritance and transfer entitlement*, where an individual is entitled to own commodities that are willingly given to the individual by another who legitimately owns them, which could be either during the lifetime of the latter if commodities are transferred in the nature of a gift, or after the latter's death, in case they are transferred by way of a will, inheritance or bequest.

5. *Endowment entitlement* is what an individual owns. It is also called *direct entitlement*.

6. *Exchange entitlement* is the set of all alternative bundles of commodities that an individual can acquire through trading (trade-based entitlement) or through production (production-based entitlement) or through some combination of the two.

7. *Usufruct entitlement* is the bundle of commodities that an individual can acquire through exchange or production or through some combination of the two, from nature.

References

Abbott, David, and Steve Pollard. 2004. *Hardship and Poverty in the Pacific*. Manila: Asian Development Bank.

ADB (Asian Development Bank). 2000. *The Growth and Sustainability of Agriculture in Asia*. Manila: ADB.

An, Shuqing, Zhongsheng Wang, Changfang Zhou, Baohua Guan, Zifa Deng, Yingbiao Zhi, Yuhong Liu, Chi Xu, Shubo Fang, Zheng Xu, Haibo Yang, Fude Liu, Jianwei Zheng and Hongli Li. 2006. 'The Headwater Loss of the Western Plateau Exacerbates China's Long Thirst'. *AMBIO: A Journal of the Human Environment*, vol. 35, no. 5, pp. 271–72.

Asia Foundation. 2007. *Understanding Bonded Child Labour in Asia: An Introduction to the Nature of the Problem and How to Address it*. Bangkok: Child Workers in Asia Foundation.

Ban, K. 2008. 'Address by the Secretary-General to the High-Level Conference on World Food Security'. UN Food and Agriculture Organization, Rome, 3 June.

Bhattamishra, Ruchira and Christopher B. Barrett. 2008. *Community-based Risk Management Arrangements: An Overview and Implications for Social Fund Programs*. Washington, D.C.: The World Bank.

Black, Richard, Claudia Natali and Jessica Skinner. 2005. 'Migration and Inequality'. Second Draft, World Development Report 2006 Background Papers, 20 January. Development Research Centre on Migration, Globalization and Poverty, University of Sussex.

Brown, Lester. 2003. 'World Creating Food Bubble Economy Based on Unsustainable Use of Water'. Earth Policy Institute Eco-Economy Update, no. 2, 13 March. http://www.earth-policy.org/Updates/Update22.htm (accessed on 19 October 2005).

———. 2008. *PLAN B 3.0 Mobilizing to Save Civilization*. New York: Earth Policy Institute, and London: W. W. Norton.

CARE International. 2009. 'Nargis Emergency Response'. Interim Report, 3 March. http://expert.care.at/uploads/media/MMR812_CARE_InterimReport_April09.pdf.

Chan, K. C., H. G. Fung and W. K. Leung. 2004. 'Daily Volatility Behavior in Chinese Futures Markets'. *Journal of International Financial Markets, Institutions and Money*, vol. 14, no. 5, pp. 491–505.

Chatrath, A., and F. Song. 1999. 'Futures Commitments and Commodity Price Jumps'. *The Financial Review*, no. 34, pp. 95–112.

Chen, S., and M. Ravallion. 2008. 'The Developing World Is Poorer Than We Thought, But No Less Successful in the Fight against Poverty'. Policy Research Working Paper 4703. World Bank, Washington, D.C.

Commission of the European Communities. 2006. 'Communication from the Commission: An EU Strategy for Biofuels' (SEC 142). Brussels, 8 February.

Dalrymple, D. 1985. 'The Development and Adoption of High-Yielding Varieties of Wheat and Rice in Developing Countries'. *American Journal of Agricultural Economics*, vol. 67, no. 5, pp. 1067–73.

David, C. C., and K. Otsuka (eds). 1993. *Modern Rice Technology and Income Distribution in Asia*. Boulder: Lynne Reinner.

Davidson, Paul. 2008. 'Crude Oil Prices: "Market fundamentals" or speculation?' *Challenge*, vol. 51, no. 4, pp. 110–18.

Deaton, Angus. 1997. *Analysis of Household Surveys: A Microeconomic Approach*. Baltimore: Johns Hopkins University Press.

Dev, Mahendra. 2009. 'Social Protection: Food Based Programme in India'. http://www.socialprotectionasia.org/Conference-Unesco-Icssr/India-Mahendra-Dev.ppt#418,36,Conclusion

Dixon, J., and A. Gulliver, with D. Gibbon. 2001. *Farming Systems and Poverty: Improving Farmers' Livelihoods in a Changing World*. Rome: Food and Agriculture Organization.

Doornbosch, R., and R. Steenblik. 2007. *Biofuels: Is the Cure Worse than the Disease?* Paris: Organization for Economic Cooperation and Development.

ECLOF (Ecumenical Church Loan Fund) International. 2008. 'Micro-Loans to 56 Rice Farmers in the Philippines'. http://www.globalgiving.com/pr/1200/proj1101a.html#progressReports (accessed on 19 January 2012).

Economist. 2008. 'India's Civil Service: Battling the Babu Raj', and 'India's Economy: What's Holding India Back?' 6 March.

ERS-USDA (Economic Research Service of the US Department of Agriculture). 2003. *International Food Consumption Patterns*. Washington, D.C. http://www.ers. usda.gov/data/InternationalFoodDemand/ (accessed on 19 January 2012).

ESCAP (Economic and Social Commission for Asia and the Pacific). 1995. *State of the Environment in the Asia-Pacific, 1995*. Bangkok: UNESCAP.

ESCAP (Economic and Social Commission for Asia and the Pacific). 2003. 'Emerging Issues and Developments at the Regional Level: Emerging Social Issues', Item 4 (c) of the Provisional Agenda—Shanghai Implementation Strategy: Regional Implementation Strategy for the Madrid International Plan of Action on Ageing 2002 and the Macao Plan of Action on Ageing for Asia and the Pacific 1999. ECOSOC, UNESCAP, E/ESCAP/1280, 27 February 2003. Fifty-Ninth Session, 24–30 April 2003, Bangkok. http://www.unescap.org/esid/psis/ageing/strategy/shanghai.pdf (accessed on 18 January 2011).

———. 2005. *State of the Environment in the Asia-Pacific, 2005*. Bangkok: UNESCAP.

———. 2008a. *2008 ESCAP Population Data Sheet*, Bangkok: UNESCAP.

———. 2008b. 'Major Issues in Transport', Item 4 (b) of the provisional agenda—Transport and Poverty: From Farm to Market—Extending the Reach Of Logistics. ECOSOC, UNESCAP, E/ESCAP/CTR/2, 10 September 2008. Committee on Transport, First session, 29–31 October 2008, Bangkok. http://www.unescap.org/ttdw/ct2008/ctr_2e.pdf (accessed on 26 April 2012).

———. 2009. *Sustainable Agriculture and Food Security in Asia and the Pacific*. Bangkok: Economic and Social Commission for Asia and the Pacific.

FAO (Food and Agriculture Organization). 1999. *Better Practices for Agricultural Lending*. Rome: FAO.

———. 2002. *FAO Production Yearbook 2002*. Rome: FAO.

———. 2003. *State of the World's Forest 2003*. Rome: FAO.

———. 2005a. *Global Forest Resources Assessment 2005*. Rome: FAO.

———. 2005b. *The State of Food and Agriculture—Agricultural Trade and Poverty: Can Trade Work for the Poor*. Rome: FAO.

———. 2005c. He Changchui, FAO Assistant Director-General and Regional Representative for Asia and the Pacific, Opening Speech at the Workshop on Reducing Food Insecurity Associated with Natural Disasters in Asia and the Pacific, 27–28 January. http://www.fao.org/world/regional/rap/speeches/2005/20050127.html (accessed on 23 December 2008).

———. 2005d. *State of the World's Forests 2005*. Rome: FAO.

———. 2008. *State of Food Insecurity 2008*. Rome: FAO.

FAO (Food and Agriculture Organization) and WFP (World Food Programme). 2008. 'Special Report: FAO/WFP Crop and Food Security Assessment Mission to the Democratic People's Republic of Korea'. Mimeograph, 8 December. Rome: FAO and WFP.

Feffer, John. 2008. 'Military Spending in East Asia'. *Asian Perspective*, vol. 32, no. 4.

Ferrett, Grant. 2007. 'Biofuels "Crime against Humanity"'. BBC News, 27 October. http://news.bbc.co.uk/2/hi/americas/7065061.stm (accessed on 19 January 2012).

Fischer, G., and G. Heilig. 1997. 'Population Momentum and the Demand on Land and Water Resources'. *Philosophical Transactions of the Royal Society*, Series B, vol. 352, no. 1356, pp. 869–89.

Garcia, P., R. M. Leuthold and H. Zapata. 1986. 'Lead-lag Relationships between Trading Volume and Price Variability: New Evidence'. *Journal of Futures Markets*, vol. 6, no. 1, pp. 1–10.

Glewwe, P., and V. Linh. 2008. 'Impacts of Rising Food Prices on Poverty and Welfare in Vietnam'. Paper presented at the Midwest Conference on International Development Economics, Madison, 1–2 May.

Goonesekere, Savitri. 1999. 'Children, Law and Justice: A South Asia Perspective'. The International Journal of Children's Rights, vol. 7, no. 3, pp. 283–98.

Government of India. 2005. 'Press Note on Indebtedness of Farmer Household (January–December, 2003)'. New Delhi: National Sample Survey Organization, Ministry of Statistics and Programme Implementation, 3 May.

———. 2006. 'Report of the National Commission for Enterprises in the Unorganised Sector (NCEUS)'. New Delhi: Government of India.

———. 2007. 'Report of the Expert Group on Agricultural Indebtedness'. New Delhi: Ministry of Finance, Banking Division, July.

GSO (General Statistics Office of the Government of Vietnam). 2005. Vietnam Migration Survey: Life Course Events. Hanoi: GSO.

Gwatkin, Davidson, Michel Guillot and Patrick Heuveline. 2000. The Burden of Disease among the Global Poor: Current Situations, Future Trends and Implications for Strategy. Washington D.C.: The World Bank.

Haigh, M., J. Hranaiova and J. Overdahl. 2007. 'Hedge Funds, Volatility, and Liquidity Provision in Energy Futures Markets'. Journal of Alternative Investments, vol. 9, no. 4, pp. 10–38.

Hayami, Y., and V. Ruttan. 1985. Agricultural Development: An International Perspective. Baltimore: Johns Hopkins University Press.

He, S. 2007. 'Government Holds Fast to Arable Land'. China.Org.Cn, 16 August. http://www.china.org.cn/english/government/221140.htm (accessed on 19 January 2012).

Holmes, R. and N. Jones. 2009. 'Putting the Social Back into Social Protection: A Framework for Understanding the Linkages between Economic and Social Risks for Poverty Reduction'. ODI Background Note, ODI London.

Holzmann, Robert, and Steen Lau Jorgensen. 2000. 'Social Rsk Management: A New Conceptual Framework for Social Protection and Beyond'. Social Protection Discussion Paper 0006, World Bank, Human Development Network, Washington DC.

Hussain, Akmal. 2000. Impact of Agricultural Growth on Changes in the Agrarian Structure of South Asia. New Delhi: Konark Publishers Pvt Ltd.

International Federation of Red Cross and Red Crescent Societies. 2005. 'World Disaster Report: Focus on Information in Disasters'. Geneva: International Federation of Red Cross and Red Crescent Societies.

———. 2007. 'World Disaster Report: Focus on Discrimination'. Geneva: International Federation of Red Cross and Red Crescent Societies.

International Food Research Institute. 2008. 'Speculation and World Food Markets'. IFPRI Forum, July.

International Rice Research Institute. 2008. World Rice Statistics. www.irri.org (accessed on 3 November 2008).

IRIN (Integrated Regional Information Network). 2008. 'Afghanistan: Food Insecurity May Cause Deaths This Winter—Government'. 27 November. http://www.irinnews.org/Report.aspx?ReportId=81692 (accessed on 8 July 2009).

Irudaya Rajan, S., U. S. Mishra and P. S. Sharma. 1999. *India's Elderly: Burden or Challenge?* New Delhi/Thousand Oaks/London: SAGE Publications.

Irwin, S. H., and S. Yoshimaru. 1999. 'Managed Futures, Positive Feedback Trading and Futures Price Volatility'. *Journal of Futures Markets*, vol. 19, no. 7, pp. 759–78.

Irwin, Scott. 2008. 'Commodity Arbitrage'. Econbrowser, 16 April. http://www.econbrowser.com/archives/2008/04/commodity_arbit.html (accessed on 6 November 2008).

Jalan, Jyotsna, and Martin Ravallion. 2002. 'Household Income Dynamics in Rural China'. Working Papers UNU-WIDER Research Paper, World Institute for Development Economic Research (UNU-WIDER).

Jazairy, Idriss. 1991. *The State of World Rural Poverty: An Inquiry into its Causes and Consequences.* Rome: International Fund for Agriculture and Development (IFAD).

Kabeer, N., and T. V. A. Tran. 2000. 'Leaving the Rice Fields, But Not the Countryside: Gender, Livelihood Diversification and Pro-poor Growth in Rural Viet Nam'. Occasional Paper No. 13, UN Research Institute for Social Development, Geneva.

Koomson, George. 2000. 'Mining Civil Wars: Peace or Plunder'. *African Agenda*, vol. 3, no. 1, pp. 9–11.

Krugman, Paul. 2008. 'Running out of Planet to Exploit'. *New York Times*, 21 April.

Kulkolkarn, K., T. Potipiti and I. Coxhead. 2007. 'Immigration and Labor Market Outcomes in Thailand'. Manuscript, Thammasat University and University of Wisconsin-Madison.

Lavoy, Peter R. 2003. 'Managing South Asia's Nuclear Rivalry: New Policy Challenges for the United States'. *The Nonproliferation Review*, Fall–Winter.

Macwhirter, Iain. 2008. 'The Trading Frenzy That Sent Prices Soaring'. *New Statesman*, 17 April.

Manning, C., and P. Bhatnagar. 2004. 'The Movement of Natural Persons in Southeast Asia: How Natural?' Working Paper in Trade and Development No. 2004/02, Division of Economics, Australian National University.

Mansfield, David. 2008. 'The Role of Opium as a Source of Informal Credit in Rural Afghanistan'. Working Paper # 4, UNODC, Afghanistan.

Mazhar, Farhad, Daniel Buckles, P.V. Satheesh and Farida Akhter. 2008. *Food Sovereignty and Uncultivated Biodiversity in South Asia: Essays on the Poverty of Food Policy and the Wealth of the Social Landscape.* New Delhi: Academy Foundation in association with International Development Research Centre (IDRC).

Millennium Ecosystem Assessment. 2005. *Ecosystems and Human Well-Being: Wetlands and Water Synthesis.* Washington, D.C.: World Resources Institute.

Millman, Gregory, J. 2008. 'Futures and Options Markets'. http://www.econlib.org/library/Enc/FuturesandOptionsMarkets.html (accessed on 23 October 2008).

Moser, C. O. N. 1998. 'The Asset Vulnerability Framework: Reassessing Urban Poverty Reduction Strategies'. *World Development*, vol. 26, no. 1, pp. 1–19.

Mukherjee, Neela. 1993. *Participatory Rural Appraisal.* New Delhi: Concept Publishing Company.

Mukherjee, Amitava, and Neela Mukherjee. 2008. 'Rural Women and Food Insecurity: A Longitudinal Study', in Neela Mukherjee, Meera Jaiswal and Bratindi Jena (eds), *Learning to Share Experiences and Reflections on PRA and Other Participatory Approaches.* New Delhi: Concept Publishing Company.

Munich Re Group. 2005. *Topics Geo Annual Review: Natural Catastrophes 2004.* Berlin: Munich Re Group.

OCHA. 2005. Indian Ocean Earthquake Tsunami, 26 December 2004. http://relief web.int/node/221627

Olimova, Saodat K. 2006. *Remittances of Labour Migrants in Khatlon Region, Tajikistan.* Dushanbe: International Organization of Migration, Sales No. E08.II.F.7.

Otsuka, K. 2005. 'The Green Revolution in Asia and Its Sustainability'. http:// www.fasid.or.jp/daigakuin/sien/kaisetsu/051125report.pdf (accessed on 3 November 2008).

Pace, Noemi, Andrew Seal and Anthony Castello. 2008. 'Has Financial Speculation in Food Commodity Markets Increased Food Prices?'. Field Exchange Issue 34, The Emergency Nutrition Network (ENN), Oxford, October 2008.

Pakistan Water Gateway. 2001. 'Water Quality Survey of Rawalpindi and Islamabad'. http://www.waterinfo.net.pk/doc1.htm (accessed on 1 October 2005).

Pandey, S. 2008. 'The True Price of Rice'. *Rice Today,* January–March, pp. 36–37. www.irri.org (accessed on 3 November 2008).

Pardey, P., and N. Beintema. 2001. *Slow Magic: Agricultural R&D a Century after Mendel.* Washington, D.C.: International Food Policy Research Institute.

Phan, D., and I. Coxhead. 2008. 'Interprovincial Migration and Inequality during Vietnam's Transition'. Staff Paper Series No. 507, Department of Agricultural and Applied Economics, University of Wisconsin, Madison.

Prasad, K., P. Belli and M. Das Gupta. 1999. 'Links Between Poverty, Exclusion and Health', World Bank Background Paper for the World Development Report 2000/2001.

Prins, Nomi. 2008. 'Oil Speculation: Why We Don't Have Answers'. CNNMoney. com, 8 July. http://money.cnn.com/2008/07/07/news/economy/oil_prins. fortune/ (accessed 23 October 2008).

Raizada, G. and G. S. Sahi. 2006. 'Commodity Futures Market Efficiency in India and Effect on Inflation'. Working Paper, Indian Institute of Management (Lucknow), India.

Rao, V. V. B. 1988. 'Income Distribution in East Asian Developing Countries'. *Asian Pacific Economic Literature,* vol. 2, no. 1, pp. 26–45.

Ravallion, M. 1987. *Markets and Famines.* Oxford: Clarendon Press.

Ridge, Milan. 1999. 'Poverty an Unhealthy Start to Ageing'. *The Sunday Nation,* Singapore, 4 July, pp. A5.

Sen, Amartya. 1981. *Poverty and Famines.* New York: Oxford University Press.

Shariff, Abusaleh. 1999. India Human Development Report. New Delhi: Oxford University Press.

Short, P. 1999. *Mao: A Life.* New York: Henry Holt.

Skons, Elisabet, Wuyi Omitoogun, Sam Perlo-Freeman and Peter Stalenheim. 2003. 'Military Expenditure'. *Stockholm International Peace Research Institute Yearbook 2003.* Oxford: Oxford University Press.

Slayton, T., and C. P. Timmer. 2008. 'Japan, China and Thailand Can Solve the Rice Crisis—But U.S. Leadership Is Needed'. Centre for Global Development Notes, Centre for Global Development, Washington, D.C., May.

Smith, L., U. Ramakrishnan, A. Ndiaye, L. Haddad and R. Martorell. 2003. 'The Importance of Women's Status for Child Nutrition in Developing Countries', Research Report 131, International Food Policy Research Institute, Washington, D.C.

Smith, Yves. 2008. 'Is the Commodity Boom Driven by Speculation?' *Naked Capitalism*, 19 May. http://www.nakedcapitalism.com/2008/05/is-commodities-boom-driven-by.html (accessed on 23 October 2008).

Stewart, Frances. 1993. 'War and Deprivation'. Developing Studies Working Papers No. 56, International Development Centre, Oxford.

Strait Times. 2004. 'Plight of Chinese Farmer', *Strait Times*, 5 September. http://www.hartfomd-hwp.com/archives/55/333.html (accessed on 19 January 2012).

Southgate, D., D. Graham and L. Tweeten. 2007. *The World Food Economy*. Malden: Blackwell Publishing.

UNDP (United Nations Development Programme). 1999. 'Human Development Report 1999'. New Delhi: Oxford University Press.

———. 2002. 'Human Development Report 2002'. New York: Oxford University Press.

———. 2004. 'North East Asian Dust and Sand Storms Growing in Scale and Intensity'. Press Release ENV/DEV/760 UNEP/216, 31 March.

———. 2007. 'Fighting Climate Change: Human Solidarity in a Divided World'. Human Development Report 2007/2008. New York: Palgrave Macmillan.

———. 2008. 'Tackling Corruption, Transforming Lives'. Asia Pacific Human Development Report. New Delhi: Macmillan India Ltd for UNDP Regional Centre, Colombo.

United Nations. 2006. 'International Migration Report: A Global Assessment'. Population Division, Department of Economic and Social Affairs. New York: United Nations.

United Nations ECE and FAO. 2005. *Forest Products Annual Market Review 2004–2005* (ECE/TIM/BULL/2005/3), Geneva, United Nations.

UNEP (United Nations Environment Programme). 2001. *Asia Pacific Environment Outlook*, Chapter 1, Bangkok.

UNICEF (United Nations Children's Fund). 1995. *The State of the World's Children*. Oxford: Oxford University Press for UNICEF.

Volunary Health Association of India. 1997. 'Report of Independent Commission on Health in India'. New Delhi: VHAI.

von Braun, Joachim. 2008. 'Global Response Needed to Rising Food Prices'. Press release, 16 May. http://www.ifpri.org/pressroom/briefing/global-response-needed-rising-food-prices (accessed on 6 March 2012).

Warford, Jeremy. 2004. 'Infrastructure Policy and Strategy in the East Asia and Pacific Region: Environmental and Social Aspects', in Asian Development Bank, Japan Bank for International Cooperation and the World Bank, *Connecting East Asia: A New Framework for Infrastructure*. Washington, D.C.: World Bank.

WHO (World Health Organization) and UNICEF (United Nations Children's Fund). 2000. 'Global Water Supply and Sanitation Assessment, 2000 Report'. Geneva and New York: Water Supply and Sanitation Collaborative Council. http://millenniumindicators.un.org/unsd/mi/mi_goals.asp (accessed on 19 January 2012).

Williams, David B. 2009. 'Merely Increasing Production Is Not the Way to Improve Food Security'. http://loopy232.webcrossing.com/?14@871.gIKOgi 2CSKy@.59c8c090/1 (accessed on 5 March 2012).

Woods, Stanley, Kate Sebastian and Sara J. Scherr. 2000. *Pilot Analysis of Global Ecosystem: Agroecosystems*. Washington, D.C.: World Resources Institute. http://www.ifpri.org/pubs/books/page.htm (accessed on 15 March 2006).

World Bank. 2000. 'World Development Report 2000'. Oxford: Oxford University Press for the World Bank.

———. 2001. 'Attacking Poverty: World Development Report 2000/2001'. Oxford etc.: Oxford University Press.

———. 2008a. 'Double Jeopardy: Responding to High Food and Fuel Prices'. G8 Hokkaido-Toyako Summit. http://siteresources.worldbank.org/NEWS/Resources/risingfoodprices_chart_apr08.pdf (accessed on 2 July 2008).

———. 2008b. *World Development Indicators Online*. http://ddp-ext.worldbank.org/ext/DDPQQ (accessed on 9 October 2008).

Worldwatch Institute. 2001. 'The Hidden Freshwater Crisis', *Earth Times*, San Diego, January. http://www.sdearthtimes.com/et0101/et0101s6.html (accessed on 10 January 2006).

Xie Yu. 2008. 'Land Erosion "Threat" to Food Supply'. *China Daily*, 22–23 November.

Yang, J., B. R. Balyeat and D. J. Leatham. 2005. 'Futures Trading Activity and Commodity Cash Price Volatility'. *Journal of Business Finance Accounting*, vol. 32, nos 1 & 2, pp. 297–323.

Zhai, F., and T. Hertel. 2004. 'Labor Market Distortions, Rural-Urban Inequality and the Opening of China's Economy'. Policy Research Working Paper 3455, World Bank, Washington, D.C.

Further Readings

ADB (Asian Development Bank). 2005. *Asian Development Outlook 2005: Promoting Competition for Long-Term Development*. Hong Kong: ADB.

Burkhard, James. 2008. 'The Price of Oil: A Reflection of the World'. Testimony before the Committee on Energy and Natural Resources, United States Senate, Cambridge Energy Research Associates, Washington D.C., 3 April.

Cassman, K. G., and P. L. Pingali. 1995. 'Intensification of Irrigated Rice Systems: Learning from the Past to Meet Future Challenges'. *GeoJournal*, vol. 35, no. 3, pp. 299–305.

Coxhead, I., and R. Øygard. 2007. 'Land Degradation', in B. Lomborg (ed.), *Solutions for the World's Biggest Problems: Costs and Benefits*. Cambridge: Cambridge University Press, pp. 146–61.

Curran, L. M., S. Trigg, A. McDonald, D. Astiani, Y. M. Hardiono, P. Siregar, I. Caniago and E. Kasischke. 2004. 'Lowland Forest Loss in Protected Areas of Indonesian Borneo'. *Science*, vol. 303, no. 5660, 13 February, pp. 1000–03.

Dursin, Richel. 2001. 'Indonesia: Shrimp Farming Destroying Mangroves'. Inter-Press Service, 16 March. http://forests.org/archive/indomalay/shfadest.htm (accessed on 23 December 2008).

Economist. 2008. 'Land Reform in China: Promises, Promises'. 16 October.

FAO (Food and Agriculture Organization). 1996. *Data of Food and Agriculture, 1996*. Rome: FAO.

———. 2008a. 'Myanmar's Food Bowl Devastated'. http://www.fao.org/news room/EN/news/2008/1000838/index.html (accessed on 22 December 2008).

International Monetary Fund (IMF). 2006. *World Economic Outlook*. Washington, D.C., April.

Irwin, Scott. 2008. 'The Misadventures of Mr. Masters: Act II'. Econbrowser, 17 September. http://www.econbrowser.com/archives/2008/09/scott_irwin_tak.html (accessed on 19 January 2012).

Molle, F., and J. Berkoff (eds). 2007. *Irrigation Water Pricing: The Gap between Theory and Practice*. Wallingford: CAB International Publishing.

Oldeman, L. R., R. T. A. Hakkeling and W. G. Sombroek. 1991. *World Map of the Status of Human-Induced Soil Degradation: An Explanatory Note*. Wageningen: International Soil Reference and Information Centre, and Nairobi: United Nations Environment Programme.

Pakistan Transparency International. 2008. 'Persistently High Corruption in Low-Income Countries Amounts to an "Ongoing Humanitarian Disaster"'. Press Release, 23 September. http://www.transparency.org.pk/PressRelease/PRE SS%20RELEASE%20CPI%202008.doc (accessed on 23 December 2008).

People's Daily On-Line. 2008. 'Survey: 90% Chinese Slate Local Officials for Inefficiency'. 18 February 18. http://english.people.com.cn/90001/90776/6356146.html (accessed on 19 January 2012).

Pingali, P., C. Marquez, A. Rola and F. Palis. 1995. 'The Impact of Long-Term Pesticide Exposure on Farmer Health: A Medical and Economic Analysis in the Philippines', in P. Pingali and P. Roger (eds), *Impact of Pesticides on Farmer Health and the Rice Environment*. Boston: Kluwer Academic Publishers.

Preston, B., R. Suppiah, I. Macadam and J. Bathols. 2006. 'Climate Change in the Asia Pacific Region'. http://www.csiro.au/resources/pfkd.html (accessed on 6 November 2008).

Quisumbing, A. E., L. R. Brown, H. Feldstein, L. Haddad and C. Peña. 1995. *Women: The Key to Food Security*. Washington, D.C.: International Food Policy Research Institute.

Ramesh, A. 1999. 'Priorities and Constraints of Post Harvest Technology in India', in Y. Nawa et al. (eds), *Post Harvest Technology in Asia*. Tokyo: Japan International Research Centre for Agricultural Sciences.

Rosegrant, M., X. Cai and S. Cline. 2002. *Global Water Outlook to 2025*. Washington, D.C.: International Food Policy Research Institute.

Roy, A. 2003. 'Fertilizer Needs to Enhance Production: Challenges Facing India', in R. Lal, D. Hansen, N. Uphoff and S. Slack (eds), *Food Security and Environmental Quality in the Developing World*. Boca Raton: Lewis Publishers.

Scherr, S. 1999. *Soil Degradation: A Threat to Developing-Country Food Security by 2020?* Washington, D.C.: International Food Policy Research Institute.

UNCTAD (United Nations Conference on Trade and Development). 2007. 'The Development Role of Commodity Exchanges', TD/B/COM.1/EM.33/2, UNCTAD, Geneva.

UN Millennium Project. 2005. 'Environment and Human Well-Being: A Practical Strategy: Report of the Task Force on Environmental Sustainability'. London: Earthscan Publications.

Winters, L. A. 2002. 'Trade Liberalization and Poverty: What Are the Links?' *The World Economy*, vol. 25, no. 9, pp. 1339–67.

Young, J. 2008. 'Speculation and World Food Markets'. *IFPRI Forum*, July.

3

Food Insecurities Faced
by Women and Girl Children

A child said, 'What is the grass?' fetching it to me with full hands,
How could I answer the child? I do not know what it is, any more than he.
I guess it must be the flag of my disposition out of hopeful green stuff
woven,
Or I guess it is the handkerchief of the Lord,
A scented gift that Remembrancer designedly dropped,
Bearing the owner's name somewhere in the corners,
That we may see and remark, and say 'whose?'

—Walt Whitman, *Leaves of Grass*

Some Conceptual Clarifications

The scented gift of the 'Remembrancer' has given the world, especially this part of the world, its principal food-grains. The grass has given rice, wheat, barley, oats, millets and fodder. They were the foods of antiquity. Two crops which have retained their importance from antiquity till today are rice and wheat. The Green Revolution in South Asia, as in other parts of the world, was aimed primarily at improving the productivity of these crops, and it did give the region the fruits of modern technology: better seeds, newer forms of fertilizers and more and newer methods of irrigation, as a consequence ultimately increasing the availability of food. Yet millions are food-insecure in South Asia. Why? Because, achieving 'food security' means not just ensuring that sufficient food is available in the system, but also that everyone has economic, social, cultural and physical access to it—failures of access to food, particularly for the most marginal subgroups of the population, are largely hidden from public view—and that food has adequate nutritive value. Achieving 'food security' also means that people have the right physiological condition to absorb and utilize food, which is contingent upon, *inter alia*, access to potable water and availability of promotive and preventive health care. Regrettably, many of these elements are missing in most of South Asia because of a wide spectrum of causes: social, economic, financial, military, natural, and 'Acts of God', as it were.[1]

How to Monitor Food Insecurity?

The simplest way of monitoring food security is to look at outcomes—to count how many people are hungry.[2] For this, there are three principal measures. The first addresses consumption, typically by estimating the proportion of the population whose food intake falls below the minimum dietary energy requirement of 1,800 calories per day (see Figure 3.1). The greatest problems are in South and South-West Asia where 21 per cent of the population is undernourished. The country with the most acute problems is Afghanistan, where the proportion is more than one-third. But levels of undernourishment are also high—between 20 and 34 per cent—in a number of other countries, including (in descending order of the proportion undernourished) Bangladesh, Pakistan, Sri Lanka and India (UNESCAP, 2009). The second principal way of monitoring food security is by weighing a sample of children to arrive at the proportion who are underweight for their age (see Figure 3.2; UNDP and FAO, 2008). More than half the world's underweight children, around 79.5 million, live in South and South-West Asia alone (United Nations, 2008), where on average 42 per cent of children are underweight—with the highest figures in Bangladesh, at 47 per cent, and India, at 46 per cent.[3] A third way

---

[1] There has been an explosion of literature on the subject in recent times.

[2] Being hungry and being food-insecure are different things. But hunger can be used as a measure of food security.

[3] However, even in South-East Asia, the majority of countries in the sub-region have more than one-quarter of their children undernourished.

FIGURE 3.1 Three Indicators of Hunger: Regional Comparison

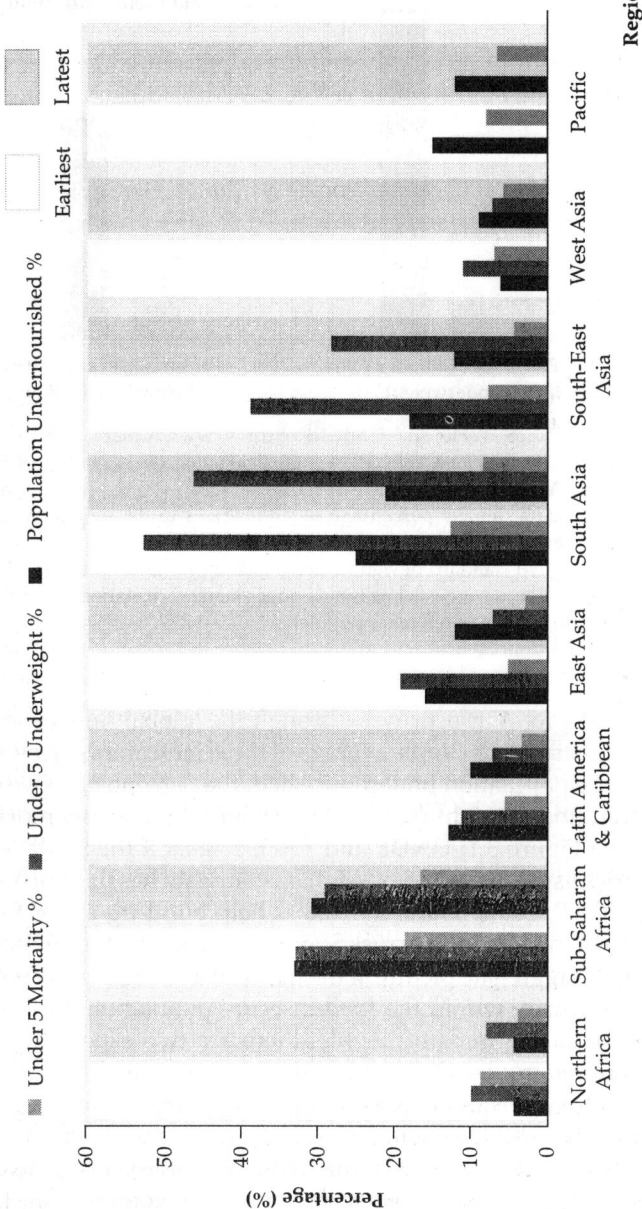

Source: UNDP and FAO (2008).

FIGURE 3.2   Status and Poverty in Asia and the Pacific

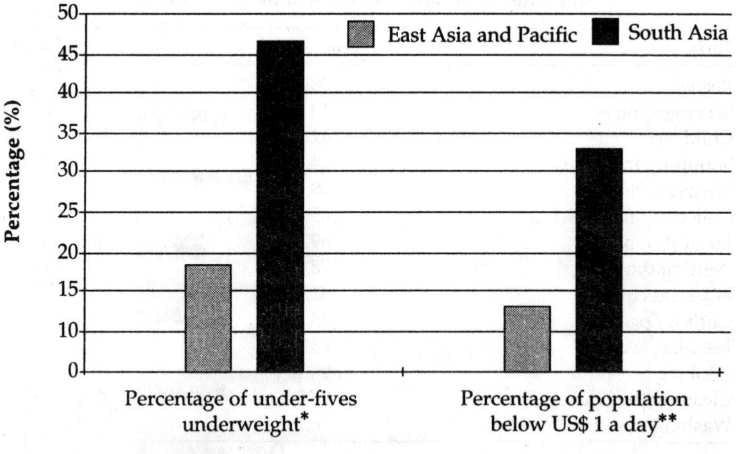

Source: DHS, MicS, WHo and UNICEF. http://www.unicef. org/progress
forchildren/2006n4/files/PFc4_statistical_table.xls as quoted in UNDP
and FAO (2008).
Notes: *The data depicts percentage of under-fives underweight (moderate
and severe) during the period 1996–2005. Moderate and severe—below
minus two standard deviations from median weight-for-age of reference
population; severe—below minus three standard deviations from median
weight-for-age of reference population.
**Data from 2004.

of tracking food insecurity is through the 'global hunger index',
which is based on a simple average of three indicators: the percentage of the population undernourished; the percentage of under-5
children underweight (see Table 3.1); and the under-5 mortality
rate (see Figure 3.1; UNDP and FAO, 2008). Of the South Asian
countries listed in this index, Afghanistan again has the worst score
(UNESCAP, 2009). These indicators help build up a fairly good
picture of malnutrition across South Asia, country by country.

But the aggregate picture at the national level may mask greater
food insecurity within the food-insecure population. Within the
food-insecure groups of people, there are two especially food-
insecure subgroups. One is rural children—who are twice as likely
to be undernourished as those living in urban areas. Some estimates
indicate that children living in rural areas are more than twice as
likely to be underweight as compared with children in urban areas.
Although in percentage point terms the greatest contrast is in Nepal,

**TABLE 3.1   Household Work during One Week
in Landless Indian Households in South India**

Tasks	Female (%)	Male (%)
Kolam*	38.5	–
Serving spouse	40.4	17.6
Childcare	42.3	15.7
Shopping in the city	44.2	47.1
Whitewashing	48.1	62.7
Collecting firewood	65.4	47.0
Local shopping	69.2	43.1
Wetting the yard	78.8	05.9
Washing clothes	82.7	25.5
Cutting/peeling	88.5	11.8
Fetching water	88.5	23.5
Cooking	90.1	11.8
Cleaning the house	94.2	15.7
Washing vessels	96.2	5.9

Source: Bread for the World Institute (1995).
Note: Rice powder or powdered chalk is used to make drawings in front of the
          home.

there are also strikingly wide gaps in other countries, including
India and Sri Lanka. Another important overall distinction is by
sex: in South Asia the rate is 44 per cent for boys while for girls it
is 47 per cent (UNESCAP, 2009).[4]

The second group is women: in most countries in South Asia,
there is persistent gender inequality within and outside households.
The situation is bleakest for rural women in South Asia. They are
often the main food producers—contributing about 65 per cent of
total food production (Inter-Press Service, 2008) and even more
(see Boxes 3.1 and 3.2). Yet, rural women find it more difficult to
get access to a range of resources such as credit,[5] land, agricultural
inputs and extension services and employment, both within the
community and the household, having obvious bearing on their
food security.

In fact, women in South Asia face the seven inequalities (Sen,
2001), having serious implications for their own food security and

---

[4] It may be noted that for most countries in Asia, the rates of underweight are the same for both sexes.

[5] Barring the honourable exceptions provided by borrowers of the
Grameen Bank in Bangladesh.

BOX 3.1   The Hunter, the Gatherer, the Shopper, the Cook

But I have been cooking all day.
Standing over a hot stove.
Slaving over a hot stove.
Cooking.

I've been shopping for groceries.
Putting them away.
Setting the table.
Cooking the food.
Making this dinner.
Wracking my brains.

I've been wracking my brains over this meal.
What to buy.
How much to pay.
I've been budgeting.
Looking for sales.
I've been feeding this family on $6.00.
Making it do. I've been wracking my brains over this meal.

I have been cooking all day.
Shopping for bargains.
Hunting for bargains.
I've been hunting all day.

I've been up and down hundreds of aisles.
Hunting.
Hunting and gathering and cooking this food.
Loading my cart.
Carrying carcasses.
I have been hunting all day.

I have gathered this food from across the land.
I've been everywhere.
I've been everywhere.
I have made this meal.
I have created this food.
This is my time.
My thought.
What you have on your plate is my blood.
My brains.
I tell you I have been cooking all day.

What do you mean you don't want it?

Source: Segal and Sklar (1987).

BOX 3.2   The Hierarchy of the Chicken

The father eats the breast.
He only likes white meat.

The kids eat the drumsticks,
the thighs and the wings.

The mother eats the neck,
the back,
the liver,
the gizzard,
the feet and the heart.

Source: Segal and Sklar (1987).

---

the food security of households, especially of children who depend on women for their food security.

Eight Kinds of Food Insecurities Faced by Women and Girl Children

Flowing from the seven inequalities faced by women in South Asia, women and girl children face eight kinds of food insecurities. These are as follows.

Mortality-based Food Insecurities

In some regions in South Asia, inequality between women and men directly involves matters of life and death, and takes the brutal form of unusually high mortality rates of women. In both absolute and relative terms, the maternal mortality rate (MMR) is highest in South Asia. With 226,077 deaths, this region accounted for more than two-thirds of maternal deaths in Asia and the Pacific in 2000. Nepal had the highest MMR, at 740, while India and Pakistan also had high levels, at 540 and 500, respectively (UNESCAP, 2007: 31–32).[6] These rates are among the highest in the world. A consequence is the preponderance of men in the total population, as

---

[6] Afghanistan also had maternal mortality rate that was amongst the highest in the world but a 2010 Afghanistan Maternal Mortality Survey found the rate to be below 500.

opposed to the preponderance of women found in societies with little or no gender bias in health care and nutrition. This has substantial bearing on the food security of women and children.

Girl children who survive their mothers dying at child birth are not only denied access to the mother's milk, with consequent undernutrition, but are additionally more vulnerable to denial of adequate nutritious food than male surviving newborns. Such newborn girl children, denied of the mother's care and love, may also suffer from ingesting food and water not fit for infants' consumption, notably water not properly purified or boiled. Additionally, such children in general, and girl children in particular, would have inadequate access to preventive health care [or health education] and elementary education, both of which depend largely on the initiative of the mothers. These deprivations impair the long-term capability of such children to access food. Thus, mortality-based inequalities faced by women, jointly and severally, contribute to higher food insecurity among girl children as also infant and girl child mortality rates.

Natality-based Food Insecurities

Given a preference for boys over girls that afflicts many male-dominated societies in Asia, gender inequality can manifest itself in the form of the parents wanting the newborn to be a boy rather than a girl. With the availability of modern techniques to determine the gender of the foetus, sex-selective abortion has become common in Asia and is beginning to emerge as a statistically significant phenomenon in India and South Asia as well, as a negative fallout of the march of medical science. Some, like the United Nations Population Fund, have called this phenomenon 'gendercide', while Sen (1990) calls it the case of the 'missing millions'.

If sex-selective abortion does not succeed or is not possible due to structural, institutional or medical conditions, the girl child starts from the womb with familial and parental environments that are hostile to her existence. This leads to infant girls and women being in a weaker position to be food-secure, by suffering from limitations to physical and social access to food. Mothers carrying female foetuses neglect their diet (poor as it is) or are forced to do so, making the foetus even more undernourished. In harsher societal environments where sex-selective abortion fails, girl children so born are fed a diet, including an excess of mother's milk or excessive salt (Burns, 1994), so that they do not become their parents' burden. Hark these words:

> If the female foetus is lucky enough to survive till her birth then she faced the peril of elimination in infancy by female infanticide. It is defined as, 'Killing of an entirely dependent girl child under one year of age by mother, parents or others in whose care she is entrusted'. (Kollor, 1990)

Historically, female infanticide has been in existence for a long time. Girl infants have been known to be killed by rubbing poison on the nursing-mother's breast, by feeding infants with 'milk of errukam flower or oleander berries, by using sap of calotropis plant, paddy grains, giving sleeping tablets. . . . Law banned this heinous practice in 1870, more than a century ago' (Walia, 2005). Nevertheless, such abuse of the girl child, which is tantamount to a violation of her right to life, continues not only in some parts of Rajasthan and Gujarat, but has recently been found in some parts of Tamil Nadu and Maharashtra in India as well (Kumari, 1995: 177–88).

All these factors taken together demonstrate the various forms that natality-based food insecurities assume for women and girl children in South Asia, ranging from deprivation of nutrition in the mother's womb, to overfeeding, to the feeding of substances that lead to food insecurity and can ultimately cause death.

Basic Facility-based Food Insecurities

Even when demographic characteristics do not show much or any anti-female bias, there are other ways in which women in South Asia can have less than a square deal. In much of South Asia, women and girl children have far less opportunities of schooling than men and boys do. It is true that among the South Asian countries, the primary school attendance of girls has improved,[7] but SAARC countries on average are still well behind other groups, such as the Association of South-East Asian Nations (ASEAN)

---

[7] For example, the ratio of girls per 100 boys at this level in India increased from 76 in 1991 to 93 by 2005, and in Nepal from 63 in 1991, one of the lowest ratios in the region, to 91 in 2005.

countries, with 96 girls per 100 boys enrolled in primary education in 2005. Two countries in the region have even less than 80 girls for every 100 boys enrolled in primary school education in that year (UNESCAP, 2007: 63–64). In some parts of South Asia, it has been reported recently that girls have been banned from attending schools, under threat of dire consequences.

There are other deficiencies in basic facilities available to women and girl children, varying from access to preventive and curative health care, encouragement to cultivate one's natural talents, to fair participation in rewarding social functions of the community: 'A girl between her first and fifth birthday in India or Pakistan has a 30–50% higher chance of dying than a boy' (Filmer et al., 1998). This neglect may take the form of poor nutrition, lack of preventive care (specifically immunization) (Pande, 2003: 395–418), and delays in seeking health care for disease' (Pande and Yazbeck, 2003: 2075–88; Fikree and Pasha, 2004). All these factors taken together lead to higher capability poverty among women and girl children, which contributes to limiting the productivity of women in producing food and opportunities for employment, in turn jeopardizing availability of food and their long-term access to purchasing power and economic access to food, respectively. Indeed, access to basic services like elementary education and primary health care, including access to reproductive health facilities, is key to attaining food security.

Food Insecurities Based on Inequality in Special Facilities

Even where women may have access to basic facilities such as primary health care and elementary education, lack of opportunities for higher education and vocational and professional training for women is proverbial in South Asia vis-à-vis the opportunities available to young men, because, *inter alia*, 'the culture does not see this as "feminine"' (Sen, 2001). Girls may be discouraged from studying subjects that are deemed to be 'the province of men'. This includes agricultural sciences and training in techniques for improving agricultural productivity, post-harvest processing and marketing. Such inequality prevents women from growing more food and/or achieving improved physical and economic access to food. In both cases, the food security of women and girl children is put at risk either by reducing food availability or truncating economic access to food.

The number of women in engineering and medical schools is less than 15 per cent in most South Asian countries. Even in developed countries, women account for only 14 per cent of those studying engineering at university, and that abysmal proportion shrinks further when it comes to the percentage of women pursuing a career as a professional engineer (Mclellan, 2005). In India, for example, the percentage of women in the higher echelon of the civil services is less than 10 per cent.

Many women in South Asia face barriers to the use of modern technology, which include lack of adaptation to local conditions and needs, discriminatory practices, beliefs and stereotypes.

These findings imply that women are underrepresented in high-paying jobs, and to that extent, their economic capacity to access food is lower than that of men. In the longer run, their capacity to withstand idiosyncratic shocks to themselves and their children is so much less.

Profession-based Food Insecurities

In terms of employment as well as promotion in work and occupation, women in some countries in Asia often face a greater handicap than men, notwithstanding the fact that the region has three countries which have produced women prime ministers and presidents. The percentage of women in the higher level of the 'power–status structure' in South Asia varies between 2 and 8, except in Pakistan where it is 20 per cent, mostly due to reservations (Ghimire, 2006). Moreover, there is the 'glass ceiling', as it were, which curtails the movement of women into more senior managerial positions through institutional barriers including norms and attitudes. For example, in the armed forces in this part of the world, women cannot be in combat positions. This is the case even in occupations dominated by women, where men are likely to occupy the more skilled, 'responsible and higher paid positions' (ILO, 2004).

Even where a country may be quite egalitarian in matters of demography or basic facilities, and even to a great extent, in higher education, progress to the elevated levels of employment and occupation seems to be much more problematic for women than for men. Women's income-earning potential is therefore, hindered, which in turn reduces their ability to access food.

Then there is the issue of wage differentials. Throughout South Asia, women's wages range from half to two-thirds of the wages received by men. In Pakistan, for example, women receive, on an average, just about one-third of the wages paid to men, and in India, the wage differential is as high as 38 per cent (Ramachandran, 2006). The Maldives and Nepal have the least disparity in the wage structure, but everywhere in South Asia, women receive less than or equal to 60 per cent of the wages earned by men. In the case of India, gender-specific wage rates for both agricultural and non-agricultural work has been found to be at least 20 to 30 per cent lower for women workers than those earned by men for the same activity. In non-farm activities, not only women are paid less than half the wages earned by their male counterparts but also have lower job security (Rustagi, 2004). In Bangladesh, for example, the female wage rate is so low that a day's wage cannot maintain a family of three, even if the female worker is employed full time (UNESCAP, 2009).

Women's low status in employment is not only a denial of basic rights, it is a very costly economic mistake.[8] The women-to-men employment ratio in South Asia is about 40 per cent, reiterating the low level of employment among women. The limited data available also show that women are underrepresented in non-agricultural activities, which tend to have higher returns and labour productivity, with few exceptions. The difference is much more striking in the women-to-men employer ratio, which went down to 1.6 per cent in Pakistan in 2005 (UNESCAP, 2007; UNESCAP et al., 2008: 22). In the last analysis, what all this boils down to is that women as a class have lower economic access to food and are more vulnerable to food insecurity.

Moreover, even where women are working in the region, they are employed predominantly in informal, labour-intensive, low-value-added manufacturing and service-sector jobs—and thereby vulnerable to domestic and global shifts in demand. 'South Asia has not only relatively high share of vulnerable employment in

---

[8] Figures for South Asia are not available, but for Asia-Pacific the cost of continuing gender discrimination could reach $80 billion per year.

total employment but also relatively large gender gap in vulnerable employments (exceeding 10 percentage points)' (ILO, 2009, quoted in UNDESA, 2009). This is inimical to food security, as 'women in informal work have no access to social security or protection and have limited potential to organise to ensure the enforcement of international labour standards and human rights'. In consequence, such women are extremely vulnerable to both transitory food insecurity and food insecurity caused by idiosyncratic shocks.

Ownership-based Food Insecurities

The important role of women in food production and processing in our part of the world 'underscores the need to provide them with security of tenure of land they cultivate as well as access to and control over resources necessary to increase agricultural productivity and food security' (UNDESA, 2009). Access to land rights, housing credit, technology, markets and extension services determines sustainable livelihood options for women.

However, regrettably, in many societies, especially in rural areas in Asia, the ownership of property is very unequal. 'Asian rural women as small farmers and/or as rural labourers represent the most vulnerable rural poor. This is mainly due to lack of access to assets and unequal control over monetary and non monetary resources at household level' (Polman, 2002). Critical assets such as homes and land are very asymmetrically owned. For example, the laws of inheritance are biased against women and girl children. The absence of claims to property not only reduces the voice of women, but also makes it harder for them to enter and flourish in commercial, economic and even some social activities.

A serious constraint for women farmers is their lack of access to security of tenure or ownership of land. 'In largely agrarian economies, arable land is the most valued form of property and productive resource. It is a wealth creating and livelihood-sustaining asset' (Agarwal, 2002a). Agriculture in South Asia can be classified as falling in the male farming system, dominated by patriarchy and extreme forms of gender discrimination (Centre for Policy Dialogue, 2000). This includes the right to ownership of land.

Traditionally, women have been denied equal inheritance rights to property under both the Hindu and the Islamic systems of law.

In India, for example, under both the Dayabhaga and Mitakshara Schools of Hindu law, a woman could inherit ancestral property only in the absence of four generations of men in the male line of descent. Even then, her rights were limited to a lifetime of interest without the right to demise the property in any way by way of mortgage, sale or lease, except under exceptional circumstances. The Hindu Succession Act of 1956 attempted to rectify some of these infirmities by way of investing equal shares in sons, daughters and widows in a man's own property and in his share of joint family property, but, curiously enough, it kept agricultural land beyond the purview of the act. It was only in 2005, by an amendment of the Hindu Succession Act, that the equal inheritance right has been granted to sons and daughters in all forms of property, including agricultural land (Agarwal, 2005).

Similarly, under Muslim law in India, daughters were allowed only half the share of sons in the property bequeathed by their father. Like the Hindu Succession Act of 1956, the Muslim Personal Law Shariat (Application) Act of 1937 had attempted, much earlier, to enhance the property rights of Muslim women, but excluded all agricultural land, both tenanted and owned, from the Act's ambit, barring some states of south India. The gender-discriminatory pattern of inheritance is also common in Nepal, Bangladesh and Sri Lanka, where 'personal laws govern inheritance, marriage and other social contracts and are in most cases inherently discriminatory. In Bangladesh, inheritance rights are governed by religion and under all religious laws, women have a lesser share than men' (Ramachandran, 2006). In Sri Lanka, the situation is similar in all parallel systems of law such as the general law, the Kandyan law, the Muslim law and the Thesawalamai (Chulani, 2003), although considerable reform has been initiated for gender equality in the general law.

Even under the much-touted land reform and resettlement programmes in the late 1970s and early 1980s called 'Operation Barga', implemented by the communist government in West Bengal to secure the rights of tenants and fair returns to them, few women received land. For example, a village study showed that 98 per cent of the 107 holdings of land distributed went to men, and in nine out of 10 female-headed households, it went to the sons. Married women did not even receive joint titles (Ramachandran, 2006).

Ownership of land has assumed a critical dimension for women in agriculture, because of the rapid feminization of agriculture (Kelkar, 2007) and the increasing migration of men from rural to urban areas in search of a living. An increasing number of *de facto* woman-headed households struggle to eke out a living and ensure the food security of their families without access to credit, technology or extension services. This is because what is involved is not just land ownership, but also all that goes with it, such as collateral to access institutional credit, and training and the ability to deal with extension facilities on an equal basis. (Such inequality has long existed in other parts of South Asia, as well as in other parts of the world, but there are also local variations.)[9] Equal property rights for women are relevant for developing production. They are relevant even for matters like raising wages, since the reservation wage is sticky. Moreover, ownership inequality reduces women's ability and incentives to invest in agricultural land, which has a bearing on food availability and hence food security. Moreover, because of these and other factors, women often become dependent on others for food, especially as widows, or when divorced or abandoned. All of these factors together contribute to the greater food insecurity of women.

Household-based Food Insecurities

There are basic inequalities in most parts of Asia in gender relations within the family or the household which can take many different forms, often based on the intra-familial distribution of power.

Even where no overt signs exist of anti-female bias in, say, survival or son-preference or education, or even in promotion to higher positions in the job market, family arrangements can be quite unequal in terms of sharing food and the burden of housework and childcare, apart from limiting women's opportunities for earning an income, thereby making inroads into women's economic

---

[9] For example, even though traditional property rights have favoured men in the bulk of India, matrilineal inheritance has existed in the state of Kerala for a long time among an influential part of the community, namely, the Nairs, as also in the north-eastern state of Meghalaya. In the Kingdom of Bhutan, land is inherited in the female line.

access to food. Such arrangements may also include girls being fed less food and food of lower nutritional value than boys (see Figure 3.3; N. Mukherjee, 2003). Aspects of intra-familial distribution of resources, including food, such as women being expected to eat the least, often the leftovers, and only after all others in the family have eaten, make women vulnerable to food insecurity. Thus,

> gender, in particular, is noted to be an important signifier of differences in interests and preferences, incomes are not necessarily pooled and self-interest resides as much within the home as in the market place, with bargaining power affecting the allocation of who gets what and who does what. (Agarwal, 2002b)

Intra-household power–status structures directly impact women's food security and indirectly impact the food security of others in the family, particularly children. Indirect evidence in terms of gender-specific malnutrition levels points to existing disparities. 'In poor households, in particular, the incidence of severe malnutrition is greater among girls. In fact, gender has been found to be the most statistically significant determinant of malnutrition among young children' (Ramachandran, 2006). In Punjab, a sharp difference in calorie intake among adult men and women has been found in some studies, with women consuming approximately 1,000 fewer calories than men (Development Gateway, 2004; Mukhopadhyay, 2007). It has also been found that, in some cases, though boys and girls were treated similarly in terms of total calorie intake, the nutritional value of the food was biased against the girl children: girls were given more carbohydrates in the form of cereals, while boys were given more milk and fats with their cereal.

Intra-household gender bias in favour of male children, both in terms of feeding and seeking health care, has been noted in Pakistan (Nazli and Hamid, 1999). Women in Bangladesh are seemingly a residual category in intra-household food distribution, eating after the men and the children, and eating what is left over after all others have eaten (Rahman, 2002). A similar pattern prevails in most Asian countries. The structure of the family also plays a role in women's food security. In nuclear families where the woman herself is responsible for food distribution, she gives preference to her husband and children at the cost of her own needs (Mondal, 2003).

FIGURE 3.3 Food Discrimination

Seasonal Food Calendar, Girl Child/Male Child, 1 Year Old

MONTHS / ITEMS	CHAITRA		BAISHAR		JYAISTHA		ASHAR		SRABON		BHADRA		ASWIN		KARTIK		AGHRAN		POUSH		MAGH		PHALGUN	
	M	F	M	F	M	F	M	F	M	F	M	F	M	F	M	F	M	F	M	F	M	F	M	F
RICE	2	1	3	2	5	3	6	4	6	5	7	6	8	7	6	5	7	6	8	7	9	8	6	5
PULSES	6	6	6	5	4	3	6	4	5	4	7	6	7	6	6	5	6	5	6	5	5	4	7	6
VEGETABLES	4	2	4	3	3	2	4	4	4	3	6	5	7	4	5	3	4	3	7	6	6	5	5	4
FISH	2	1	2	1	2	1	3	2	3	2	4	3	5	4	4	3	4	3	5	4	6	5	4	3
EGG	6	4	4	3	3	2	3	2	4	3	5	4	6	5	4	3	7	5	8	7	4	3	10	9
MILK	9	7	4	3	5	3	4	3	4	3	5	4	6	5	6	5	6	3	8	7	6	5	7	6
BANANA	7	3	6	4	4	3	3	2	4	3	5	4	7	6	6	5	4	4	6	5	5	4	8	7
ORANGE	4	3	4	2	3	2	2	1	3	2	2	1	4	3	4	3	4	4	5	4	–	–	–	–
MANGO	–	–	–	–	–	–	3	2	–	–	–	–	–	–	–	–	–	–	–	–	–	–	–	–
JACKFRUIT	–	–	–	–	–	–	4	3	–	–	–	–	–	–	–	–	–	–	–	–	–	–	–	–
PINEAPPLE	–	–	–	–	–	–	–	–	–	–	2	–	1	1	3	2	4	3	5	4	–	–	–	–
APPLE	–	–	–	–	–	–	–	–	–	–	–	–	2	1	–	–	–	–	–	–	5	4	–	–
PAPAYA	–	–	–	–	–	–	–	–	–	–	4	3	–	–	–	–	–	–	–	–	–	–	6	5
BISCUIT	3	1	2	1	2	1	3	2	3	2	2	1	5	4	5	4	5	4	6	5	4	3	7	6

Source: Neela Mukherjee (2003).

A study in India, for example, documented the practice of 'maternal buffering'—as mothers deliberately eat less to allow men, particularly younger men, and children in their households to get enough to eat (UNESCAP, 2009). While pre-school children are best fed within the family with meat, fish and dairy products in their diets, boys are favoured over girls. These gender disparities among pre-school children tend to disappear in the middle- and higher-income groups, but the neglect of the adult female persists across all groups (Rahman, 2002). Among adults, the female is the most neglected, with adult and even elderly males receiving more nutritious food (see Figure 3.4; Mukherjee, 2003).

Therefore, household-based food insecurity confronting women and girl children takes many forms, from lower quantity of food in terms of reduced intake of food, to less nutritious food in terms of poorer quality of food, or reduction in women's purchasing power leading to reduced access. Some of these deprivations are based on misconceived social constructs.

Seasonality-based Food Insecurities

The multi-tasking of women as food growers, gatherers, hunters, processors and custodians of food is well documented. Yet women, even when they may not suffer from chronic or transitory food security, suffer from seasonal food insecurity (Mukherjee and Mukherjee, 1994; Smith and Wiesmann, 2007). There are certain seasons or months when food security is routinely jeopardized for women. It has been found in studies in India that generally during the summer months of May and June and during the cultivation season, June to September, women are particularly vulnerable to food insecurity (Mukherjee, 2004; Mukherjee and Mukherjee, 2001). And these are the months when their calorific needs are at their highest.

Men, too, are affected by the seasonality dimension of food security but, with the feminization of agriculture (Kelkar, 2007; Vepa, 2007), it affects women disproportionately more. During the leans months, men migrate to cities and towns, as mentioned earlier; usually, the out-migration from the villages starts after the harvesting season. The emigrants return just before the following agricultural season. For example, in large parts of Bangladesh (Afsar, 2000) and in Bihar, Jharkhand and West Bengal in India,

FIGURE 3.4 Food Discrimination Once Again

MONTHS ITEMS	CHAITRA		BAISHAR		JYAISTHA		ASHAR		SRABON		BHADRA		ASWIN		KARTIK		AGHRAN		POUSH		MAGH		PHALGUN	
Seasonal Food Calendar, Male Adult/Female Adult																								
	M	F	M	F	M	F	M	F	M	F	M	F	M	F	M	F	M	F	M	F	M	F	M	F
RICE	2	2	6	5	7	6	9	8	9	8	7	6	6	5	5	4	8	7	6	3	6	5	6	5
CHAPPATI	5	4	–	–	–	–	–	–	–	–	–	–	–	–	2	2	–	–	–	–	3	2	3	2
VEGETABLE	1	1	2	2	6	6	10	10	7	7	5	5	4	3	6	4	3	3	3	2	2	2	2	2
PULSES	10	10	5	5	4	4	2	2	3	2	6	6	7	7	7	6	7	6	9	7	8	8	9	9
FISH	1	1	5	2	4	3	2	3	5	2	3	3	8	3	8	2	5	2	4	2	1	2	2	2
EGG	3	2	4	2	3	2	3	2	4	1	4	2	8	3	10	4	9	9	10	9	8	7	7	6
MILK	11	10	5	5	4	4	3	3	3	2	3	3	4	3	2	2	2	2	2	2	2	2	8	8
MANGO	–	–	–	–	4	4	–	–	–	–	–	–	–	–	–	–	–	–	–	–	–	–	–	–
JACKFRUIT	–	–	–	–	9	6	8	5	–	–	VILLAGE ANALYSIS—REHANA PARVIN AND FERBUSI BEGUY VILLAGE BANSTAOL, DISTRICT TANGAIL, BANGLADESH														

Source: Mukherjee (2003).

Note: The figures show the number of seeds used in scoring; M stands for male, F for female.

men migrate to urban areas around January–February and return with the onset of the monsoons in May–June. As a result of this cyclical migration of men, women who stay back at home to hold on to agricultural operations, however small, and conservation of bio-resources, face lower availability of food, as food from their own farms lasts six to nine months at best, and there are fewer employment opportunities, lower wages if employment is available at all, and increased vulnerability to diseases that inhibit food utilization. Hence, seasonality impinges on women's food security from different perspectives.

The Future

Thus, over the long term, there are serious impending challenges to the future capacity of the region to ensure food security for all women and children.[10] Policy actions thus have to be led on four broad fronts. First, expansion in systemic food availability for women and children (i.e., meeting the challenges of higher food production, water scarcity, energy security, climate change, industrial agriculture and trade, and promoting community-based responses). Second, improving economic, physical and social access to food for women and children (through transport and social protection). Third, improving utilization of food by women and children (through vigorous expansion of promotive and preventive health care as also access to reproductive health care and potable water). And fourth, improvements in personal and social hygiene.

The Framework for Food Security for Women and Children

Given this scenario, a successful strategy to deal with the present and future food security needs of women and children in the South Asia region would aim at:

---

[10] Not least because of changes in demographics, the rapid depletion of water resources, the impact of climate change and erratic weather events, and anticipated proliferation of natural and human-induced disasters.

- Ensuring sustainable supply of appropriate food in adequate quantity for all women and children.
- Protecting women and children against shocks of both the covariate and the idiosyncratic varieties.
- Meeting the challenges of water scarcity, energy security and climate change.
- Meeting the challenge of making trade and transportation work for the food security of women and children.
- Providing women and children with economic access to food.
- Ensuring that women and children have physical and social access to food.
- Ensuring that women and children utilize and absorb the food that is consumed.

In order to achieve these goals, countries in Asia, irrespective of whether they are self-sufficient or self-reliant in food or suffer net food deficits, will need to establish a set of policies that would ensure, among other things:

1. Increase and diversification of production especially in agriculture to expand employment opportunities for women, and availability of food on a sustainable basis for women and children.
2. Enhancement of general economic growth especially in agriculture, expansion of employment for women, and guaranteeing decent rewards for work to them to ensure increased availability of purchasing power for women on a sustainable basis.
3. Protecting large sections of the population comprising women and children from both idiosyncratic and covariate shocks that impinge on food security, and providing special social protection to vulnerable subpopulations *within* women and children (including small women farmers, women and children with disabilities, women and children living with HIV/AIDS, elderly women, and infirm women and children) to guarantee economic, physical and social access to, and utilization of, food for women and children, based on justice and equity.

4. Reduction in gender-based inequalities that lead to the eight kinds of food insecurities faced by women.
5. Enhancement of literacy and health care for women and children, which includes supply of potable water and enhanced systemic social and personal hygiene.
6. Strengthening gender-sensitive governance and institutions, including gender-sensitive news media and civil society organizations.

The Call for a Second Green Revolution

Having said all of that, we shall present a partial set of options addressing five of the six elements of the framework. Enhancing literacy and health care for women and children requires a full discussion, and is reserved as an issue for future discussion.

Options at the Country Level

Start a Gendered Green Revolution

The first Green Revolution in the 20th century achieved significant yield increases in the South Asian region through promotion of high external input agriculture (HEIA), involving irrigated water, chemical fertilizer, chemical pesticides and insecticides and energy use, with male farmers at the centre. However, it also brought with it several attendant problems, including additional workload for women. Now, a gendered Green Revolution is needed, one that will increase yields even further, but that moves agriculture from high external input-intensive agriculture to 'high tacit-and-explicit knowledge-intensive agriculture'. Since a large body of tacit knowledge in food production rests with women, they will occupy the centre stage. The proposed Green Revolution must integrate traditional knowledge and technology with advances in modern-day science and agricultural engineering,[11] including plant genetics,

---

[11] Genetically modified (GM) crops may have a role in this regard but the risks are not all known, and so harnessing their power in agriculture should be preceded by a robust testing, regulatory and safety regime. Germany recently became the sixth EU country to ban a type of genetically

plant pathology and information technology and encompassing ecologically integrated approaches, like intergraded pest and soil fertility management, minimum tillage and drip irrigation. Given that large chunks of these inputs relating to ecologically integrated approaches will be internal to the faming household system itself, where women are key players, it will be an empowering process for women. A high tacit-and-explicit knowledge-intensive agriculture thus commends itself on grounds of both resilience and equity, as it will attempt to return the 'power to produce' to the women in farming households, rather than investing the whole of it in corporate boardrooms. It also commends itself on grounds of environmental sustainability.

Get the Fundamentals Right

To make the *gendered Green Revolution* happen, there will be a need to focus on setting the *factor inputs* right for women. Access to key factor inputs for women, namely, access to assets (land, tools, machinery, water[12] and 'energy') without demur or recourse; efficient credit for women (fair interest rates and timely availability); gender-sensitive knowledge system (technology serves women's need best and a robust agricultural extension service to transfer technology of food production from the laboratory to women[13] in the farms, especially the small and marginal farms); women's access to information communication technology as applicable to farming (such as precision agriculture); and risk management for women in farming households, especially the small and marginal ones, need to be made available on a sustainable basis. An economic Sherlock Holmes may well say: 'Elementary, my dear Watson', but Alas the

---

modified maize, the only GM crop permitted until now in the country. See EUbusiness, 'Agriculture in EU'. http://www.eubusiness.com/agri (accessed on 15 April 2009).

[12] Including time-tested irrigation structures like wells, canals, percolation tanks and ground water extraction, to which traditionally women have access as well as in water harvesting and ground-water recharges systems that are gender sensitive.

[13] Women's needs tend to be ignored, even in agricultural research and technological innovations. Worldwide, only 5 per cent of extension services have been addressed to rural women. See FAO (1996).

elementary things often matter and are often overlooked for more glamorous but may be less efficient solutions.

Institute Universal Social Protection

Many countries in the Asia-Pacific region provide protection to vulnerable groups through subsidies, outright grants (like old-age pension or widow pensions), price support or price control. These instruments of social protection are often gender-insensitive and/ or inefficient because they lack in *range, reach* and *depth*, or some combination of these. For example, the widow pension scheme in India gives widows only ₹200 or so per month,[14] if they get it at all. The scheme covers a very small percentage of widows,[15] and is riddled with informal taxation (see Box 3.3). And even these inefficient instruments come with a high cost: subsidies, outright grants (like pensions) and price support can create havoc in government finances, while price control, though beneficial to food consumers, concomitantly carries the unintended effect of reducing farmers' incentives to produce more food.

A universal social protection represents a better alternative, given that targeting is always a problem and riddled with inefficiencies and corrupt practices. That is the reason that there is a growing demand for universal rather than targeted social protection (ESCAP, 2011). Given the diversity of Asia, the challenge is to devise innovative ways of providing universal social protection

---

[14] The scheme is called the Indira Gandhi National Widow Pension Scheme (IGNWPS), implemented by the Ministry of Rural Development, Government of India. The fund is financed 50:50 between central and state governments. The pension is given to widows between 45 and 64 years of age, belonging to households below the poverty line as per criteria prescribed by the Government of India. The pension amounts to ₹200 per month per beneficiary, and the concerned state government is also urged to provide an equal amount to the person. These funds are to be credited into a post office or public-sector bank account of the beneficiary.

[15] There are similar examples in other regions as well. For example, China's basic health insurance covers only 30 per cent of its 1.3 billion population. The system is being revamped and the amount that each person covered supposed to get as subsidy was about $17 per year starting in 2010.

BOX 3.3   Crooks Wait on the Road to Widows' Pension

Vrinda Gopinath, Vrindavan, 6 February

The bundle of papers wrapped in cloth and kept behind a curtain in Swami Vivekananda School is forever mocking at the bitter sacrifices made by the widows and abandoned women in Vrindavan.

The 50-odd old-age pension forms with photographs, thumb impressions and official signatures have been lying for over a year, forgotten and dusty. For the women of Vrindavan, however, the papers carry a fervent hope that one day they will receive their meagre pension of ₹1,500 a year, which they have been entitled to for a decade and more. Kamala Ghosh, the school's principal and mother to the *mais*, can only offer solace as she comforts the women with some optimistic cheer. Says Ghosh, as she unwraps and displays the fraying papers, 'These forms were filled after we organized a camp for the women to come and take what is rightfully theirs. . . . they were not even aware that they were entitled to a pension. I got my students to help in filling forms and completing formalities but the officials have not bothered to forward them to the district headquarters.' Ghosh adds that work gets done if the district magistrate is sympathetic, as some of the previous efforts have been quite fruitful.

The procedure is quite straightforward if it is carried out wilfully—a health officer from the State Social Welfare Ministry certifies the age, the SJM and DM attest their signatures, it goes back to headquarters where a cheque is drawn and deposited in the bank. What could be an easier task—as there are about 2,000 aged widows in this temple town—but a whole corrupt network has thrown a ring around them, picking on their drying bones. There are several cases of cheques being issued in the names of landlords, account books which are wrongly tabulated, pension forms which are suspiciously lost and, last year, 250 cheques that were returned because the beneficiaries could not be traced due to incomplete forms.

Bhanu Ghosh, a 70-year-old widow, still clutches on to a cheque which has come in the name of Premlal, her landlord, despite the fact that her husband's name is Gurudas Ghosh. Says the frail, old lady, 'We have to bribe the Patwari [a local revenue official] to get our money from the bank, even the postman to receive the money order from home.'

Source: The Indian Express, New Delhi, Monday, 7 February 2000.

of appropriate depth and adequate reach to a range of people. It could include the following:

- Undertaking *ex ante* management of covariate[16] shocks to food security by boosting the coping strategies of women at risk of covariate shocks through installation of insurance and insurance-like programmes for women with flexible targeting, flexible financing and flexible implementation arrangements, before the onset of natural disasters.
- Provisioning for *de jure* and *de facto* insurance for idiosyncratic shocks,[17] including: (*a*) more effective, ubiquitous and continuing insurance programmes, whether through financial innovations such as micro insurance for women or index insurance schemes that are especially designed for women farmers, and community-based health insurance programmes for women and girl children; and (*b*) *de facto* insurance, via, for example, a robust system of protecting common property resources with special usufructuary rights for women guaranteed, public employment guarantee schemes like the National Rural Employment Programme in India, underpinned by food-for-work or cash-for-work projects as a means of protecting vulnerable women from idiosyncratic shocks like sudden loss of valuable and productive assets, unexpected loss of means of livelihood through desertion, sudden illness, or other adverse effects.
- Eliminating the seven gender-based food inequalities: (*a*) through a multi-sectoral programme involving, *inter alia*, social security, affirmative actions, changing laws relating to inheritance and ownership of productive resources and making the rights to food, education, health care and information for women justiciable rights; (*b*) by adopting an agent-oriented approach to the women's agenda and regarding women as potentially active agents of major social change rather than

---

[16] Covariate shocks are shocks that affect everyone in a community or area (see Chapter 2 for more details).

[17] Idiosyncratic shocks are shocks that affect a household or an individual (see Chapter 2 for more details).

as solicitors of social equity; (c) by creating an enabling environment, to use Amartya Sen's phraseology, for 'cooperative conflict' between genders and devising ways and means for amicable resolutions; and (d) taking affirmative action including reservation of seats for women in all legislatures and parliaments as a fair outcome and realization of the benefits of law.

- Making guaranteed employment for 100 days a legal right, for especially marginalized subgroups among the population of women (like women having dependents with disabilities, elderly women, widows, women small farmers, women migrants, women categorized as internally displaced persons and women-headed households) and women facing discrimination on the basis of race, religion, caste, ethnicity and communicable diseases, who are often among the poorest of the poor, commensurate with their needs, noting that employment guarantees are among the best forms of *de facto* insurance for such groups.

Get Gender-sensitive Institutions in Place

Gender-sensitive institutions that help women farmers should be put in place. Especially important among these are extension education accessible to women of farming households, institutions for post-harvest facilities, purveyors of credit for women farmers, markets where women farmers can operate, adequate infrastructure that is gender-sensitive, communication networks, connectivity, including mobile phones, and building the capacity of women in farming households to meet international standards (especially on food safety).[18]

A key element in this picture is the system that can help small and marginal women farmers reap the benefits of economies of scale in marketing outputs, and enable them to meet the standards of

---

[18] One reason why the benefits of higher prices did not accrue to farmers in many countries of South Asia in 2007 and 2008 was the paucity of institutions which allow farmers to take advantage of the higher prices of food in the international market. Where institutions existed, farmers have reaped benefits; in Vietnam, small farmers have benefited from high food prices by accessing export markets.

TABLE 3.2 Underweight Children in South Asia

	Proportion of underweight children (%)		Number of underweight children (Thousands)	
Country	(Oldest observation since 1990)	(Most recent observation)	(Oldest observation since 1990)	(Most recent observation)
China*	19.1 k (90)	6.9 k (05)	22,703	5,885
Myanmar	32.4 h (90)	31.8 k (03)	1,625	1,327
Afghanistan	48.0 o (97)	39.3 f, r (04)	1,691	1,830
Bangladesh	67.4 r (92)	47.5 k (04)	11,569	8,985
Bhutan		18.7 r (99)		13
India	53.4 c, i (93)	45.9 g (05)	67,775	58,244
Maldives	38.9 l (94)	30.4 k (01)	15	10
Nepal	48.7 p (95)	38.6 k (06)	1,695	1,394
Pakistan	40.4 b, k (91)	37.8 e, k (02)	8,337	7,720
Sri Lanka	37.7 n (93)	29.4 n (00)	662	475

Sources: As quoted in UNESCAP (2009); proportion of children underweight from United Nations (2008). Number of children age 0–4 years from United Nations (2007).

Notes: b. Data refer to 1990–91.
c. Data refer to 1992–93.
e. Data refer to 2001–02.
f. Data refer to 2003–04.
g. Age group is 0–35 months.
i. Age group is 0–47 months.
k. Age group is 0–59 months.
l. Age group is 0–60 months.
n. Age group is 3–59 months.
p. Age group is 6–35 months.
r. Age group is 6–59 months.
* Projected here for benchmarking purposes only.

up-market stores. In the past, this role was often played very inef-
ficiently by government-run parastatals, but even these have since
been dismantled under the onslaught of the global programme
of liberalization, privatization and deregulation. A system of
incentives has to be instituted for private companies, community-
based women's organizations (such as self-help groups in India),
gender-sensitive grassroots non-governmental organizations, and
women farmers' organizations that have the capability to enable
and empower small and marginal women farmers.

References

Afsar, Rita. 2000. *Rural Urban Migration in Bangladesh: Causes, Consequences and
Challenges*. Dhaka: University Press Limited.
Agarwal, Bina. 2002a. 'Are We Not Peasants Too? Land Rights and Women's Claims
in India'. *Seeds*, no. 21. New York: Population Council.
———. 2002b. 'Gender, Inequality, Cooperation and Environmental Sustainability'.
www.santafe.edu/research/publications/workingpapers/02-10-058.pdf
(accessed on 23 April 2009).
———. 2005. 'A Landmark Step to Gender Equality'. *Hindu*, 25 September.
Bread for the World Institute. 1995. *Causes of Hunger*. Maryland: Silver Spring.
Burns, John F. 1994. 'India Fights Abortion of Female Fetuses.' *New York Times*, 27
August. http://www2.hu-berlin.de/sexology/IES/india.html (accessed on
19 April 2009).
Centre for Policy Dialogue. 2000. 'Gender, Land and Livelihood in South Asia',
Report No. 30, Centre for Policy Dialogue, Dhaka.
Chulani, Kodikara. 2003. 'Engaging with Muslim Personal Law in Sri Lanka: The
Experience of MWRAF'. *Lines Magazine*, August 2003. http://issues.lines-
magazine.org/Art_Aug03/Chulani.htm (accessed on 22 January 2012).
———. 2008. 'The National Machinery for the Protection and Promotion of Women's
Rights in Sri Lanka', Dossier 29, Women Living Under Personal Law, London.
http://www.wluml.org/english/pubs/pdf/dossier29/dossier29-en.pdf
(accessed on 26 April 2009).
Development Gateway. 2004. 'News on Food Security, Intra Household Gender
Disparities and Access to Food'. Quoted in Anish Kumar Mukhopadhyay,
'Gender Inequality and Child Nutritional Status: A Cross Country Analysis'.
Kolkata: Indian Council of Social Science Research, Centre for Studies in Social
Sciences. http://www.isical.ac.in/~wemp/Papers/PaperAnishKumarMukh
opadhyay.doc (accessed on 9 March 2012).
ESCAP (Economic and Social Commission for Asia and the Pacific). 2011. *The Promise
of Protection*. Bangkok: United Nations.
FAO (Food and Agricultural Organization). 1996. 'FAO Focus: Women and Food
Security: Women Hold the Key to FOOD Security'. www.fao.org/FOCUS/E/
Women/WoHm-e.htm (accessed on 23 April 2009).

Fikree, Fariyal, and Omrana Pasha. 2004. 'Role of Gender in Health Disparity: The South Asian Context'. *British Medical Journal*, vol. 328, no. 7443, pp. 823–26. http://www.popcouncil.org/publications/articles/823.html (accessed on 22 April 2009).

Filmer, D., E. M. King and L. Pritchett. 1998. 'Gender Disparity in South Asia: Comparison Between and Within Countries', Policy Research Group Working Papers 1867, Washington D.C., World Bank.

Ghimire, Durga. 2006. 'South Asian Situation on Women in Politics'. Paper presented at the 6th Asia Pacific Congress on Political Empowerment of Women, Centre for Asia-Pacific Women in Politics, Manila, 10–12 February.

International Labour Organization (ILO). 2004. 'Breaking through the Glass Ceiling: Women in Management', 2004 Update. Geneva: ILO.

Inter-Press Service. 2008. 'Q&A: Women Do Most, with Least Assistance', Interview with Lennart Båge, President of the International Fund for Agricultural Development. http://ipsnews.net/news.asp?idnews=43554 (accessed on 22 January 2012).

Kelkar, Govind. 2007. 'The Feminization of Agriculture in Asia: Implications for Women's Agency and Productivity'. Food and Fertilizer Technology Centre, Taiwan. http://www.agnet.org/library/eb/594/ (accessed on 26 April 2009).

Kollor, T. M. 1990. 'Female Infanticide: A Psychological Analysis', in *Grass Roots Action*, Special issue on the Girl Child, 3 April.

Kumari, Ranjana. 1995. 'Rural Female Adolescence: Indian Scenario'. *Social Change*, vol. 25, no. 2, pp. 177–88.

Mclellan, Amy. 2005. 'The Battle for Hearts and Minds: The Number of Women in Engineering Remains Low'. *The Independent, London*, 24 February.

Mondal, S. K. 2003. *Health, Nutrition and Morbidity: A Study of Maternal Behavior*. New Delhi: Bookwell.

Mukherjee, Amitava. 2004. *Hunger: Theory, Perspectives and Reality*. London: Kings College, University of London, and Ashgate: Aldershot.

Mukherjee, Amitava, and Neela Mukherjee. 2001. 'Rural Women and Food Insecurity: What Food Calendars Reveal—A "Longitudinal Study"', in Neela Mukherjee and Bratindi Jena (eds), *Learning to Share: Experiences and Reflections on PRA and Other Participatory Approaches*, vol. 2. New Delhi: Concept.

Mukherjee, Neela. 2003. *Participatory Rural Appraisal, Methodology and Applications* (reprint). New Delhi: Concept.

Mukherjee, Neela, and Amitava Mukherjee. 1994. 'Rural Women and Food Insecurity: What a Food Calendar Reveals'. *Economic and Political Weekly*, vol. 29, no. 11, pp. 597–99.

Mukhopadhyay, Anish Kumar. 2007. 'Gender Inequality and Child Nutritional Status: A Cross Country Analysis'. Paper read at the seminar on Gender Issues and Empowerment of Women, Platinum Jubilee of Indian Statistical Institute, Kolkata, 1–2 February.

Nazli, H., and S. Hamid. 1999. 'Concerns of Food Security, Role of Gender and Intra Household Dynamics in Pakistan'. Pakistan Institute of Development Economics (PIDE), Islamabad. www.pide.org.pk/research/report175.pdf.

Pande, R. P. 2003. 'Selective Gender Differences in Childhood Nutrition and Immunization in Rural India: The Role of Siblings'. *Demography*, vol. 40, no. 3, pp. 395–418.

Pande, R. P., and A. S. Yazbeck. 2003. 'What's in a Country Average? Wealth, Gender, and Regional Inequalities in Immunization in India'. *Social Science Medicine*, vol. 57, no. 11, pp. 2075–88.

Polman, Wim. 2002. 'Role of Government Institutions for Promotion of Agriculture and Rural Development in Asia and the Pacific Region: Dimensions & Issues'. Food and Agriculture Organization, Asia-Pacific Regional Office, Bangkok.

Rahman, Aminur. 2002. 'On Measuring Intra-household Inequality in Food Distribution—Is the Conventional Calorie Intake Enough to Understand Individual Wellbeing within the Household?' Mimeograph, Department of Economics, University College, London.

Ramachandran, Nira. 2006. 'Women and Food Security in South Asia: Current Issues and Emerging Concerns', Research Paper No. 2006/131, WIDER, UN University, Helsinki.

Rustagi, Preet. 2004. 'Women and Development in South Asia'. *South Asian Journal*, vol. 4, April–June.

Segal, Sondra, and Roberta Sklar. 1987. *Women's Body and Other Natural Resources.* http://theater2.nytimes.com/eme/theater/treview.html?res=9D06E4D6143 BF933A25756C (accessed on 16 April 2009).

Sen, Amartya. 1990. 'More Than a Hundred Million Women Are Missing', *New York Review of Books*, vol. 37, no. 20, 20 December.

———. 2001. 'Many Faces of Gender Inequality', *Frontline*, vol. 18, no. 22, 27 October–9 November.

Smith, Lisa C., and Doris Wiesmann. 2007. 'Is Food Insecurity More Severe in South Asia or Sub-Saharan Africa? A Comparative Analysis Using Household Expenditure Survey Data', Discussion Paper 00712, International Food Policy Research Institute, Washington, D.C.

UNDESA. 2009. *World Survey on the Role of Women in Development.* New York: United Nations.

UNDP and FAO. 2008. *Combating Hunger: A Seven Point Agenda.* Colombo and Bangkok: UNDP Regional Centre in Colombo and FAO Regional Office for Asia and Pacific.

UNESCAP. 2007. *UNESCAP Statistical Yearbook for Asia and the Pacific 2007.* Bangkok: UNESCAP.

———. 2009. *Sustainable Agriculture and Food Security in Asia and the Pacific.* Bangkok: UNESCAP.

UNESCAP, UNDP and ADB. 2008. *A Future Within Reach.* Bangkok: UNESCAP.

United Nations. 2007. 'World Population Prospects: The 2006 Revision. Population Database'. http://esa.un.org/unpp/index.asp?panel=2 (accessed in 9 March 2012).

———. 2008. Millennium Development Goals Indicators: The official United Nations site for the MDG indicators. http://mdgs.un.org/unsd/mdg/Data.aspx (accessed in November 2008).

Vepa, Swarna Sadasivam. 2007. 'The Feminization of Agriculture and the Marginalization of Women's Economic Stake', in Maithreyi Krishnaraj (ed.), *Gender, Food Security and Rural Livelihoods.* Kolkata: Stree.

Walia, Ajinder. 2005. 'Female Foeticide in Punjab: Exploring the Socio-economic and Cultural Dimensions'. *IDEA*, vol. 10, no. 1.

4

A Select Suite of Government Response to Hunger

There are two things you don't want to see being made: Sausage and legislation.

—Bismarck

The problem of food security has confronted the countries of the Asia-Pacific region for centuries. Major famines have afflicted all the major countries of the region, on an almost continuous basis, though for different sets of reasons, including state policy.[1] Take the case of China. Malthus (1798) saw China as one of the most terribly afflicted of all countries. It was a country where in 'times of famine' which 'are here but too frequent', 'millions of people' perished of hunger. There is a list of the 1,282 Chinese famines in the 2,019 years preceding 1911. In China, it is estimated that there was a drought or flood-induced famine in at least one province almost every year from 108 B.C. to A.D. 1911. In the 17th century, for instance, famines became common, especially in north China, worsened by unusually cold and dry weather. It is estimated that some nine million fatalities were caused by the famine in north China from 1876 to 1879. Famine continued in China until very recently. Reportedly, the 'Great Leap Forward' led to 'famine on a gigantic scale', a famine that according to one estimate claimed 20 million lives or more between 1959 and 1962. China has only just escaped from famine. Similarly, there were famines in other parts of East and South-East Asia (see Table 4.1).

---

[1] See Mukherjee (2010), where the deliberate policy of Churchill is located as an important cause for the great Bengal famine of the early 1940s.

TABLE 4.1 Famine in East Asia, 1700–2000

Country	Incidence	Excess death (million)	Population	As % of total population	Nature of disaster
Japan	1782–87 (Temmei Famine)	0.2–0.9	26,010,600 (est. 1780)	0.77–3.46	Drought, flood, cold wind and the eruption of various volcanoes
	1832/33–1836/39 (Tempo Famine)	'Worse than the Temmei famine in 1782'	27,063,907 (est. 1834)	–	Flood and cold weather
North Korea	1997–99	0.2–3.5	25,904,124 (est. 1996)	0.77–13.51	'Unprecedented floods' in 1994–96
China	1877–78	9.5–13	308,803,939	3.08–4.21	Drought across north China plains
	1892–94	1	335,134,795	0.3	Drought
	1920–21	0.5	456,200,000	0.11	Drought
	1928–29	3	446,679,832	0.67	Drought
	1938	1	479,084,651	0.21	Levee breach of Yellow River
	1958–61	16.5–30	659,940,000	2.50–4.55	Drought and floods in different parts of the country

Source: Compiled from various sources.

142   FOOD SECURITY IN ASIA

South Asia is another area where massive famines occurred until very recently. Even Malthus (quoted in William Godwin, 1820) saw it as one of the great famine areas of the world: 'India . . . has in all ages been subject to the most dreadful famines.' According to him, famines seem to recur in India at periodical intervals. Every 5 or 10 years, the annual scarcity widens its area and becomes a recognized famine; every 50 or 100 years, whole provinces are involved, loss of life becomes widespread, and a great famine is recorded. In the 140 years since Warren Hastings initiated British rule in India, there have been 19 famines and five severe scarcities. Braudel (1984: 41) refers to the 'terrible and almost general famine in India in 1630–31'. In 1769–70, it is estimated that about 10 million people died of starvation. Famines continued throughout the 19th century. Seavoy lists some 12 major peace-time famines between 1812 and 1901. The closing decade of the 19th century 'was distinguished by the occurrence of terrible famines'(MacFarlane, 2002); indirect calculations suggest that 19 million persons perished. Others estimate that 10 famines occurred in India between 1860 and 1900. A census report of 1901 stated that 'in ancient times the occurrence of a severe famine was marked by the disappearance of a third or a fourth of the population of the area afflicted'. After 1901, there was a lull in famines, but then they recurred. The Bengal famine of 1943–44 has been studied extensively. It is reckoned that it took between 1.5 and 3 million lives, which was 'due in part to the complications of war and the (British) administration's incompetence, but little to do with crop failure'(MacFarlane, 2002). Even in 1966–67, 'only a major inflow of food . . . saved India from famines in which hundreds of thousands, if not millions, might have died'(MacFarlane, 2002). The Bangladesh–Bengal–Assam food shortage of 1974–75 developed into a famine that claimed an estimated 1.8 million lives.

These famines[2] were caused due to a variety of reasons. In order to deal with them, both the government and the community responded in various ways over the years. To put the discussion in this book in perspective and in order to better appreciate the agenda proposed in Chapter 7, a select suite of policies pursued by different governments is discussed in what follows. The policies discussed in this chapter are by no means, exhaustive; neither space nor the scope of this work permits such a discussion, important though it

[2] Without going into the polemics of the debate, famines here have been taken as synonymous with food insecurity.

may be. This chapter does not provide a detailed typology of all programmes across Asia, of all the food security programmes in the countries in Asia, but offers a glimpse (a summary, to make it more readable) of what governments in various countries have tried to do, without being judgemental.

Increasing Food Production: Technology

In the past, when tackling hunger was thought of as being synonymous with producing more food (read guaranteeing food availability), it was common to see greater reliance on what came to be known as the 'new technology' (use of high-yielding varieties of seeds, chemical fertilizers and irrigation), or what some rechristened as the Green Revolution, in agriculture to boost food availability. For example, the Government of Indonesia combined price interventions and economic incentives through subsidized inputs, substantial investment in irrigation and rice marketing activities in the outer islands to encourage agricultural production, especially of the staple crops (Bautista et al., 1997; Piggot et al., 1993). Until 1998, policies included intensification programmes, Bimbingan Massal Swa Sembada Bahan Makanan (BIMAS) or mass guidance for food security, for rice, field crops and livestock (combination of subsidized inputs and guaranteed prices for output), and 'nucleus estate' programmes aimed at integrating smallholders into large plantation production.[3] During the late 1980s and 1990s, one of the major domestic reforms affecting agriculture was the phasing out of input subsidies. The subsidies on pesticides were removed in 1989. Fertilizers subsidies, by far the largest input subsidy, were eliminated in December 1998 (Barichello, 2003), but reinstated in 2003.

In China, after 1985, technological advances were the main engine of productivity growth (Huang and Rozelle, 1996). China was one of the first countries to develop and extend Green Revolution technology in the 1960s, 1970s and 1980s. It developed hybrid rice in the late 1970s, and until the mid-1990s it was the only country to have commercialized the technology of hybrid rice production.

---

[3] These programmes promoted high-yield varieties together with subsidized fertilizer, pesticides and credit, and offered technical assistance to farmers on the new cultivation techniques (Fuglie, 2001).

However, China's agricultural research faced major challenges by the late 1980s (Pray et al., 1997), and it was apparently not responding to many demands for new agricultural technologies; moreover, the system of providing extension services was chaotic. Thus, in the mid-1980s, China launched a nationwide reform of agricultural research to increase research productivity. Funding for institutional support was shifted to competitive grants, supporting agricultural research that was useful for economic development. Applied research institutes were encouraged to become self-financing by selling the technology that they were able to develop. In the late 1980s and early 1990s, new horticultural seeds, improved breeding livestock (Rae et al., 2006) and new dairy technologies were imported (Ma et al., 2006).

After the mid-1990s, investment in R&D began to rise. Funding for plant biotechnology was increased. China is now a global leader in agricultural biotechnology: in the late 1990s, China invested more in agricultural biotechnology research than all other developing countries combined, and its public spending on agricultural biotechnology was second only to the USA. Investment in government-sponsored R&D increased by 5.5 per cent annually between 1995 and 2000 and by 15 per cent per year after 2000 (Hu et al., 2007).

The investment in R&D has paid off. During China's early reform period, yields of major food crops rose steadily. Although there was concern about the effect of the slow-down on R&D spending during the 1980s and early 1990s, the growth of output continued to outpace the growth of inputs; productivity trends continued to rise. During the 1980s and early 1990s, China's total factor productivity (TFP) increased at 2 per cent per annum in all provinces and for all crops. This must have increased the incomes of all farmers, regardless of whether the crop was being protected or taxed, thereby contributing to better food security. What is significant is that when new technologies are released, poor farmers have been just as likely to adopt them as wealthier farmers (Huang et al., 2002), and that TFP in poorer areas also rose very fast (Jin et al., 2002). There was apparently no measurable negative impact of the extension of new agricultural technologies on the poor in China.

The Green Revolution substantially increased rice production in Indonesia, where rice is the main staple crop. Similarly, in India, from the mid-1960s to the late 1980s, the Green Revolution did give the country the fruits of new technology: rice production increased from 53.6 million tons in 1980 to 100 million tons in 2007–08; wheat production increased from 54.5 million tons in 1990 to 78 million tons in 2010–2011. Over 30 per cent of the yield growth of wheat during the 1970s and 1980s can be attributed to the provision of irrigation and electricity (Kumar, 2001). But, hunger could not be eliminated because merely increased food availability was not enough. Moreover, since then, productivity gains have slowed down, leading some experts to call it 'the yellowing of the green revolution' (Shiva, 1991), in part because of falling soil nutrition and partly because of the neglect of investment in rural infrastructure, especially after the 1990s following the initiation of the structural adjustment programme. There have also been other serious externalities. For example, because the government provided substantial power subsidies, there was overuse of groundwater, waterlogging, soil salination and falling water tables in the Green Revolution belt (FAO, 2006). All said and done, the Green Revolution did bring a lot of people out of hunger, including through enabling the government to run programmes based on availability of food, to which we shall return later. 'On balance, the Green Revolution probably saved millions from famine' (The Unfinished Campaign against Hunger, 2009).

Given that supply of land is fairly inelastic, a critical option for many governments even today is to raise levels of food production through new technology in agriculture, among other things. GM crops and biotechnology are now at the forefront of technological advancements for increasing food supply. Unfortunately, as of now, only one variety of rice, the 'Golden Rice', has been produced as a GM food crop.[4] The transgenic 'Golden Rice' contains two 'transgenes' from daffodils and one from a bacterium, which, taken together, have the effect of accumulating pro-vitamin A in the rice endosperm and, hopefully, in the final polished rice grains. Experiments are also under way on several species of fish. In fact, scientists are reportedly trying to insert the 'Bt' gene into several crops, not knowing whether this is desirable or not. Most importantly, the risks associated with these technologies are not fully

---

[4] Besides cotton, genetic engineering experiments are being conducted on maize, mustard, sugarcane, sorghum, pigeon-pea, chickpea, rice, tomato, brinjal, potato, banana, papaya, cauliflower, oilseeds, castor, soya bean and medicinal plants (Sharma and Kapoor, 2004).

known, the consequences for hunger are uncertain, and cultivating GM crops requires huge additional investment.[5] Thus, governments in the region are eager to enhance food availability to deal with incipient hunger, but are proceeding cautiously. The Government of India has decided to put in place robust regulatory mechanisms, like the Genetic Engineering Approval Committee (GEAC) under the Ministry of Environment and Forests, to approve the results of GM crops.

Increasing Food Production: Land Reforms

Many countries that face severe structural constraints to increasing food production, such as inequitable or inefficient ownership of land, have undertaken land reforms to increase food production.

China, for example, has been through a series of land reforms. In the 1940s, the Sino-American Joint Commission on Rural Reconstruction, with the support of the national government, carried out land reform and community action programmes in several provinces. Then a thorough land reform programme was launched by the Communist Party of China in 1946 (made famous in the west by William Hinton's book *Fanshen*), under which all the lands of landlords were expropriated and redistributed, so that each household in a rural village would have a comparable holding. In the mid-1950s, a third land reforms programme made individual farmers join collectives, which, in turn, were grouped into people's communes with centrally controlled property rights and an egalitarian principle of land distribution. This policy was generally a failure in terms of production, and was therefore reversed in 1962 through the proclamation of the Sixty Articles. As a result, the ownership of the basic means of production was divided over three levels with collective land ownership vested in the production team. The fourth set of land reforms in the 1980s, one of the most significant developments in improving land tenure systems in Asia, was the establishment of the household responsibility system (HRS) in China between 1979 and 1984, triggered by the belligerent action of a group of farmers in Xiaogang village, Anhui province, in 1979 (see Box 4.1 for details).

---

[5] These technologies may lead to patented seeds that keep the control of production in the grip of multinational companies (Southgate et al., 2007).

**BOX 4.1   Production Response from China:
Ushering in Reform of Global Proportions**

In a village where it all began!!

Like other villages, village Xiaogang, Anhui province, China, went under collective ownership of land in 1956, later becoming the people's commune. Everyone earned from 1 to 10 work points per day according to the quantity and quality of their labour. A boy could earn up to 2 work points—worth about 1 fen ($0.001), herding the commune's cattle (Li Honggu, 2009). Farmers were forbidden from leaving the commune to work outside, but had little interest in agriculture. 'Begging could feed them better' (ibid.).

The situation was so bad that, in 1962, among the 34 households in Xiaogang village, 60 people had died, succumbing to starvation. Into the 1970s, life barely improved, with the final straw being a severe drought in 1978. That drought pushed 18 famished farmers in the remote village of Xiaogang, taking matters into their own hands, to do away with the collective farming that had suppressed grain production and left them hungry for years, 'even expecting death for it' (*Economist*, 2009). Recalling that fateful night in December 1978, Mr Guan Youjiang (now aged 62), one of the 18 men who pressed their red-inked thumbs onto a crumpled piece of paper to do away with collective farming, said: 'Many of us could not even understand all the words on that piece of paper. But we knew we just had to do it. We had to fill our stomachs, feed our families' (Sim Chi Yin, 2008).

Thus in 1978, the starving farmers of Xiaogang began parcelling out their hitherto communal land, cattle and farming tools to individual family units which eventually became known as the household responsibility system (*dabaogan*). At first, families were allotted equal plots of land with common access to ponds, through draw of lots (Li Honggu, 2009). But in years of drought, disputes arose over water usage, prompting a second land distribution in 1983, which guaranteed a pond for every family. Thus, rural reforms began in late 1978 in the central province of Anhui. In consequence, productivity and grain production rose dramatically as the farmers found a new incentive to work hard to better their own lot. Though local officials in Xiaogang initially did not approve of this move, it later won the 'backing from a provincial leader and Deng ally, Mr Wan Li. Others gradually followed suit' (*Economist*, 2008). Indeed, it was at this critical juncture that Den Xiaoping himself declared that farmers' decisions should be respected, hailing the new system as 'a great creation of Chinese farmers' (*China Daily*, 2009). Though the communes were officially dismantled in 1984, by the end of 1982, more than 90 per cent of China's agricultural households had returned to some sort of family farming.

Source: Sim Chi Yin (2008).

In 1979, HRS radically altered the organization of production in agriculture and the incentives for rural households (Rozelle et al., 2008). The HRS reforms dismantled agricultural collectives and contracted agricultural land to households, mainly on the basis of family size and the number of labourers in each household. Significantly, control and income rights belonged to individuals after the HRS reforms. Land was not privatized, however: ownership remained with villages of about 300 households or small groups of 15–30 households. Even if they did not own the land, farmers were able to keep all of the grain earnings. Under this system, land that had been farmed collectively was contracted to individual households. Initially, the land use right was contracted for 15 years, but later this increased to 30 years. By 1984, 99 per cent of agricultural land was contracted to individual households for 15 years. The impact of the HRS reforms could not have been more dramatic: productivity, output and incomes rose. The effect on hunger was also dramatic. By 1984, China had a bumper harvest of 407 million tons of food, making it the first time in years that the country had enough to feed itself. China expects a harvest of 528.2 million tons of grain in 2009 (*China Daily*, 2008: 1). In October 2008, the government agreed to pass further legislation to permit contract holders to transfer their land use rights or to lease the land (Government of China, 2008). On the whole, land reforms had a very positive role in reducing hunger in PRC. It is often thought that this rise in the vibrancy of the rural economy was one of the triggers of the rest of the economic reforms in China (Rozelle et al., 2008).

Brandt et al. (2002) demonstrate that the system of land rights initiated by the HRS reforms was mainly beneficial to farmers, and that the cost of insecure tenure was not too serious for agricultural output, at least in the short term. After several years of policy debate, one of the most important changes in recent years has been the renewal of land use contracts for an additional 30 years. Cultivated land is not private, but the right to use land has been granted until 2028.

A major new policy was established by the Rural Land Contract Law (RLCL) approved in 2006 by the Standing Committee of the National People's Congress, according to which ownership of land remains with the collectives but other rights are given to contract holders, such as they would have under a private property system. In particular, RLCL clarifies the rights of transfer and exchange of contracted land: this may be taking effect already, because researchers are finding that more land in China is being rented. The law also allows family members to inherit land during the contract period. The aim is to encourage farmers to use their land to increase short-term and long-term productivity.

China's central leadership has begun to increase the rights of rural families over their cultivated land. The recent pronouncements at the Third Plenary Session of the 17th Central Committee of the Communist Party of China, in October 2008, try to bring forward the implementation of RLCL. There is a perception that, despite RLCL, tenure security is still weak, and as a result, farm size and the quality of investments in land are limited. Without secure tenure, rural residents do not have the asset base to access finance that would permit them to move to cities, improve their land or expand off-farm businesses. The debate in China is now whether or not the rural economy is ready for indefinite, titled land security. Fully secure tenure will probably not occur immediately, but with the continued effort of reformers, the demand for it will gradually become stronger.

Land reforms were also started in India. Due to the taxation and regulation policies under the British, at the time of independence in 1947, India inherited a semi-feudal agrarian system, with ownership of land concentrated in a few individual landlords (also called zamindars). Within a national agenda of rural reconstruction since 1947, land reforms in India had the major objectives of:

- reordering of agrarian relations in order to achieve a democratic social structure;
- elimination of exploitation in land relations and enlarging the land base of the rural poor; and
- increasing agricultural productivity and infusing an element of equality in local institutions.

Land reforms involved three interrelated steps: recording the tenants so that they could not be evicted at will; putting legal limits to the rent that an absentee landlord could exact from the tenants; and land ceiling, which implied distribution of surplus land to the landless. Land reforms were successful in West Bengal in so far as they succeeded in recording one million of the 1.8 million tenants as sharecroppers, and distributed surplus land to landless households. The experiment also succeeded in Kerala. Another successful land reform programme was launched in Jammu and

Kashmir after 1947. However, this success was not replicated in other states like Andhra Pradesh and Madhya Pradesh. In the state of Bihar, tensions between land owners' militias, villagers and Maoists have resulted in numerous massacres. In the state of Uttar Pradesh, former feudal lords still own hundreds of acres of land, either by exploiting legal loopholes or through illegal stratagems. The number of the absolutely landless and the near-landless (those with less than half an acre of land) make up 43 per cent of rural households in India.[6] All in all, land reforms have been successful only in pockets of the country, as people have often found loopholes in the laws setting limits on the maximum area of land held by any one person. The impact of land reforms on hunger in India has also not been assessed, but West Bengal, which undertook vigorous land reforms, has witnessed the highest rate of growth in food production over the last decade in India. By implication, the reforms played a substantive role in reducing hunger.

Land reforms were started in many other countries in Asia with mixed success, such as in the Philippines. In September 1972, the entire Philippines became a land reform area, though only rice and corn lands were included. Holdings of more than 7 hectares were to be purchased and parcelled out to individual tenants (up to 3 hectares of irrigated or 5 hectares of unirrigated land), who would then pay off the value of the land over a 15-year period. Sharecroppers on holdings of less than 7 hectares were to be converted to leaseholders paying fixed rents.

The Marcos land reform programme succeeded in breaking down many of the large haciendas in Central Luzon, but in the country as a whole, the programme was generally not very successful. For example, reportedly only 20 per cent of rice and corn land, or 10 per cent of total farmland, was covered by the programme. Thus, in 1988, the country undertook the Comprehensive Agrarian Reform Programme (CARP) under which land owners were allowed to retain up to 5 hectares of land plus 3 hectares for each heir at least 15 years of age. The amount of land that could be retained was to be gradually decreased, and a non-land-transfer, profit-sharing programme to be used as an alternative to actual

---

[6] NaukriHub, 'Land Reforms in India and Their Implications'. http://www.naukrihub.com/hr-today/land-reforms.html (accessed on 1 September 2009).

land transfer. The government claims that in the first three years of its implementation, CARP met with considerable success, but this is open to question. Though between July 1987 and March 1990, 430,730 hectares were distributed, about 80 per cent of this was from the continuation of the Marcos land reform programme. Distribution of privately owned lands other than land growing rice and corn, of 3,470 hectares, was insignificant in absolute terms, and was only 2 per cent of what had been targeted. As CARP faced its possible second extension in 2007, a review of the programme commissioned by German Technical Assistance (GTZ) said: 'Despite a total of 34 years of agrarian reform—agrarian reform failed to increase agricultural productivity, reduce rural poverty, and land-owners' investment in rural-based industries' (Ombion, 2007).

In the 1980s, Vietnam embarked on the transition from a socialist command economy towards a market-oriented economy. An important (and catalytic) part of the transition was major reforms in the agricultural sector, which would lead Vietnam's graduation from the socialist mode of agricultural production to a market-oriented one (Ravallion and van de Walle, 2008). In the years after World War II, even before the formal division of Vietnam, land reform was initiated in North Vietnam. This land reform programme (1953–56) redistributed land to more than two million poor peasants, though at considerable cost. South Vietnam made several further attempts in the post-Diem years, the most ambitious being the Land to the Tiller programme instituted in 1970 by President Nguyen Van Thieu. Under this programme, the landholdings of individuals were limited to 15 hectares. The owners of expropriated tracts were compensated, and legal title was extended to peasants in areas under the control of the South Vietnamese government, to whom land had previously been distributed by the Viet Cong. While it was implemented effectively only in some parts of the country, 'in the Mekong Delta and the provinces around Saigon, the programme worked extremely well. . . . It reduced the percentage of total cropland cultivated by tenants from sixty percent to ten percent in three years' (Moyar, 1996).[7] Vietnam's story of increasing availability of food has been well documented.

---

[7] Land distribution and tenure were also major issues in many other countries—often with gender implications, since, despite women being responsible for much of the food production, the land titles are typically granted to men (Agarwal, 2003).

Increasing Food Availability: Trade Policies

China has been very successful in its efforts to liberalize agricultural trade. Trade barriers have fallen. Rights to import have been extended in the case of most commodities to thousands of private traders and trading enterprises. Non-tariff barriers have been reduced. On the one hand, there is evidence that China has responded to signals from world markets and made adjustments in its production structure to reflect its comparative advantage; on the other, China has tried to minimize the impact on those hurt by trade liberalization (see Huang et al., 2004, for more details).

Increasing Food Availability: Fiscal Policy

In the past, countries concerned about low agricultural prices set minimum support prices to offer some incentives to farmers and reduce uncertainty (Timmer and Dawe, 2007). The agricultural price policy in India in the late 1960s sought to provide adequate incentives to farmers for sustainable increase in food production, to evolve an agricultural production system in sync with the Indian economy, and increase the economic and physical access of the masses to food. The policy was modified in 1980 and in 1991. The policy includes a minimum support price offered to farmers in India for some of the more important crops like rice, wheat and sugarcane (Government of India, 2009). The minimum support price is the guaranteed support price (declared every year) at which the government buys specified food-grains from farmers. This provides incentives for farmers to grow more food without being vulnerable to the vagaries of volatile food prices. And the food procured by the government is distributed to the hungry (and the potentially hungry) at predetermined prices (usually lower than the market price) through what is called the public distribution system.

The price policy in India has contributed much to issues pertaining to food security. According to one estimate, price policy interventions contributed 53 per cent to yield growth and 51 per cent to output growth in rice during the 1990s. Their contribution to the growth of productivity and production of wheat was 70 and 48 per cent, respectively, during the same period. However, the price policy also created certain distortions in the food economy.

More recently, several countries like Afghanistan, Azerbaijan, Bangladesh, Bhutan, China, Fiji, India, Indonesia, Kazakhstan, Kyrgyzstan, Maldives, Mongolia, Pakistan and Sri Lanka, faced with volatile and rapidly rising food prices, have used reduction in domestic taxes, such as the value added tax (VAT), on basic food commodities to stabilize food prices (World Bank, 2008). Fiji went even further to introduce zero-rating for VAT on locally produced eggs to lower its prices. Such measures have the benefit of lowering food prices and increasing the economic access to food for the hungry, but their impact depends on the commodities that are targeted and whether they are consumed by the hungry. This is particularly important, since the consequent reduction in government revenue is significant. For example, loss in revenue averages from 0.15 per cent of GDP in Bangladesh, India and Indonesia, to 0.6 per cent of GDP in the Solomon Islands (IMF, 2009).

The economic crisis and the drought in 1998 resulted in severe food shortages in various parts of Indonesia. In August 1998, the Government of Indonesia introduced a targeted rice subsidy programme (called the Operasi Pasar Khusus [OPK] or the Special Market Operation) under which eligible households were allowed to buy up to 20 kg of rice per month at a subsidized price of Rupiah 1,000/per kg, to help the poor have access to rice, their main food staple. The programme reached 10.5 million households in 2000. The OPK was an in-kind income transfer. In its first year of operation, this transfer was equivalent to 11 per cent and 17 per cent, respectively, of the monthly income expenditure of an average poor household and of the lowest income decile. Additionally, since 2001, the Government of Indonesia has allocated Rupiah 279.9 billion to support a similar programme as a compensation for reduction of fuel subsidies. Though far from perfect, both these programmes have been considered the most successful so far in bringing food directly to the poor (Dillon and Rantetana, 2005).

The Government of Indonesia also has vast experience with fiscal policy on various other items of basic needs. Policies and programmes implemented by the Government of Indonesia have included safety nets, such as health insurance for the poor, a school operational assistance programme, free elementary and junior high-school education, and financial support for the poor including subsidies on fuel, rice and cooking oil (ESCAP, 2008). In 2008, Indonesia implemented non-targeted price controls and consumer subsidies, reduced import duties and VAT on basic food commodities, and increased domestic supply using food-grain stocks (World Bank, 2008). More specifically, this included:

- Increased rice subsidies
- Subsidies on cooking oil for poor households
- Price subsidies for small-scale producers of processed soybeans
- Removal of import tariffs on flour and soybeans
- Increased export tax on palm oil
- Exemption of cooking oil from VAT
- Increased supply of rice from government stocks

On the other hand, over the same period, food subsidies were decreased in India, Indonesia and Turkmenistan (World Bank, 2008). Governments have also offered subsidies on fuel to reduce transport costs, which contributes to reducing the price of food and increasing the economic access to food for the hungry. For example, since 2001, the Government of Indonesia has undertaken a programme to provide fuel to the poor at subsidized rates as compensation for reduction in fuel subsidies (Dillon and Rantetana, 2005). However, subsidy on transport also benefits many non-hungry households. Subsidies on food, by comparison, tend to be more pro-poor and also less expensive.

The Government of China launched a major programme of direct subsidies to farmers in 2004 and is currently debating the extent to which these should be increased. The national grain subsidy system, which is designed to increase the production of grain for national food self-sufficiency and to be a rural income-transfer programme, is a combination of four elements:

- Subsidies for farmers in areas that grow grain
- Nationwide agricultural seed subsidies
- Input subsidies—payments to help farmers cope with the rising costs of fertilizer and other inputs
- A general transfer programme

Nearly 80 per cent of farm households receive subsidies. Payments were relatively small in the first year of the programme, but by the second year many farmers were receiving CNY 20–30 (US$3.50, or US$11–17 in PPP terms) per mu (15 mu = 1 ha).

The Government of China has also eliminated almost all taxes and fees in villages. In 2001 and 2002, these fees were all converted into a single agricultural tax that was limited to 8.5 per cent of the gross value of agricultural output of a household or village. Subsequently, even this tax was eliminated altogether, and by 2007 farmers were paying almost no taxes.

In summary, taken together, the recent policy innovations in rural infrastructure, free nutrition in rural schools, agricultural subsidies, tax reductions and health insurance subsidies have been substantial in China: they have contributed significantly to the observed improvements in household incomes in rural areas, better economic access to food and food security.

Thailand has a similar approach to rice, called the government's income guarantee policy. Under this policy, rice farmers receive the difference between the guaranteed price (currently at Bhat 10,000 per ton) and the reference price of rice. The government sets the reference price depending on the market price. For example, in the second week of March 2010, the reference price for rice was set at Bhat 8,718 per ton, down from the previous week's Bhat 9,074 per ton. The market price for white rice during the same period was quoted at Bhat 8,700–9,100 per ton. Thus, on 10 March 2010, to shore up domestic prices and farmers' income, the government hiked its compensation to farmers from Bhat 926 to Bhat 1,282 per ton (Pratruangkrai, 2010).

Investment in agriculture has also been resorted to by several countries to increase food availability. Though growth in food output may have been held back by government policy, which has tended to tax agriculture at the expense of urban consumers, in recent years this trend has started to reverse (Anderson, 2008). Notably, China and India in Asia have rapidly increased their spending on agricultural R&D over the past few decades. For example, these two countries together with Brazil increased their spending on agricultural R&D from a third for all developing countries in 1981 to half in 2000 (World Bank, 2008).

Stocks and Reserves: Stabilizing Supply and Insurance against Risks

Governments in Asia and the Pacific have traditionally held national stocks of rice and other staple foods to serve as buffers to meet food shortages during calamities and periods of volatile food prices and to combat speculative activities. In addition to acting as a kind of insurance against covariate risks, building these stocks has entailed purchases from farmers, which contributes towards ensuring that farmers receive the minimum prices for their crop and have the right incentives to grow more food. India, for example, built a reserve of 50 million tons of rice and wheat in 2009 (Dev, 2009), and prevented millions from falling into hunger due to the failed monsoon in 2009. This building of food reserves goes hand in hand with the minimum support price policies of the governments mentioned earlier, and often with the public distribution system, to which we shall come later. At times of shortage, food held in stocks and reserves is released to those in need of food assistance, along with creating employment towards increasing the economic access of the hungry to food through such programmes as the 'food-for-work' programme in Bangladesh, Cambodia and Laos and employment guarantee programmes such as the National Rural Employment Guarantee Scheme (NREGS) in India.

China set up a grain reserve system in 1990. Reserves were divided based on their importance into four categories: central, local, national temporary and commodity.

> Last April, Premier Wen Jiabao said China had grain reserves of 150 to 200 million tons. That's equal to about 30 to 40 per cent of China's annual grain consumption or double the 17–18 per cent level regarded by the UN Food and Agriculture Organization (UNFAO) as a safe minimum for global stocks. (Xinhua, 2009)

These reserves helped China deal with impending hunger in 2007, which it could not do in earlier shocks, like that of 1959–61, when China was hit by a severe famine, and 30 million Chinese starved to death as the country had no food in reserve to tide over the difficulty (Brown, 1995).

Indonesia has operated a logistic agency, Badan Urusan Logistik (BULOG), since 1968, with the objective of, *inter alia*, undertaking rice procurement to guarantee a minimum price for farmers, distributing the food-grains so procured to guarantee food availability, and maintaining national food stocks to shield the domestic prices of food-grains from international price volatility. The agency has been by and large, successful in meeting the needs of the urban consumers (Dillon and Rantetana, 2005). However, in the rural areas, the picture is different.

Inter-governmental agencies have also held food reserves. In 1988, SAARC established a food security reserve, consisting of 241,580 tons of food-grains, including rice and wheat. However, despite a number of crises, this reserve had not been drawn upon by 2004, partly because in order to draw upon it, governments would have to declare a national food emergency, which they were reluctant to do. At a SAARC Summit in 2004, therefore, government representatives recommended that the reserve be re-established as a more flexible regional food bank. Each country would contribute a certain amount to this bank, which it could draw upon as needed, but it could also borrow from food held in storage in a neighbouring country. The borrowing country would not be charged for this, but would need to replace the stock when in a position to do so. The SAARC Colombo Summit in 2008 called for revamping the SAARC food security reserve. Governments which use food stock reserves to avert hunger generally do so by combining these with trade and other measures. All these measures can have different effects for producers and consumers, some positive, some negative. For some, it can be difficult to gauge the net impact.

Under the ASEAN Emergency Rice Reserve, the five original ASEAN member countries (Indonesia, Malaysia, Philippines, Singapore and Thailand) made a commitment to voluntarily pool rice into a common regional stockpile, with the main objective of meeting emergency requirements from extreme and unexpected natural or man-made calamities. The agreement was not intended to fill continuing deficits of individual member countries, which were usually met through imports. The total initial rice reserve was 50,000 tons. When Brunei Darussalam and Vietnam joined ASEAN, the reserve ballooned to 67,000 tons in 1997. However, in spite of more than 27 years of existence, the reserve did not increase significantly, reaching only 87,000 tons. In fact, it was not even enough to meet a half-day ration for the rice-consuming populations of the

10 ASEAN member countries. Considering that the ASEAN emergency rice reserve has never been operationalized, and since the small stock could not meet emergency needs, it is difficult to assess if it did contribute to reducing hunger in ASEAN countries.

Protecting Special Categories of People from Hunger: Food Transfers

Food from reserves and elsewhere can be released into national food markets—sold by the government to traders who in turn sell it to needy people at a predetermined price, usually lower than the market price. But food from food reserves, along with food purchased locally or internationally, can also be distributed to those in need. In some cases, food from food reserves (like subsidized wheat) is also provided unconditionally, as was done recently in Bangladesh, Fiji, India, Indonesia, the Philippines, Pakistan and Sri Lanka, through what is called a 'ration card' system (World Bank, 2008), where everyone is issued a card indicating the maximum quantity of food that s/he can access at predetermined prices over a specified period (usually a month).

India's public distribution system (PDS) is the largest food distribution system in the world, reaching 600 million people (del Ninno et al., 2005). The system is based on food procured from the farmers by a state agency (called the Food Corporation of India) at the declared procurement prices and distributed through what are called the ration shops. Households below the poverty line can purchase 35 kilograms of rice/wheat per month at 48 per cent of the cost to the government.[8] Households above the poverty line are obliged to pay 70 per cent of the cost. One of the most important considerations in these programmes is targeting—since distributing food to large numbers of people can be prohibitively expensive.

Currently, there is a debate going on in India regarding the Right to Food Act. Under the proposed Food Security Act, the government plans to give a legal entitlement to cheaper food-grains at the rate of 4 kilograms of food-grain per person each month to nearly 70 per cent of the country's population. Seeking the universalization of the PDS entitlement, the campaign said that the problems

---

[8] These figures keep changing.

related to identification of below-poverty-line households and exclusion errors were well known.

Sri Lanka had a programme to provide food at subsidized rates more than 60 years ago during World War II, when the government subsidized a rationed amount of rice to all consumers. Gradually, the subsidies were targeted more precisely: by 1978 this measure had reduced the proportion to 19 per cent of total cost of food grains and by 1984 to 4 per cent (Foster and Leathers, 1999).

Food-for-work is a conditional food programme that serves as an alternative to selecting beneficiaries for receiving subsidized food based on income. The food-for-work programme is a system of self-targeting, directed to those who provide unskilled or semi-skilled labour to build social overheads such as roads, bridges and culverts or undertake social forestry, on the assumption that the non-poor would not opt for such menial work in exchange for the kind of food on offer. Afghanistan, Bangladesh, Cambodia, India and Nepal have been implementing food-for-work programmes for a fairly long time. In fact, Bangladesh expanded its food-for-work programme in early 2008 in response to disasters and the increased price of food. The impact of the food-for-work programmes has been assessed extensively and, by all accounts, both during normal times and in emergencies, they have been efficacious in reducing hunger (Dev et al., 2004).

Conditional programmes for providing supplementary nutrition have also been devised, such as food-for-education, or the mid-day meal scheme in schools. Distribution of food through schools encourages school attendance, and the food not only provides school meals, but also offers rations for the rest of the family to reduce their hunger. In Bangladesh, the Food-for-Education programme, changed to Cash-for-Education in 2002, has provided cash transfers to households with children in poor areas on the condition that children are enrolled at school and maintain a minimum attendance level.[9] These cash transfers add to the purchasing power that poor households use to access food. Governments in Cambodia, China, Bhutan, India, Maldives, Pakistan and Sri Lanka,

---

[9] The main welfare outcomes show a 9–17 percentage point rise in the school enrolment rate (from a base of 55 per cent), and nearly full attendance among beneficiaries, with improvement in long-term opportunities for children.

among other countries, have also used school feeding programmes (World Bank, 2008). The food-for-work programme has also been used in Bangladesh for capability development of the people, which has had the effect of enhancing their economic access to food in the longer term. In its Income Generation for Vulnerable Group Development Programme, the Government of Bangladesh provides nearly 300,000 rural women with 30 kilograms of wheat per month, as long as they participate in savings groups and receive training for entrepreneurial skills and enterprise development (Ahmed, 2005). These programmes make a significant contribution to reducing hunger in the country.

The Integrated Child Development Scheme in India seeks to integrate efforts at improving child health and nutrition, non-formal education as well as maternal health and nutrition into a single service delivery window, reaching out to 4.8 million expectant and nursing mothers and 22.9 million children under 6 years of age, through a network of 4,200 projects, covering nearly 75 per cent of the development blocks and 273 urban slum pockets in the country. All ICDS services are provided through childcare centres (called Anganwadi). Under its supplementary nutrition component, food supplements are provided to the daily diet of the beneficiaries, and are meant for consumption at the childcare centre itself. Severely malnourished children are given double the daily supplement provided to the other children. In addition to calories and proteins, specific micronutrients are also provided in accordance with the requirements of various age groups. The success of the scheme varies across states. A concurrent evaluation of the ICDS reveals that in terms of beneficiaries' satisfaction, more than 80 per cent of the surveyed households were satisfied with the functioning of the nutrition component. However, state-wise variations of coverage do exist, ranging from about 90 per cent of households in some states to only 30 per cent in others.

It may be added that all schemes for food transfer need to be carefully designed so as not to distort local food markets, to prevent both a collapse of food-grain prices that hurt the local farmers and also a surge in food prices by procuring food locally, hurting poor consumers. In many cases, procuring a basket of food locally and targeting the distribution to those who cannot access food helps to keep prices unharmed (Barrett and Maxwell, 2005), while opening markets for local food producers. In Bangladesh, for example,

---

BOX 4.2   Recent Policy Developments in Asia

Bangladesh

23 October 2009: The government removed the ban on rice exports imposed in November 2008, and allowed private traders to export 10,000 tons of aromatic rice from 1 September to 31 December 2009.

China

12 October 2009: The State Council raised the minimum purchase price of wheat by CNY 60 per ton to CNY 1,720–1,800 (US$252–US$264 per ton). The minimum purchase price of rice was also to be increased. The government would also continue buying other major crops, including maize, soya beans and rapeseed, for state reserves to stabilize domestic output.

23 October 2009: Export taxes were levied on wheat and rice (3 per cent), wheat flour, wheat starch and rice flour (8 per cent). Taxes on soybean (5 per cent) and soy flour (10 per cent) were removed. Maize export taxes had already been removed in 2008.

India

20 August 2009: The government increased the minimum support price for rice by 5.40 per cent to INR 950 per kilogram (US$198 per ton).

22 October 2009: The government lowered by 18 per cent the minimum export price for basmati rice from US$1,100 per ton fixed in January 2009 to US$900 per ton.

22 October 2009: Duty-free sugar imports were extended until March 2010 for raw sugar and up to November 2009 for white sugar.

22 October 2009: The government announced that it would continue the ban on export of non-basmati rice introduced in 2008, in view of an expected historical-low rice crop in the 2009–10 crop year. According to the new arrangements, the ban would remain in place until the middle of 2010.

(BOX 4.2 *Continued*)

(BOX 4.2 *Continued*)

22 October 2009: The government announced the sale of a further 1 million tons of wheat from strategic reserves under the open market sale scheme (OMSS), after the open market sale of 3 million tons of wheat and 2.5 million tons of rice from state reserves on 18 August 2009.

27 October 2009: The government announced the removal of the 70 per cent import tax on certain varieties of rice to boost supplies, after late and uneven monsoon rains led to a significant reduction in the main (Kharif) crop plantings and production. Duty-free imports of semi- and wholly milled rice would now be permitted until 30 September 2010.

Indonesia

22 October 2009: The Logistics Agency (BULOG) planned to release 2,250 tons of rice through a market operation to avoid price spikes before the harvest of the second season.

Japan

22 October 2009: Japan cut the price at which it sold imported wheat to domestic flour millers by an average 23 per cent to JPY 49,820 (US$549) per ton.

Pakistan

23 October 2009: To counter rising food prices, particularly for sugar, the Lahore High Court ordered traders to ensure a retail price of PKR 40/kilogramme (US$0.50), 27 per cent lower than the peak of the previous month.

23 October 2009: The government removed a 35 per cent export duty on wheat products. The ban was imposed in 2007 because of shortages and high domestic prices.

Philippines

28 October 2009: The National Food Authority announced that it would allow private-sector traders to import up to 563,000 tons of rice annually. The measure aimed at enhancing market participation ahead of liberalization of the sector, including the removal of quantitative restrictions on imports in 2012.

(BOX 4.2 *Continued*)

(BOX 4.2 *Continued*)

Sri Lanka

23 October 2009: Fertilizer subsidies in Sri Lanka continued. The government was supplying to each farmer 5 kg of fertilizer worth LKR 9,000
(US$78.60) at LKR 350 (US$7.50), to support rice cultivation.

Thailand

21 October 2009: On 17 July, the National Rice Policy Committee agreed
to release 763,920 tons of intervention rice stocks from the marketing
year 2009/09, including 300,000 tons of fragrant rice, through tenders
for domestic and export markets.

22 October 2009: The government rice intervention scheme due to
end on 20 July 2009 was reinstated in September for a month following protests. The price of US$535 per ton for benchmark 200 per cent
B-grade white rice was maintained.

Vietnam

8 September 2009: The Vietnam Food Association (VFA) confirmed
the purchase of 400,000 tons of husked rice for state reserves under
the first phase of the procurement plan announced by the government
in mid-June. Under the plan, the VFA was instructed to buy 2 million
tons of summer–autumn rice to prevent a fall in domestic prices at the
peak of the harvest, when export demand is low.

22 October 2009: Vietnam decided to ban rice export to destinations
where sales would be in competition with government contracts
handled by the top two state-controlled exporters. According to the
Vietnam Food Association, from 10 August, exporters would not be
allowed to sell rice to foreign companies that had signed government-
backed deals, or to foreign traders who competed in the markets where
Vietnam aimed at signing such contracts.

Kazakhstan

28 October 2009: The Agriculture Minister announced that, in an effort
to improve the competitiveness of its grain exports, it would spend
US$33 million in subsidizing shipments to Baltic and Black Sea ports.

Source: FAO (2009).

biscuits provided in the school feeding programme opened a new market opportunity for local wheat farmers (Caldes and Ahmed, 2004). During the economic crisis in the 1990s, the Government of Indonesia had a countrywide school feeding scheme which excluded the local staple.[10] Meals were prepared by local women organized through local women's associations. In a survey, 72 per cent of farmers said that the school feeding scheme had given them more opportunities to sell produce from their fields and vegetable gardens (Studdert et al., 2004).

A new low-income programme is also being launched across China with a view to developing a social safety net system for people in the rural economy. The current annual payments—CNY 200 (US$26.3, or US$111 in PPP terms)—are low, but coverage is broad. It has been found that 6 per cent of rural households and 10 per cent of households in poor rural areas are receiving these transfers. It is possible that future increases in the annual amounts would eliminate much of the remaining absolute poverty and undernutrition in China.

Availability of Food at the National Level: Seeking International Food Aid

The question of availability of food is also impacted by international food aid. The United States and other countries have generated food surpluses. Food aid flows globally reach between 125 and 250 million people (Lowder and Raney, 2008), and around one-third of these flows go to countries in Asia and the Pacific. In 2007, the region's top five recipients were, in descending order: the Democratic People's Republic of Korea, 752,352 metric tons; Afghanistan, 224,237; Bangladesh, 220,988; India, 118,586; and the Philippines, 110,204 metric tons (WFP INTERFAIS data). Of the total amount of food aid flowing to Asia and the Pacific, more than half is used for emergency purposes. Around 20 per cent is used for projects such as food-for-work or school feeding. The rest is programme food

---

[10] In this case, partly to avoid meal substitution at home, it was stipulated that the programme could use only locally grown commodities.

aid, through which a donor government sells food at a discount to a recipient government, which can then sell it and use the revenue for any number of purposes.

References

Agarwal, B. 2003. 'Gender and Land Rights Revisited: Exploring New Prospects via the State, Family and Market'. *Journal of Agrarian Change*, vol. 3, nos. 1 and 2, pp. 184–224.

Ahmed, S. S. 2005. 'Delivery Mechanisms of Cash Transfer Programs to the Poor in Bangladesh'. Working Paper 32751, Social Protection Discussion Paper Series 0520, World Bank, Washington, D.C.

Anderson, K. 2008. 'Distorted Agricultural Incentives and Economic Development: Asia Experience'. CEPR Discussion Paper No. DP6914, Centre for Economic Policy Research, London, July 2008.

Barichello, R. 2003. 'Taxation, Expenditure and Policy Situation Facing the Indonesian Agricultural Sector: Factors Influencing the Demand for and Supply of Irrigation'. Report submitted as part of ADB RETA 5866 'Irrigation Investment, Fiscal Policy, and Water Resource Allocation in Indonesia and Vietnam', International Food Policy Research Institute, Washington, D.C.

Barrett, C. B. and D. G. Maxwell. 2005. *Food Aid after Fifty Years: Recasting Its Role*. New York: Routledge.

Bautista, R., Nu Nu San, D. Swastika, S. Bachri and Hermanto. 1997. 'Evaluating the Effects of Domestic Policies and External Factors on the Price Competitiveness of Indonesia Crops: Cassava, Soybean, Maize, and Sugarcane'. Trade and Macroeconomics Discussion Paper 18, International Food Policy Research Institute, Washington, D.C.

Brandt, L., J. Huang, G. Li and S. Rozelle 2002. 'Land Rights in China: Fact, Fiction, and Issues'. *China Journal*, vol. 47, no. 1, pp. 67–97.

Braudel, Fernand. 1984. *Civilization and Capitalism*. New York: Harper and Row.

Brown, Lester R. 1995. *Who Will Feed China?* Washington D. C.: Worldwatch Institute and W. W. Norton.

Caldes, N. and A. Ahmed. 2004. *Food-for-Education: A Review of Program Impacts*. Washington, D.C.: International Food Policy Research Institute.

China Daily. 2008. 'Bumper Harvests Lift Farm Income'. 29 December, p. 1.

———. 2009. 'Enlighten Policies Spread the Wealth', 13 January, p. 18.

del Ninno, C., P. A. Dorosh and K. Subbarao. 2005. 'Food Aid and Food Security in the Short and Long Run: Country Experience from Asia and Sub-Saharan Africa', Social Protection Discussion Paper Series 0538, World Bank, Washington, D.C.

Dev, S. Mahendra. 2009. 'Inclusive Growth in India: Policies, Prospects and Challenges'. Presentation at UNESCAP, MPDD Development Seminar Series, Bangkok, 2 September 2009.

Dev, S. Mahendra, C. Ravi, Brinda Viswanathan, Ashok Gulati and Sanghamitra Ramachander. 2004. *Economic Liberalisation, Targeted Programmes and Household Food Security: A Case Study of India*. Washington, D.C.: International Food Policy Research Institute.

Dillon, H. S. and Marcellus Rantetana. 2005. 'Food Security in Indonesia', in V. S. Vyas (ed.), *Food Security in Asian Countries in the Context of the Millennium Goals*. New Delhi: Academic Foundation and Asian Development Research Forum, and Bangkok: Thailand Research Fund, in cooperation with International Development Research Centre, Canada.

ESCAP (Economic and Social Commission for Asia and the Pacific). 2008. Statements Made at the Sixty-Fourth Commission Session, Bangkok, 28 April 2008.

———. 2009. *Sustainable Agriculture and Food Security in Asia and the Pacific*. Bangkok: United Nations.

FAO (Food and Agriculture Organization). 2005. *The State of Food and Agriculture: Agricultural Trade and Poverty—Can Trade Work for the Poor*. Rome: FAO.

———. 2006. *Rapid Growth of Selected Asian Economies: Lessons and Implications for Agriculture and Food Security, China and India*. Bangkok: FAO Regional Office for Asia and the Pacific.

———. 2009. *Crop Prospects and Food Situation*, No. 4, FAO, Rome, November 2009.

Foster, P. and H. D. Leathers. 1999. *The World Food Problem* (Second Edition). Boulder and London: Lynne Rienner.

Godwin, William. 1820. *Of Population. An Enquiry Concerning the Power of Increase in the Numbers of Mankind, Being an Answer to Mr. Malthus's Essay on That Subject*. London: Longman, Hurst, Rees, Orme and Brown.

Government of China. 2008. 'China Liberalizes Farmers' Land Use Right to Boost Rural Development'. Press release, 19 October 2008. www.gov.cn (accessed in October 2008).

Government of India. 2009. *The Economic Survey 2009*. New Delhi: Ministry of Finance, Government of India.

'Great Leap Forward And The Great Famine'. http://factsanddetails.com/china.php?itemid=69 (accessed on 6 March 2012).

He, S. 2007. 'Government Holds Fast to Arable Land'. China.Org.Cn, 16 August 2007. http://www.china.org.cn/english/government/221140.htm (accessed on 23 January 2012).

Hu, R., K. Shi, Y. Cui and J. Huang. 2007. 'China's Agricultural Research Investment and International Comparison'. Working Paper, Center for Chinese Agricultural Policy, Institute of Geographical Sciences and Natural Resource Research, Chinese Academy of Sciences, Beijing.

Huang, Jikun and Scott Rozelle. 1996. 'Technological Change: Rediscovering the Engine of Productivity Growth in China's Rural Economy'. *Journal of Development Economics*, vol. 49, pp. 337–69.

Huang, J., S. Rozelle, C. Pray and Q. Wang. 2002. 'Plant Biotechnology in China'. *Science*, vol. 295, no. 25, pp. 674–77.

Huang, J., S. Rozelle and M. Chang. 2004. 'Tracking Distortions in Agriculture: China and Its Accession to the World Trade Organization'. *World Bank Economic Review*, vol. 18, no. 1, pp. 59–84.

IMF (International Monetary Fund). 2009. 'World Economic and Financial Surveys: World Economic Outlook Database'. Washington: IMF.

Indian Census Commissioner. 1901. *Census of India 1901 (1902)*. Calcutta: Superintendent of Government Printing.

Jin, S., J. Huang, R. Hu and S. Rozelle. 2002. 'The Creation and Spread of Technology and Total Factor Productivity in China's Agriculture'. *American Journal of Agricultural Economics*, vol. 84, no. 4, pp. 916–30.

Kumar, P. 2001. 'Agricultural Performance and Productivity', in S. S. Acharya and D.P. Chaudhri (eds), *Indian Agricultural Policy at the Crossroads*. New Delhi: Rawat Publications.

Li, Honggu. 2009. 'Land of Opportunity'. *China Daily*, 13 January, p. 18.

Lowder, Sarah and T. Raney. 2008. 'Food Aid: A Primer'. FAO Working Paper No. 05–05, Agricultural Development Economics Division (ESA), FAO, Rome. http://ideas.repec.org/p/fao/wpaper/0505.html (accessed on 23 January 2012).

Ma, H., A. Rae, J. Huang and S. Rozelle. 2006. 'Enhancing Productivity on Sub-urban Dairy Farms in China'. Working Paper, Freeman Spogli Institute for International Studies, Stanford University, Palo Alto, CA, USA.

MacFarlane, Alan. 2002. *The Savage Wars of Peace: England, Japan and the Malthusian Trap*. New York: Palgrave Macmillan.

Malthus, Thomas Robert. 1798. *An Essay on the Principle of Population*. London: J. Johnson, in St. Paul's Church-yard.

Moyar, Mark. 1996. 'Villager Attitudes during the Final Decade of the Vietnam War'. Presented at the Vietnam Symposium 'After the Cold War: Reassessing Vietnam', 18–20 April 1996, Lubbok, Texas, USA.

Mukherjee, Madhushri. 2010. *Churchill's Secret War: The British Empire and the Ravaging of India during World War II*. New York: Basic Books.

Ombion, Karl G. 2007. 'Philippines's Land Reform Program a Failure, Study Shows'. PinoyPress, 3 December 2007. http://www.pinoypress.net/2007/12/03/philippiness-land-reform-program-a-failure-study-shows/ (accessed on 1 September 2009).

Piggot, R. R., K. A. Parton, E. M. Treadgold and B. Hutabarat. 1993. 'Food Price Policy in Indonesia'. ACIAR Monograph Series 22, Australian Centre for International Agricultural Research, Canberra.

Pratruangkrai, Petchanet. 2010. 'Government Hikes Compensation Price to Shore Up Rice Price'. *Nation*, 11 March, p. 4A.

Pray C., S. Rozelle and J. Huang. 1997. 'Can China's Agricultural Research System Feed China?' Working Paper, Department of Agricultural Economics, Rutgers University, New Brunswick, NJ, USA.

Ravallion, Martin and Dominique van de Walle. 2008. *Land in Transition: Reform and Poverty in Rural Vietnam*. Basingstoke: Palgrave Macmillan and World Bank.

Rae, A. N., H. Ma, J. Huang and S. Rozelle. 2006. 'Livestock in China: Commodity-Specific Total Factor Productivity Decomposition Using New Panel Data'. *American Journal of Agricultural Economics*, vol. 88, no. 3, pp. 680–95.

Rozelle, S., J. Huang and K. Otsuka. 2008. 'Agriculture in China's Development: Past Disappointments, Recent Successes and Future Challenges', in L. Brandt and T. Rawski (eds), *China's Great Economic Transformation*. Cambridge: Cambridge University Press.

Sharma, Devinder, and Aditi Kapoor. 2004. 'Genetically Modified Crops in India'. http://www.countercurrents.org/en-sharma050404.htm (accessed on 1 September 2009).

Shiva, Vandana. 1991. *The Violence of the Green Revolution*. Penang: Third World Network.

Sim Chi Yin. 2008. 'The Village Where It All Began'. *Straits Times*, Singapore, 6 December.

Slayton, T. and C. P. Timmer. 2008. 'Japan, China and Thailand Can Solve the Rice Crisis—But U.S. Leadership Is Needed'. Centre for Global Development Notes, May. Centre for Global Development, Washington, D.C.

Southgate, D., D. H. Graham and L. Tweeten. 2007. *The World Food Economy*. Malden, Oxford and Carlton: Blackwell.

Studdert, L., K. Soekirman and J. Habicht. 2004. 'Community-based School Feeding during Indonesia's Economic Crisis: Implementation, Benefits and Sustainability'. *Food and Nutrition Bulletin*, vol. 25, no. 2, pp. 156–65.

The Unfinished Campaign against Hunger. 2009. *Green Revolution*. http://www. silentkillerfilm.org/green_revolution.html (accessed on 4 September 2009).

Timmer, C. P. and D. Dawe. 2007. 'Managing Food Price Instability in Asia: A Macro Food Security Perspective'. *Asian Economic Journal*, vol. 27, no. 1, pp. 1–18.

WFP (World Food Programme) INTERFAIS data. http://wfp.org/ (accessed in September 2009).

World Bank. 2008. 'Rising Food Prices: Policy Options and World Bank Response'. http://sitere-sources.worldbank.org/NEWS/Resources/risingfoodprices_backgroundnote_apr08.pdf.

Xinhua. 2009. 'How Much Grain Does China Really Have?' 14 April 2009, as reported in *China Daily*, 4 September 2009. http://www.chinadaily.com.cn/china/2009-04/14/content_7676241.htm (accessed on 4 September 2009).

Further Readings

Bhattamishra, Ruchira and Christopher B. Barrett. 2007. 'Community-based Risk Management Arrangements: An Overview and Implications for Social Fund Program Design'. Working Draft, Cornell University, Ithaca, 1 October.

Bread for the World. 2004. *Hunger Report 2004*. Washington, D.C.: Bread for the World.

CARE. 2009. 'One Year On from Cyclone Nargis and Another Emergency is Brewing', 1 May 2009. http://www.reliefweb.int/rw/rwb.nsf/db900sid/FJTC-7RMBVE?

Deaton, Angus. 1997. *The Analysis of Household Surveys: A Microeconomic Approach*. Baltimore: Johns Hopkins University Press.

Economist. 2009. 'How to Feed the World' and 'The Parable of the Sower'. 21–27 November.

FAO (Food and Agriculture Organization). 1983. *World Food Security: A Reappraisal of the Concepts and Approaches*. Rome: FAO.

———. 1996. *Report of the World Food Summit*. Rome: FAO.

———. 2008. 'Briefing Paper: Hunger on the Rise—Soaring Food Prices Add 75 Million People to Global Hunger Rolls', 17 September. http://www.fao.org/newsroom/common/ecg/1000923/en/hungerfigs.pdf (accessed on 12 January 2009).

Gangadharan, S. 2009, 'Don't Reduce Drought Impact to a Statistic'. *DNA Read the World*, 22 August 2009. http://www.dnaindia.com/money/analysis_don-t-reduce-drought-impact-to-a-statistic_1284393 (accessed 3 September 2009).

Gwatkin, Davidson R., Shea Rustein, Kuetsten Johnson, Rohini Pande and Adam Wagstaff. 2000. *Socio Economic Differences in Health, Nutrition and Population*. Washington, D.C.: World Bank.

Holmes, Rebecca and Nicola Jones. 2009. *Putting the Social Back into Social Protection*. London: Overseas Development Institute.

Holzman, Robert and Steen Lau Jorgensen. 1999. 'Social Protection as Social Risk Management: Conceptual Understandings for Social Protection Sector Strategy Paper'. Social Protection Discussion Paper No. 9904, World Bank, Washington, D.C.

IMF (International Monetary Fund). 'Food and Fuel Prices: Recent Developments, Macroeconomic Impact and Policy Responses'. http://ww/imf.org/external/np/pp/eng/2008/091908.pdf (accessed in July 2008).

Moser, Caroline. 1998. 'The Asset Vulnerability Framework: Reassessing Urban Poverty Reduction Strategies'. *World Development*, vol. 26, no. 1, pp. 1–19.

OCHA-29. 2005. 'OCHA Situation Report No. 29 Earthquake and Tsunami Indonesia, Sri Lanka, Maldives, & Thailand', 25 February 2005. http://www.cidi.org:8080/disaster/tsunami/ixl85.html (accessed on 2 September 2009).

Prasad, K., Paolo Belli and Monica Das Gupta. 1999. 'Links between Poverty, Exclusion and Health'. Background Paper for World Development Report 2000/2001, World Bank, Washington, D.C.

Sen, Amartya. 1982. *Poverty and Famines: An Essay on Entitlement and Deprivation*. Oxford: Oxford University Press.

———. 1997: 'Hunger in the Contemporary World'. Discussion Paper DEDPS/8, STICERD, London School of Economics, London, November.

World Bank. 1986. *Poverty and Hunger*. World Bank Policy Study. Washington, D.C.: World Bank.

———. 1998. 'Reducing Poverty in India: Options for More Effective Public Services'. Report No. 17881-IN, World Bank, Washington, D.C.

———. 2000. *Engendering Development*. New York: Oxford University Press.

———. 2001. *World Development Report 2000/2001*. New York: Oxford University Press.

5

Community-based Responses to Food Security

A journey of a thousand miles begins with a single step.

—Confucius

This chapter begins with a brief reference to the conceptual framework within which to analyze community responses, and a clarification of the special terms used. It then discusses how communities associate well-being with food security. The chapter then presents a set of community responses to prevent food insecurity or to deal with it once it sets in. The responses described here are set out in a particular order: production responses; increasing food availability; increasing economic access to food; consumption responses; food storage and protection responses; multiple coping responses; and finally distress measures. Although the distress measures are essentially individual responses, given their import in the overall social fabric, they are mentioned briefly here. In some cases, communities respond to a given food insecurity situation in more than one way. Therefore, the chapter also includes a few examples of integrated community responses to deal with food insecurity. Policy conclusions are drawn based on these experiences. In all cases of community-based response, an attempt has been made to contextualize the contribution of the response to food security (as defined in this study) within the framework of the entitlement and deprivation thesis. It is underscored here that community responses are not necessarily the most effective basis for responses to food insecurity. Further, in many instances, community responses can be facilitated or impaired by the policy environment created by the state.

Some Concepts

This chapter does not evaluate community-based responses to food insecurity, except to draw broad conclusions. It chronicles community-based responses to food insecurity that could inform policy making, either in the form of new ideas or through taking corrective action where warranted. It also recognizes the need to allow communities to explore ways and means to tackle their problems, to supplement state action.[1] A caveat is in order. All community-based responses are not autonomous; some are led or facilitated by civil society organizations.

Community-based responses to achieve food security are inextricably linked to risks, household responses to risks, and shocks that lead to food insecurity (Bhattamishra and Barrett, 2007). They are also linked to seasonal forces and cultural practices. Shocks are either idiosyncratic (meaning that one household's experience is only remotely related, if at all, to that of neighbouring households),[2] or covariate (meaning that many households in the same locality suffer similar shocks that lead to food insecurity).[3] Covariate shocks causing food insecurity are difficult to insure within a community, and thus require some sort of coordinated external response from the state, financial markets or civil society organizations. Idiosyncratic shocks, however, can be effectively managed within and by

---

[1] This is particularly true in a decentralized system of governance and in a political environment that encourages alternative service providers (ESCAP, UNDP and ADB, 2005).

[2] Examples of idiosyncratic shocks are those caused, for example, by crop failures associated with microclimatic variation or localized wildlife damage or pest infestation, illness, disablement and/or death in the household, and one-off events (like loss of property due to fire, theft or burglary). See Chapter 2 for more details.

[3] Covariate shocks occur because of natural disasters, war, price instability and financial crises to which (virtually) everyone in a community is vulnerable. See Chapter 2 for more details.

a community. Towards that end, communities have developed norms and institutions to reduce risk leading, *inter alia*, to food insecurity, and individuals adopt livelihood strategies that reduce the probability of suffering the consequences of food insecurity. Community-based responses, however, only partially reduce the overall risks leading to food insecurity. When food insecurity is caused by shock(s), communities usually have in place some 'mechanisms' for providing community-based food insurance, as it were, and individuals often enjoy some capacity of self-food-insurance. But these mechanisms may at times fail to prevent people from falling into situations of food insecurity.

Apart from the range of community-based responses, states have also instituted various insurance systems (like crop insurance and food loans in China), price stabilization (such as the public distribution system in India) and safety net programmes (such as cash transfers in the Philippines) to deal with food insecurity. Such interventions focus overwhelmingly on tackling food insecurity arising from covariate risks, such as the price volatility that gripped almost all countries in Asia in early 2008; droughts such as those that afflicted Vietnam and Australia in 2007–08; and earthquakes such as those that hit China and Indonesia in 2008. However, a growing body of empirical evidence suggests that idiosyncratic risk may be as important, indeed may even dominate covariate risk in rural Asia (Bhattamishra and Barrett, 2007). How communities respond to idiosyncratic risks, therefore, deserves attention.

In this study, a broad definition of 'community-based responses' for food security is adopted, to include all coordinated actions put in place and managed by groups for their protection against food insecurity (and its effects). The word 'community' is also used in its broader sense, to include individuals and institutions whose relations have an informal and non-market character, who may be linked by lineage, ethnicity, religion, occupation, historical reasons, location of habitations and the like, the key criteria being that a common interest to prevent or circumvent food insecurity binds them together and that their strategies to do so are coordinated. Additionally, the phrase 'community-based responses' implies responses adopted by groups whose management rests with the members of the groups themselves.

This chapter considers formal, semi-formal and 'informal' community-based responses to food insecurity that help food security situations within extended families, ethnic groups, neighbourhood groups and professional networks. Many community-based responses could be membership-based, but are often facilitated by civil society organizations. These community-based responses have also been considered. Since the importance of community relations in both formal and semi-formal community-based responses is the key defining feature, and because many semi-formal institutions are actually based on traditional, informal responses, this chapter takes cognizance of such organizations.[4]

The understanding of community-based responses to food security may be facilitated by an understanding of community perspectives on food security and coping mechanisms, noting that the concepts of food security,[5] well-being or ill-being and poverty and livelihood have strong linkages among themselves. Quite often they are context-specific. Communities define food security, well-being or ill-being and poverty in multidimensional terms. The criteria for these definitions often vary from rural areas to urban areas and within rural areas *inter se*. Food security for communities, therefore, has to be looked at from a holistic point of view. The example of well-being ranking depicted in Table 5.1 shows the characteristics of well-being categories compiled from well-being analysis across the Federally Administered Tribal Areas (FATA) in Pakistan during a participatory poverty assessment (PPA) exercise carried out in September 2003.

In the perception of the community under reference, then, food security coexists with health and physique in different well-being categories. Those with greater availability of food and access to resources for growing food, like land, crops and livestock, are perceived to belong to higher well-being categories relative to those having limited or no access to such resources. The quality (and, by implication, the nutritional content) of food, food availability and consumption are closely linked by the community in defining well-being. The well off are those who have a surplus of expensive food like clarified butter, meat and milk. In the perception of this community, the better off have enough food-grain and

---

[4] For example, grain banks in India and Bangladesh (discussed later) are offshoots of traditional hunger insurance systems found in many villages in South Asia.

[5] Concepts here connote the community's own concepts.

TABLE 5.1 Well-being Ranking from Pakistan

Well off	Better off	Poor	Very poor
• Good physique	• Some land	• Drinks black tea	• Hungry
• Land	• 50 sheep	• Often hungry	• Physically weak
• Crops	• Good health	• Many dependents	• Landless
• 100–150 sheep	• Enough food-grain and bread	• Bad health	• No livestock
• Surplus food (ghee, meat, milk)	• Eats two meals/day	• Very little land	• No food
		• 1–2 livestock	• Low-quality food
		• Insufficient food	• Eats dry bread
			• Visits others to obtain food
			• Depends on *zakats*
			• Begs

Source: Adapted from FATA PPA Team (2003).

Note: 'Ghee' is clarified butter; *zakat* refers to voluntary donations for the welfare of less fortunate people.

bread, and eat two meals a day. The very poor are characterized by the prevalence of food insecurity in terms of having no food, or having to eat dry bread (hence food of poor nutritive value), and dependence on charity and begging (what in the terminology of the entitlement and deprivation thesis is referred to as 'transfer entitlements').

A Suite of Community-based Responses to Chronic Food Insecurity

Communities respond to food security needs or tackle food insecurity in many ways. Since food availability, especially in subsistence economies in rural areas, appears to be most critical for the community, production responses towards increasing food availability come first into the picture. This section contains some examples of production, consumption and storage responses to chronic food insecurity from communities across the region.

Production Response from China: Ushering Reform of Global Proportions[6]

As noted in Chapter 4, like other villages, Village Xiaogang, Anhui province, China, went under collective ownership of land in 1956, later becoming the People's Commune. Everyone earned from 1 to 10 work points per day according to the quantity and quality of their labour. Farmers could not leave the commune to work outside though they earned little in agriculture, so much so that in 1962, among the 34 households in Xiaogang village 60 people had died of starvation. The final straw being a severe drought in 1978 which pushed 18 famished farmers in the village to do away with the collective farming that had suppressed grain production and left them hungry for years, knowing that it could mean death. The starving farmers under reference began parceling out their hitherto communal land, cattle and farming tools to individual family units which eventually became known as the household responsibility system (*dabaogan*). Thus, rural reforms began in the late 1978 in the

---

[6] Sim Chi Yin (2008).

central province of Anhui and productivity and grain production rose dramatically as the farmers found new incentive to work hard to better their own lot. Though local officials in Xiaogang initially did not approve of this move, it later won the backing even from Den Xiaoping himself who declared that farmers' decisions should be respected. In four years from that date, more than 90 per cent of China's agricultural households had returned to some sort of family farming and 1984, China had a bumper harvest of 407 million tons of food, making it the first time in years that the country had enough to feed itself. That decision of the farmers in village to dismantle the commune system by not only spurred a tectonic shift in the Chinese countryside, but also silently became the first chapter in China's chronicle of its dramatic economic growth and prosperity. The steps that the farmers of Xiaogang took clearly increased the food production and hence food availability in the system. In the absence of enough data, it is difficult to establish what kind of entitlement the increase of food production led to but given the fact that the farmers who were engaged in the process all had land, an increase in food production would have, from first principles, led to an increase in their endowment entitlement. By 1984, China had a bumper harvest of 407 million tons of food, making it the first time in years that the country had enough to feed itself. China expected a harvest of 528.2 million tons of grain in 2008 (*China Daily*, 2008: 1).

'It was like throwing fish back into water,' said Mr Yan Jincang, now 65 and one of the Xiaogang pioneers. As official accounts of history tell it, that desperate pact by Xiaogang's villagers not only spurred a tectonic shift in the Chinese countryside, but also, silently and unwittingly, became the first chapter in China's chronicle of its dramatic economic growth and prosperity. Unknown to the Xiaogang villagers, the winds of change were also blowing in Beijing, with a watershed meeting of the Chinese Communist Party (CCP) that same month opting for 'socialist modernization'.

The steps that the farmers of Xiaogang took clearly increased food production, and hence food availability in the system. In the absence of enough data, it is difficult to establish definitively what kind of entitlement the increase of food production led to. But given that the farmers who were engaged in the process all had land, an increase in food production would have, from first principles, led to an increase in their endowment entitlement.

Vertically Integrated Cropping: Production Response from Central Asia

Karatal village, Eskeldinskiy region, Almaty oblast in Kazakhstan has a population of about 4,000 people, dependent on agriculture for their livelihood. Villagers have devised means to maintain food security through multilayered farming. The village has seen several vicissitudes, including inflation during 2007–08, which was reminiscent of the 2,000 per cent inflation when Kazakhstan was separated from the erstwhile USSR. In 2008, the price of wheat surged 67 per cent in just three months, while that of flour jumped 58 per cent, leading to a temporary suspension in the country's bread supply (Xinhua, 2008). The price of rice of different varieties increased from 80–100 Tenge per kg to 160–200 per kg between January and June 2008.[7]

But people in Karatal are not food-insecure, at least apparently. Table 5.2 provides a seasonality diagram of food consumption.

TABLE 5.2   Seasonality of Food in Village Karatal, Kazakhstan

Months	Food	Meat (Beef)	Potatoes	Bread	Fruits	Rice	Macaroni	Salat
January		*****	*****	*	***	*	**	***
February		****	****	*	**	*	****	**
March		****	****	*	**	*	**	***
April		****	***	*	***	*	**	**
May		***	**	*	***	*	***	***
June		**	**	*	****	*	*	*****
July		**	**	*	*****	*	*	*****
August		**	*	*	*****	*	*	*****
September		***	*****	*	****	*	***	*****
October		****	*****	*	***	*	***	****
November		****	*****	*	**	*	***	****
December		*****	*****	*	**	*	**	***

Source: Prepared by students: Rustani, Olzhas, Zhandos, Arman, Omar, Kostya, Ruslan, Aida, Tumur, and Kirill, and Teacher of Physical Training Mr Amvar. Facilitated by Ms Aiman Kenzhebekova and Amitava Mukherjee. Dated 12 June 2008.

---

[7] Field notes of Dr Amitava Mukherjee and Mr Eugene Gherman from their visit to a market in the neighbouring town of Tekeli on 11 June 2008.

Consumption of meat, potatoes, fruits, macaroni and *salat*, meaning vegetables of different kind, varies over the months, while consumption of bread and rice, as staple foods, is constant. A clear substitution effect is noticeable: during the winter months, the relative consumption of meat and potatoes goes up, whereas the consumption of fruits and *salat* goes down accordingly, as the availability and nutritional needs of the season vary. The villagers have three meals all through the year. Generally, the villagers are food-secure, both in terms of quantity and nutritional content.

This result has been achieved through multilayered agriculture (see Figure 5.1). The agriculturists in the village generally have a two-track agricultural production system: first, *farming*, to produce wheat, vegetables, fruits and tubers; and second, *allied agriculture*,

FIGURE 5.1   Multilayered Agriculture in Kazakhstan

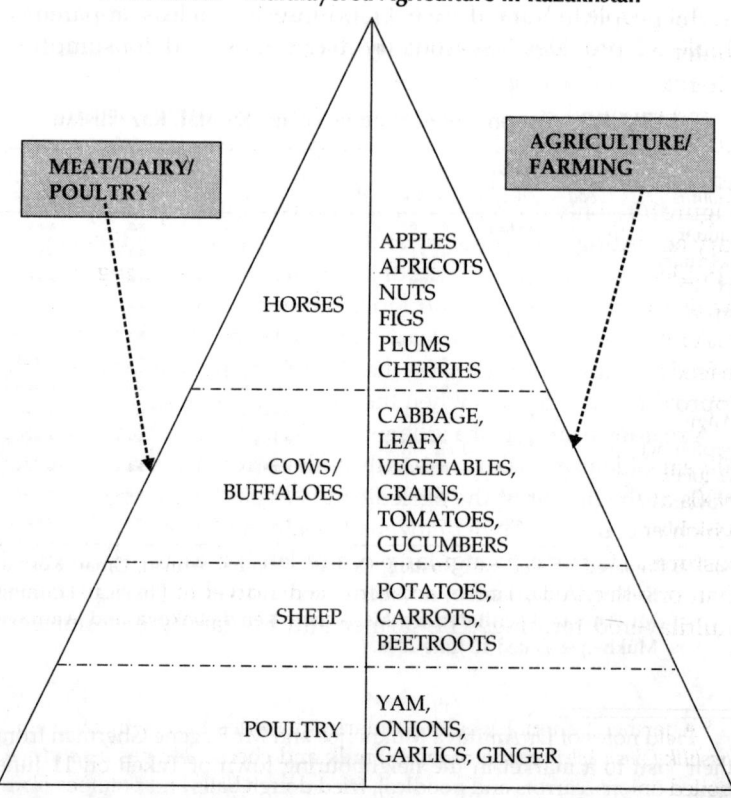

Source: Author.

for producing meat, dairy products and wool. In both systems, they undertake multilayered farming and production. In farming, they grow crops under the ground, on the surface and above the surface. The first layer is the underground layer, allowing farmers the space to grow tubers (ginger, onions, carrots, turnips, potatoes). The second layer is the surface, or the ground layer, on which generally cereals and 'other' vegetables like cabbage and green leafy vegetables are cultivated. The third layer is above ground level, providing people fruits and nuts of different varieties. Similarly, in the case of allied agriculture, sheep constitute the first layer, cows the second, and horses the third layer. The general pattern is that sheep are most numerous, followed by cows, with the number of horses being lowest. Milk is provided by cows and horses. Meat is obtained from sheep, cows and horses.[8]

Both systems of production allow the farmers to hold their produce to tide over the difficult months, usually the severe winter months. Meat and milk are saved in terms of live animals, whereas non-meat agricultural produce is 'saved', within limits, in underground storages that last them the year round. Fruits are all preserved as dried fruits. This explains why the food basket remains the same all through the year in the seasonality diagram (Figure 5.1); only the relative proportions of different components vary according to availability and the dietary needs of the season. Significantly enough, because of the multilayered farming system, the size of the farmland required is not very large.

A relatively small plot of land (say half an acre) would be enough to feed a family round the year, the effective cultivable surface being approximately 1.5 acres when the three tiers are taken together.

A real-life example of a villager would be useful. Alexandar is a 70-year-old farmer who came to the village from Siberia during the 1950s at the height of the agricultural 'virgin lands' programme, which encouraged Soviet citizens to help cultivate Kazakhstan's pastures. He has a triangular piece of land, measuring no more than one-third of an acre, adjacent to his dwelling. He practises multilayered farming. The farmer and his family encountered

---

[8] A farmer's animal wealth primarily provides him with meat and *shashlik* (*shish kebab* made of lamb), milk and cheese. He has *beshbarmak* (boiled onions, carrots and noodles), fried dough balls and potatoes from his cropping.

few problems, if any, in satisfying their food requirements and generated some cash by selling the produce from the land itself. Significantly, Alexandar obtains most of his food not from endowment entitlement, but from usufruct entitlement. This is because the land he has been cultivating for the last 50 years or so (he could not recall the exact years) does not belong to him but to the state.[9] He neither pays land-related taxes, nor is his possession disturbed by the agencies of the state. He is enjoying the usufruct rights, not the title to land; no right to cultivate either. Similarly, the grazing land used by his animal stock belongs to the state.

Horizontal Integration of Crops: The Circular Economy of the Dyke–Pond System in China

The dyke–pond system of crop production is a classic example of horizontal integration of agricultural operations. The mulberry dyke (or sugarcane) fish pond complex is a system developed by farmers in the Pearl River delta region of China to make full use of available land and water resources to grow food. It is an interrelated ecosystem that brings into full play the productive potential of humans and their environment and promotes the development of different branches of agriculture at the same time. There are six operational elements in the system, which is horizontally integrated.

The Pond

The pond is the heart of the system. To produce a pond, soil is excavated and used to build or repair the dykes surrounding the pond. The pond is prepared by clearing, cleaning and fertilization with quicklime, tea-seed cake and organic manure from livestock kept on the dykes. Most ponds are rectangular, 0.4 to 0.6 ha in area and 2 to 3 m deep. The dykes are usually 6 to 10 m wide, and extend 0.5 to 1.0 m above the surface of the pond. The pond is filled with river water. Water also enters directly as rain and through run-offs from the dykes. Water leaves the pond via the pond drainage outlet

---

[9] Alexandar started cultivating a small piece and parcel of land many years ago, and gradually expanded it to its present limits.

in controlled discharges, evaporation and transpiration, seepage into the dykes, and when water is drained out at regular intervals over a one-year period.

Mulberry and Silkworms

The co-production of mulberry (or sugarcane) and sericulture with fish farming is achieved by converting the banks of the fish ponds into mulberry dykes. Mulberry leaves are picked to rear silkworms, from which silk cocoons are harvested, while the wastes of silkworms are used to fertilize the pond to feed the fish.

The Pond Mud, Much Enriched in Nutrients, as Fertilizer for Crops

Ponds are drained two or three times a year, and the mud at the bottom is dredged out and put on the dykes. This operation raises the height of the dykes and repairs them as well, while at the same time the depth of the pond is restored and siltation prevented. The mud excavated from its pond is also used for mushroom cultivation; mushrooms are often cultivated on the floor of the silkworm shed in winter, the off-season for silkworm production. After the final crop of mushrooms has been harvested, the mud-bed is used to grow vegetables, fruit trees and grasses.

The Fishes

Various fish species which live at different pond depths, and have different feeding habits, are reared in the pond, to make full use of the water and the pond ecology. The typical poly-culture reared in the pond is a combination of the 'four big family fish': grass carp, silver carp, big head carp and common carp, requiring little or no external input.

Livestock

Livestock is an important link in the circular economy. Pigs, chickens and ducks are reared on the dykes, to provide manure to fertilize the fish ponds, and to encourage the growth of plankton that feeds the fish. Many of the dyke crops are fed directly to the fish, such as elephant grass for the grass carp, or else to the livestock, such as forage crops for pigs.

Harvests

With a tropical to subtropical climate, the dyke–pond area is well endowed with sunshine and rainfall, and hence extremely productive, especially with a system that recycles and transforms all the 'wastes' into nutrient resources. There are many harvests, fishes, silk cocoons and vegetables being the major ones for the system. Rearing of pigs would be a minor harvest along with livestock such as chickens and ducks as well as mushrooms. But for the farmers, fish sales contribute the largest source of income.

The external energy input in the dyke–pond system is minimal, and consists mainly of labour and the energy expended to make farming implements, housing and equipment for rearing silkworms, and machinery and energy to aerate the fish pond and to dredge it. The major energy input by far, consisting of more than 99 per cent, is sunlight, and it is renewable and free. However, since the late 1970s, the traditional dyke–pond system of the Pearl River delta has been undergoing dramatic changes due to the shift away from collectivist to household production, as discussed earlier, and also due to the intensification of production following the Green Revolution.

The horizontal integration of crops increases the endowment and exchange as also the trade-based entitlements of the farmers.

Ducklings and Rice Cultivation: A Case from Japan[10]

The Aigamo method!!

The organic farming method dubbed the Aigamo (the name of a crossbreed of domestic and wild ducks) method uses ducklings to provide pest and weed control, oxygenate the rice fields and fertilize rice seedlings, apart from maturing themselves. It is a hybrid of the traditional farming practices of Japanese rice farmers and experimentation by the Furano family in Fukuoka, Japan. Stated simply, the agricultural operations simultaneously raise Aigamo

---

[10] Mae-Wan Ho et al. (2008).

ducklings, loaches (a species of fish), rice and Azolla—a nitrate-fixing species of aquatic fern. The ducks are left in the fields 24 hours a day, and do not need to be herded back to a protected pen.

Ducklings are released into paddy fields soon after the seedlings are planted. Ducklings do not eat rice seedlings, as the seedlings have too much silica. Generally, about 20 ducklings are released per tenth of a hectare. It seems ducklings genuinely enjoy getting into the water, paddling between rows of rice seedlings, ducking their heads underneath the surface and then raising their heads to swallow something. These movements are crucial for rice plants, producing mechanical stimulation that makes plant stems thicker and stronger. The ducks also eat up insects, pests and the golden snail, which attack rice plants. They also eat the seeds and seed-lings of weeds, using their feet to dig up the weed seedlings, in the process oxygenating the water and encouraging the roots of rice to grow. The ducks are so good at weeding that farmers who have adopted the duckling method now save up to 240 person-hours per hectare in manual weeding every year. Indeed, the Furano family has left their fields to ducklings for plying and naturally managing their rice paddy cultivation.

The Fukuoka plot also has a patch of dry land amid the paddy fields for the ducks to rest and eat waste grain from rice-polishing factory operations. The ducks stay in the field until the rice plants form ears of grain, and then are removed before they can eat the grains. At this stage, the ducks are confined to a shed and fed waste grain exclusively. The ducks in turn fatten up, lay eggs and provide meat, and are sold in the market to generate market revenue.

The ducks are not the only inhabitants of the paddy field. The aquatic fern Azolla, or duckweed, which harbours a blue-green bacterium as symbiont, is also grown on the surface of the water. The Azolla is very efficient in fixing nitrogen, attracting insects for the ducks and also provides food for the ducks. Loaches use the Azolla as a hiding place from ducks and the plant provides protection for fish that swim in the flooded paddy. The fish feed on duck faeces, on daphnia and other worms, which in turn feed on the plankton. The fish and ducks provide manure to fertilize the rice plants all through the growing season.

In short, the paddy field with ducks is really a complex, well-balanced, self-maintaining and self-propagating ecosystem. The only external input is the small amount of waste grain for the ducks, and the output from the visibly more robust rice plants (as compared to rice plants not cultivated under the duck method) is a healthy harvest of organic rice, duck and loach. The rice-farming method also shows that organic farming does not have to be labour-intensive.[11]

Nurturing ducklings in rice fields has the twin effects of increasing rice production and the maturing of ducks that lay eggs and provide meat. In addition, the ducks are sold to provide income to the farmers. Thus, ducks increase endowment and exchange entitlements as also trade-based entitlements of the farmers, increase their economic access to food and improve the nutritional value of food for farmers. The method also contributes to better environmental management as it obviates the need to use chemical fertilizers.

Integrated Production Responses: A Case from India[12]

Often, food production is linked to availability of seeds, fodder, tools and machineries. These factors taken together delineate the technology used in agriculture. Thus, fodder banks, grain banks, seed banks and machinery and tool libraries have been established to make agriculture more sustainable for 40,000 small and marginal farmers in 400 villages in the dry desert districts of Kutch, Patan and Surendranagar, India.

Grain Bank

Community members typically experience difficulty in saving grains by themselves (Bhattamishra and Barrett, 2007), probably due to commitment problems (which have been shown to impact savings abilities in empirical studies from Asia).[13] To circumvent

---

[11] Today, this successful farming method has spread to many parts of Asia including the Republic of Korea, Cambodia, Lao PDR, Vietnam, Malaysia, the Philippines and Thailand. Farmers in these countries have increased their rice yield 20 to 50 per cent or more in the first year, with one farmer from Lao PDR increasing his income three-fold.

[12] Drawn from Nanavaty (n.d.).

[13] Ashraf et al. (2006) provide evidence from an experiment in rural Philippines that individuals with time-inconsistent preferences have a higher demand for a commitment savings product, and that by use of this commitment savings product, they are able to increase both short-term and longer-term savings.

this problem, community members resort to the setting up of grain banks. The main function of grain banks, a semi-formal arrangement,[14] is to enable households to save grains to smooth consumption over different periods of time, usually a 12-month period.[15] The grain bank as a grain storage centre concept, however, is an old concept in the Asia-Pacific region.[16]

There was a traditional practice in Gujarat of each family contributing a certain amount of grain to a common pool during good times, to be stored in a sealed, underground storage system.[17] In the lean season, this stored grain would be retrieved and used to ensure food security. The Self-Employed Women's Association (SEWA)[18] in Gujarat has facilitated grain banks apparently in an attempt to formalize this traditional practice. The grain bank as a grain storage centre is built up through local procurement of different grains and through contributions from the community in cash or kind (of a predefined amount), and the contributors become members. During difficult times such as droughts or years of poor harvest, the stored grain will be available in the form of loans, at nominal rates of interest, to households in need and to the poor at

---

[14] Grain banks flourish both as a civil society–facilitated programme and as a government-sponsored programme.

[15] To some extent, grain banks also provide credit as insurance, in that households that are not able to return loans after facing negative shocks leading to food insecurity find their repayment periods extended.

[16] Parbati Sankar Roy Choudhury pioneered the *dharmagola* system of cooperative grain banking to avoid famine and scarcity as early as the late 19th century in Joyganj (Dinajpur), Goalundo in Faridpur, and Nathpur and Shaitghar in Dacca, Bangladesh. And the system worked really well. These grain banks were subsequently registered officially as formal cooperative societies in the second decade of the 20th century. http://en.wikipedia.org/wiki/Parbati_Sankar_Roy_Choudhury (accessed on 24 January 2012).

[17] This practice is still followed in some Central Asian countries such as Kazakhstan. The author found this practice still alive in June 2008 in Karatal village, Eskeldinskiy region of Almaty oblast.

[18] A registered trade union since 1972, SEWA is an organization of poor, self-employed women workers, who earn a living through their own labour or small businesses, without regular salaried employment and without welfare benefits like workers in the organized sector.

cost (even during the lean season). The grain bank is administered and managed by the local community, ensuring local decision making and control. The grain bank thus, helps increase availability of food throughout the year, acquired by the farmers through either endowment or exchange entitlement,[19] especially at times when systemic availability of food is low.

The grain bank also serves to supplement the public distribution system of the government.[20] The fair price shops (FPS) under the PDS that provide the poor with essential food-grains and other necessities at controlled prices are at times too far from some of the villages or human habitations. In such cases, the grain bank acts as a source of food for those who have no physical access to food-grains. Additionally, the bank can cover shortfalls when FPS shops are either not open or are not well stocked when actual need arises.[21] Cases of women resorting to money-lenders to acquire food because of the failure of the PDS to deliver are not rare. These loans often carry exorbitant rates of interest. Once this happens, the borrower-women find it difficult to come out of their debt and their economic access to food (and other items in their consumption basket) is compromised.

Seed Bank

High-yielding varieties of seeds are an important input for ensuring higher production, pest control, crop disease resistance and higher food production. Under the conditions of rain-fed agriculture in many parts of India, availability of improved seeds (read high-yielding-variety seeds) is the first step in improving food production. Supplying information on the quality of seed to farmers is

---

[19] The grain bank concept accepted by SAARC countries at the SAARC Summit in Colombo in August 2008 is an extension of this basic idea. This has been discussed earlier in Chapters 1 and 3.

[20] Chapter 4 gives an account of the state-run public distribution system in India.

[21] However, SEWA understood that ensuring availability of food alone was not enough for food security. In order that the poor get maximum benefit from the grain banks, SEWA has also initiated livelihood activities at the village level to improve the economic access of the poor to food. These activities are specifically for the poor, and women helping them earn a regular income, thus ensuring their food security.

also vital as it may reduce the dependence of many farmers on unscrupulous distribution agents of seeds, pesticides and insecticides. Seed banks managed by the communities themselves have been set up in villages in selected districts of Kutch, Patan and Surendranagar in Gujarat. Seed banks store good-quality seeds, which farmers are allowed to borrow on payment of a small interest at a rate determined by the community. For example, in a village of the Patan district, in 2007, cumin seeds were distributed to 200 farmers at 8 kg of seeds for each farmer. Seeds stored in the seed bank are certified by the Gujarat State Seed Corporation, thus guaranteeing seed quality. The seed banks offer the advantage of repayment in seeds over several seasons, and the stock of seed in the seed bank gradually builds up with the interest payments (see Box 5.1).

BOX 5.1   Testimony of a Nepali Farmer

A Nepalese woman farmer Hemmaya Bhandari, 22, told UCA News that through the visit to Mymensingh in Bangladesh, she learned the benefits of digging the earth twice before planting and of a community seed bank. What impressed her most was the seed bank, where 64 indigenous varieties of superior-quality paddy seeds were stored in mud jars with paddy straw and banana leaves. She learned that the community farmers took paddy seeds from the seed bank for cultivation and then, after the harvest, returned double the amount they took.

Source: UCANEWS (2008).

The seeds banks aid in augmenting the endowment entitlement of subsistence-level and sub-subsistence-level farmers to food-grains, in that these banks contribute towards increasing total food production for the farmers, and in any case increase food availability in the community. For the above-subsistence-level farmers, a seed bank augments exchange-based and trade-based entitlements, as they may have more food-grains to exchange and trade.[22]

---

[22] The Svalbard Global Seed Vault, preserving crop varieties from over 100 countries, opened on 26 February 2008 in Svalbard, Norway, to preserve seed samples of the world's crops, guarding against the loss of the planet's biodiversity. The vault is essentially an extension of the seed

Fodder Bank

The importance of livestock rearing and dairy in the livelihood of people in general and those in disaster- and drought-prone areas like Patan district, India, cannot be overemphasized. Livestock is not only the source of draught power,[23] but can also be an all-weather source of secondary employment and income that improves the economic access of people to food and other items in their consumption basket. However, the quantity and quality of bovine wealth and of the milk they yield are directly dependent on the availability of fodder. In the dry and semi-arid districts of Gujarat, fodder is a scarce commodity especially in the lean season, and its price is unaffordable for poor farmers. And with many farmers shifting to cash crops, the availability of fodder per se on the whole is also on the decline. Dairy as a livelihood option for the poor is, thus, receding into the background and forcing many milk-folk to emigrate, and even to resort to distress sales of their bovine wealth, making inroads into their economic access to food. To circumvent their problems, SEWA decided to ensure fodder security for its members engaged in dairy activity through fodder banks in various districts.

Fodder banks collect fodder during years of good fodder harvest and store the same for use in years of drought. Each member of the dairy cooperatives (set up earlier) initially deposits an amount of ₹200[24] with the fodder bank, which is managed by poor women community members.[25]

Funds generated are then used to purchase and stock fodder when it is available. At times when fodder is needed, but unavailable in the vicinity of the village where the fodder bank is located,

---

bank idea on a global scale. The vault has the capacity to preserve 4.5 million different samples of seeds (up to two billion seeds in number) for thousands of years. Contributions were received from gene banks around the world, including rice from the Philippines.

[23] Draught power provided by nearly 82 million animals, plough 100 million hectares or 60 per cent of the area cultivated in India (Bansil, 2003).

[24] US$1 is equal to ₹48 approximately.

[25] Various kinds of training, such as how to distribute, weigh and transport fodder, and how to manage the administrative process associated with distribution, were also provided by SEWA for numerous villagers.

fodder is procured from other blocks or districts where it is available in greater quantities. Fodder so procured, is stocked and distributed to members during lean periods at cost, making dairy a sustainable livelihood activity, with members using their own resources. For example, in 2003, almost 542 metric tons of fodder worth about ₹1.9 million were distributed to members, including green fodder, dry fodder and cattle feed. Not insignificantly, the dung generated by bovine wealth is a source of manure and fuel for small and marginal farmers in that part of the world.

Since dairy contributes to increasing the food availability for farmers through higher endowment entitlements as well as increasing economic access for those in off-farm employment, fodder banks have a vital role in the agricultural life of small and marginal farmers.

Tools and Equipment Library (Read Bank)

Small, marginal and landless farmers most often have little or no access to the tools and equipment needed for agriculture. Their inability to buy the necessary tools hampers their ability to earn a sustainable livelihood in agriculture. To tackle the problem, a tools and equipment library has been started. The idea of the library is to house different kinds of agricultural equipment such as ploughs, masonry tools, water-lifting machines and 'heavy' commercial vehicles including tractors to meet the needs and requirements of a farming community. Participation in the tools and equipment library programme is based on a nominal membership fee, which is then used to buy additional tools and equipment, cover repairs and maintenance of the existing tools, and purchase new equipment as the membership grows. Thus, the library is self-sustaining, with management in the hands of local women's cooperatives. Like the seed bank, the library increases the endowment entitlements of the farming community.

Community-based Irrigation Systems

Community-based water delivery systems have been pervasive in Asia, and even today they serve a third or more of the total irrigated area. They have generally developed in mountainous or hilly areas based on the diversion of small and medium streams, most especially in the Himalayas, the Philippines, Northern Thailand and Laos, China and Japan.

Small-scale Irrigation: Zanjeras, Philippines—Equity the Key

In the Ilocos provinces of the northern Philippines, a typical form of community-based irrigation system is the *zanjeras*. These are small-scale irrigation societies all of whose infrastructure is built, maintained and managed by communities of farmers—the water users. Land is allocated among the farmers in parcels of equal size, the holdings being dispersed between upstream and downstream areas. The even land distribution ensures that the allocation of water as also maintenance obligations are fairly equitable. This has assisted the *zanjera* system to persist successfully with minimal state interference for over two centuries.

Water is allocated proportionally to the size of the area of the land cultivated, as is the size of the input of labour and material that each user must contribute to the operation and maintenance of the system.

State control was introduced in the form of the Water Code of 1976, which asserted state ownership of water sources, and required that all rights to divert water, whether for irrigation or for other purposes, be legitimized via possession of a water permit. This intervention has been successful in so far as it left intact the traditional, informal systems through which coordination and conflict resolution among users of particular irrigation systems are achieved. In some cases, it even provided new opportunities for coordination among different *zanjeras* along the same river system, helping in greater efficiency of water allocation. However, there is no recognition of traditional and ancestral claims to usufructuary rights over water that have not been officially registered, leading to instances of over-allocation and consequent conflicts between registered and unregistered users. Furthermore, the application procedure for water permits puts the wealthy and powerful individuals in an advantageous position vis-à-vis irrigation associations, despite the fact that the wording of the water codes indicates an explicit favouring of collective over private ownership (Cruz, 1989).

Farmer-managed Irrigation Systems, Thailand: The Local Bureaucrat as the Pivot

Before the establishment of Royal Irrigation Department (RID), Farmer-managed Irrigation Systems (FMIS) in Thailand were typically developed, operated and maintained by groups of farmers (Shivakoti, 2008). The systems were mostly of small scale, the optimum size for maintenance (100–1,000 rai), for providing water for a good harvest of the main rice crop, for making water available during the dry season to reap a second harvest of rice or other crops, and for expanding cultivated areas. The district head was the agent provocateur, who arranged meetings with different sub-district heads and discussed the most efficient, feasible way of supplying water to different sub-districts and villages. The sub-district head then discussed the matter with the village head and finally planned work strategies. In order to operate and maintain the system, farmers were organized and they formed water user committees, based on canal networks.

In the FMIS, the irrigation systems were constructed communally using available local materials such as bamboo, logs and stones. When the need for irrigation water arose, the community, along with the community head, identified the source of irrigation water and constructed a weir on a river to hold back water. Since the weir as well as the entire irrigation system was considered common property, households had common rights and responsibilities for repair and maintenance and hence were governed by customary rules and regulations. Each household was obliged to contribute labour, construction materials and tools based on the size of their landholding and economic status. Given traditionally oriented beliefs, the households had to collect some funds for annual ritual offerings to the weir spirit for the protection of entire irrigation systems.

The water users developed their own rules and regulations. Typically, for the maintenance of these irrigation systems, labour contribution was guided by the principle of 'more land, more labour contribution'. Since the people-managed irrigation systems were fully constructed and repaired by local materials, farmers were obliged to bring necessary tools and materials during maintenance time. The amount and types of tools and materials brought were either decided by the weir headman or, in most cases, based on already-agreed-upon rules of contribution proportionate to land-holding size. As any traditional society is bound by several belief systems, the mutual understanding of members and community solidarity have often been reflected in some ritual form. Therefore, each household had to contribute towards ritual offerings either in cash or in kind, whatever was convenient.

Weirs and Subaks, Bali, Indonesia: The Temple as the Kingpin

The irrigation systems of Balinese rice farmers achieve coordination of large numbers of individual users covering a wide geographical area without the imposition of central control. Farmers are organized into associations known as *subaks*, whose members share a common irrigation infrastructure built around weirs (small dams) into which water is diverted from a river. From the weir, water is channelled into canals and aqueducts, which carry it to the fields. Each of these weirs may be used by one or several *subaks*.

Coordination of planting schedules within and among *subaks* is necessary for two reasons. Planting schedules are staggered so that the peak demand for water from all users from a single watercourse is avoided. However, synchronization among fields in the same area is an important part of the pest control strategy. The optimum balance between these conflicting demands depends on specific local conditions, and is achieved in a surprising fashion.

Associated with the physical infrastructure for irrigation is a complex network of shrines and temples dedicated to the worship of the water goddess and various agricultural deities, which provides the basis for coordination of planting cycles within and between *subaks*. Each temple is associated with a single weir, and its congregation is composed of the members of the *subaks* serviced by the weir. The temple hosts annual meetings of representatives of all these farmers, at which a common planting cycle for the year ahead is decided. The temple's ritual calendar provides the basis for the succeeding sequence of agricultural tasks.

Coordination among users of weirs along the same river is achieved via regular exchange of delegates among temples, in fulfil-ment of obligations associated with ritual ties among the deities to which they are dedicated. This system is finely tuned to local conditions and allows farmers to achieve an optimum cropping pattern. Mathematical modelling of the network's ecological properties suggests that it allows yields to be maintained in the face of crises such as low rainfall or large increases in pest populations, and that these properties depend upon a high degree of coordination of farmers' activities. Interestingly, the system appears to have evolved as the outcome of a gradual process of experimentation on the part of individuals over a period of several centuries, rather than the imposition of any central political or religious authority. It is thus an example of coordination resulting from bottom–up organization. Attempts to replace the role of the temples by a bureaucratic system of control following the Green Revolution of the 1970s led to a loss of the coordinating function, resulting in a series of ecological crises (Lansing, 1991; Lansing and Kremer, 1993).

All the three community-based irrigation systems discussed contribute to increasing food production. They thus contribute to increasing the endowment entitlement of farmers and, to the extent that they sell a part of their produce, to their trade-based or exchange entitlements as well.

Alternative Public Distribution System

Different models for community grain banks have emerged in many communities in the region. In one case, a community grain bank in conjunction with the reclamation of fallow land and the creation of a community grain fund facilitated by an NGO (Deccan Development Society in Andhra Pradesh) has evolved into what is called an alternative public distribution system (APDS) based in Zaheerabad mandal,[26] Andhra Pradesh, India. Here, food-grain production, procurement, storage and distribution are done at the community (village) level and are entirely managed by women sangham[27] members in 11 villages.

The target groups, drawn from marginal and small farmers mostly belonging to the Scheduled Castes and the Backward

---

[26] A mandal is a sub-district-level administrative unit.

[27] A self-help group is called a sangham in the local language.

Castes,[28] were organized into self-help groups or *sanghams*. The programme implementation started, among other things, by advancing
loans to the beneficiary farmers for agricultural operations on fallow
lands over a three-year period. The total investment for bringing
fallow lands under cultivation for all the three years was fixed at
₹4,200 per acre. In the second stage, the responsibilities of collecting and disbursing loan amounts and timely implementation of all
seasonal agricultural activities at the village level were vested in the
women committee members in each village. Loan repayment was
in the form of grain from the partner farmers after the harvesting
of crops. Grains collected were stored using traditional storage
methods as parts of the community grain fund (CGF). The loan
repayments by partner farmers were spread over a five-year period
in the form of grain in pre-fixed quantities at pre-fixed prices. In
the third stage, the grain thus collected was stored in the village
for distribution during the scarcity months (usually the monsoon
season) among the poor households on a five-point poverty scale
(read level), identified by the villagers themselves based on their
criteria evolved through a participatory wealth ranking. Finally,
households thus identified were issued a sorghum card by the
sangham, entitling them to a fixed quantity of jowar at a subsidized
price of ₹3.50 per kg, the proceeds of which were deposited in a
community grain fund account held with a bank. The subsidy of
one Rupee, being the difference between the issue price and procurement price was made up from the interest payments, accruing
from CGF Bank deposited in five years' fixed deposits. Individual
village groups hold the CGF account, and the fund is used year
after year for reclaiming more fallow lands.

The APDS has increased food availability, as it yields additional quantities of different kinds of cereals and pulses from the
fallow lands. Additional labour employment for various agricultural operations and a demand for 'bullock-pair days' created by
the APDS has increased the people's economic access to food.
The APDS has also brought back biodiversity conservation as a
source of food for local communities, and helped revive several

---

[28] Scheduled Castes are among the most marginalized groups in Indian
society. They are also called dalits. Backward Castes are also marginalized
groups in Indian society, but a step better off than the Scheduled Castes.

varieties of crops, cereals, legumes, pulses and oilseeds, helping in ensuring food security for future generations. Thus, the APDS has not only helped farmers to increase their access to food through better endowment entitlements from increased yield, but has also improved the nutritional quality of food through increasing biodiversity, which provides farmers with a whole range of foods hitherto not available.

Increasing Food Availability: Accessing Food from Common Property Resources

Common property resources (CPR) (such as lands, forests, wildlife, fisheries, water) may incorporate elements of quasi-insurance by allowing free access to forage food and fodder for both consumption and trade.[29] Common property resources that involve access on a rotating basis also have a quasi-insurance character. In these cases, community members have an equal likelihood (in expectation) of receiving fertile and infertile tracts of cultivable land (in the case of agricultural societies) or equal amounts in expected catches (in the case of fishing communities). For example, in a fishing community in Sri Lanka, access to the biggest catches, which depends both on the time of the day and the location in which nets are cast, is allotted on a strict schedule, such that the expected availability of fish for all fishermen is equalized over time. This smoothes out, to an extent, access to food or the means to acquire food among fisher folk households (Bhattamishra and Barrett, 2007). In another example, the *warabandi* system of water distribution in rural Pakistan, canal water is distributed to farmers on a rotational basis at pre-specified times during the course of the week (Murgai et al., 2002). These traditional rules help households to smooth out consumption over time, thereby reducing the risk (Platteau, 1991) of seasonal food insecurity.

Common property resources may also be seen to constitute a secondary system of food production, and hence constitute a source of food availability.[30] Food-insecure households traditionally intensify

---

[29] The possibility of damaging CPRs articulated by Hardin (1968) should be borne in mind.

[30] There are two food systems. Rice, potatoes, etc., which are foods produced by the application of technology in the economic sense, are called

the use of the natural resource base, and in particular CPRs including forests and rivers, when food production and availability are insufficient.[31] Similarly, in coastal areas and areas prone to floods, many landless households try to secure their living by fishing in rivers, canals and flooded areas during the flooding season when there are fewer job opportunities. The nature of the CPRs and the products collected also vary from urban to rural contexts. In some cases, it has been estimated that food from CPRs constitutes almost 35 per cent of the total food consumed by the poor in arid and semi-arid areas in India (Jodha, 1986). In Tangail, Bangladesh, it has been found that people belonging to different economic groupings access from 39 to 89 per cent of their food from CPRs, or what they call uncultivated sources (Mazhar et al., 2007). The fact that food from the secondary food system is available when food supply from the primary food system is not available or is scarce, has led some experts to call food from the CPRs 'hunger foods' (FAO, 1984, 2002).

The use of common property resources to smooth out the availability of food over different months is also significant. A longitudinal participatory study of food insecurity and community coping mechanisms was conducted in village Rakshit Chak (inhabited by landless families of the Lodha tribe) in Midnapore district, West Bengal, India, over the years 1993, 1995 and 1998, using food calendars with women (see Tables 5.3, 5.4 and 5.5). The study demonstrates that CPRs, and the rights of the poor to access them, also act as an insurance against a sudden fall in food availability (Mukherjee and Mukherjee, 2008a, 2008b). The study of the food calendar reveals that there are four phases of food insecurity. Phase one covers the period from mid-November to mid-February, when people eat the most. Food insecurity sets in during phase two, spread over

---

foods derived from the 'primary food system'. Fish, fruits, honey, gums, small animals, birds, tubers, snails, leaves, leafy vegetables, etc., derived from common property resources, micro-environment (Chambers, 1992) and forests, and from gleaning and collecting food from lands belonging to richer, relatively well-off farmers in neighbouring villages, are called foods from the 'secondary food system' (Mukherjee, 1994).

[31] Food gathered from the CPR can be viewed as usufruct entitlement or entitlement emanating from exchange with nature.

TABLE 5.3 Seasonal Food Calendar of Village Krishna Rakshit Chak, Midnapore, 1993

Month	Rice	Potatoes	Pulses	Vegetables	Fruits	Food from water sources	Others from the wild
1	2	3	4	5	6	7	8
Magh (mid-January to mid-February)	*****************	**************	***	Cabbage	–	–	Wild borums and wild rabbits
Phalgun (mid-February to mid-March)	**********	************	**	Spinach	–	–	Neem leaves
Chaitra (mid-March to mid-April)	****	****	**2	Pumpkin	–	Fish and wild water plants	–
Baisakh (mid-April to mid-May)	****	***	*	Pui leaves and herbs	Mango, jackfruit	Fish, snails and wild water plants	–
Jyastha (mid-May to mid-June)	****	****	**	*Lota*, leaves and herbs	Mango, jackfruit	Fish, wild water plants	–
Asardh (mid-June to mid-July)	****	***	***	*Jhinge* (nearer to Sukini), green papaya	–	–	–
Srabon (mid-July to mid-August)	****	****	******	Green papaya	–	–	–
Bhadra (mid-August to mid-September)	****	***	***	Green banana	–	Fish and snails	–
Ashwin (mid-September to mid-October)	****	***	**	–	–	–	–
Kartick (mid-October to mid-November)	**	***	**	Radish, leaves	–	–	–
Ahgrayan (mid-November to mid-December)	****	*****	*****	Tomatoes	–	–	–

Source: Mukherjee and Mukherjee (2008b).

TABLE 5.4 Seasonal Food Calendar of Village Krishna Rakshit Chak, Midnapore, 1995

Month	Rice	Potatoes	Pulses	Vegetables	Fruits	Fish	Snails	Others from the wild
1	2	3	4	5	6	7	8	9
Magh (mid-January to mid-February)	*****************	*****************	***	–	–	–	–	–
Phalgun (mid-February to mid-March)	**********	************	**	–	–	–	–	Neem leaves
Chaitra (mid-March to mid-April)	*****	***	–	–	–	–	–	–
Baisakh@ (mid-April to mid-May)	****	***	–	–	–	–	–	–
Jyastha (mid-May to mid-June)	****	***	–	–	–	–	–	–
Asardh (mid-June to mid-July)	****	**	–	–	–	–	–	–
Srabon (mid-July to mid-August)	****	***	–	–	–	–	–	–
Bhadra (mid-August to mid-September)	**	–	–	–	–	***********	–	–

Source: Mukherjee and Mukherjee (2008b).

TABLE 5.5 Seasonal Food Calendar of Village Krishna Rakshit Chak, Midnapore, 1998

Month	Rice	Potatoes	Pulses	Vegetables from CPR	Fruit from CPR	Fish from CPR	Snails from CPR	Others from the wild
1	2	3	4	5	6	7	8	9
Magh (mid-January to mid-February)	***********	*****	–	–	Some fish	–	–	–
Phalgun (mid-February to mid-March)	********	****	Yes	–	Fish, *jhinuk*	–	Neem leaves	–
Chaitra (mid-March to mid-April)	*****	****	–	–	–	–	Yes	Rabbits
Baisakh** (mid-April to mid-May)	***	***	–	Brinjals	–	–	Snails, *jal geri*	*Jhinge*
Jyastha (mid-May to mid-June)	***	***	–	Kalmi sak, susmi sak, gim sak	–	Fish	–	–
Asardh (mid-June to mid-July)	***	*	–	Wild potatoes, mushroom	–	–	–	–
Srabon (mid-July)	****	*	–	–	–	Weeds	Yes	–

Source: Mukherjee and Mukherjee (2008b).

Note: The stars in the cells represent the number of stones that the villagers placed in the relative cell, as ordinal ranking of the quantity of the item consumed. This participatory methodology is called scoring and ranking. For more details of the methodology used, see Neela Mukherjee (1993). *Participatory Rural Appraisal*, New Delhi, Concept Publishing Company for the details of the methodology used.

two months, mid-February to mid-April, when people eat less. Mid-April to mid-September marks phase three, when food consumption is very low. Phase four spreads over mid-September to mid-November, when a typical diet of poor households consists mostly of rice.

There are significant seasonal variations in the availability of foods like rice, potato and pulses, which are from the 'primary system of food' produced using technology in the economic sense. For example, the consumption of rice, the staple food, reaches its peak between mid-January and mid-February, after which it declines over mid-March to mid-April, remaining stationary through mid-September to mid-October. The consumption of rice remains low throughout the summer months. Secondary food, which comes primarily from CPRs, provides valuable additions to the food basket of the poor in the form of fibre and animal protein, adding nutrition to the diet of food-insecure villagers. In the lean period (mid-March to mid-September), when food from the primary system is low, the villagers access CPRs to cope with food shortages. Access to CPRs like water bodies can make a significant difference to the quality and quantity of food available. For example, in the village, a huge tank called Rajbandh was not auctioned in 1998 by the local body due to a dispute, which allowed local villagers access to the tank, and they could collect a number of food items. While auctioning of CPRs like tanks provides resources to the local bodies, it also means that insecure households are deprived of food in lean periods.

Food-insecure villagers, in addition to collecting, hunting and gathering food from the CPRs, also glean food from the land of richer farmers. Vegetables like cabbage, pumpkin, ridge gourd (*jhinge*), papaya, green banana, radish, tomato and brinjal that have not been harvested by rich farmers because of damage due to pests, hailstorms, etc., are picked by poor families and help to cover their food insecurity to some extent. Both systems of food play critical roles in providing food to the villagers of Krishna Rakshit Chak. Figure 5.2 also gives a glimpse of the extent of substitution of grains, etc., that takes place with food from the CPRs.

'Foraging' of food from CPRs is widely practised. For instance, for the Karen community in Thailand, foraging of food through 'fishing, hunting and gathering of food from natural habitats' constitutes one of three main pillars of food procurement

(Chotiboriboon et al., 2009). Similarly, the Ainu, an indigenous people who live in northern Japan, collect substantial amounts of food from the CPRs. In fact, of the three components of food, two came from CPRs, namely, wild vegetables and fish and wild game. It is noteworthy that the food from CPRs provides proteins and much vital nutrition that they may not get from cultivated grains (Goodman et al., 2009; see also Table 5.6).

TABLE 5.6 Collection and Preservation Methods of Some Ainu Traditional Foods

Food names	Collection and preservation methods
Wild vegetables	
Wild onion	Collected in April to mid-May, dried by the end of May
Anemone	Collected in the beginning of May to mid-May, dried by mid-June
Udo, spikenard	Collected in mid-May to mid-June
Japanese butterbur coltsfoot	Collected in beginning to mid-June, dried by the end of June
Angelica, fresh	Collected in mid-May to the end of May
Aha bean	Collected in beginning to mid-November, dried by mid-November
Ostrich fern, fiddle head fern	Collected in mid-April to mid-May, dried by mid-May
Perennial lily	Collected in mid-June to beginning of July, preserved by mid-July
Cultivated grains	
Egg millet	Seeded in mid-May, weeded in mid-June, harvested in the end of September to the end of October, dried by mid-November
Barnyard millet	Seeded in mid-May, weeded in mid-June, harvested in the end of September to te end of October, dried by mid-November
Italian millet	Seeded in mid-May, weeded in mid-June, harvested in the end of September to end of October, dried by mid-November
Frozen potatoes	Collected in mid-April in the vegetable garden, soaked in water and dried
Fish and wild game	
Freshwater pearl	Collected in summer in freshwater mussel
Dried salmon	Cut into three and dried outside in the cold
Hokkaido deer	Caught and cut in mid-January

Source: Goodman et al. (2009).

It is noteworthy that one study deems the phenomenon of poor people's scavenging food from leftovers in markets in urban areas, as akin to the rural poor foraging in the CPRs (Brocklesby and Hinshelwood, 2001). The fact is that 'leftovers' partake of the nature of food from CPRs because they can be accessed by anybody without demur and recourse. This is not an uncommon sight in many parts of South Asia, Cambodia, Lao PDR and Vietnam. Box 5.2 provides a brief description of this phenomenon.

BOX 5.2 Use of Proxy Common Property Resources in Urban Vietnam, Bangladesh and Nepal

The participants from urban areas of a participatory poverty assessment carried out by the Centre of Development Studies in Vietnam pointed out that scavenging in markets for leftovers was the urban equivalent of relying on CPRs in rural areas. Analysis across participatory poverty assessments (PPAs) showed that the urban poor rely on all sorts of CPRs for their survival, from scouring rubbish and garbage dumps to growing food on scraps of roadside lands and near railway tracks.

The social differentiation in the use of and reliance on CPRs for survival was visible here also. Use of and reliance on CPRs was considered to be predominantly a female option. Any decline in the condition of and availability of resources from the CPRs directly and adversely affects women. The study also found that, apart from women, the same applies to other very poor community members, for example, 'untouchables' in Nepal or ethnic minority groups in Bangladesh.

Source: Brocklesby and Hinshelwood (2001).

Increasing the Nutritive Value of Food: Use of NTFP to Diversify the Food Base

In most developing countries of Asia and the Pacific, rural and poor people depend on non-timber forest produce (NTFP), *inter alia* for food, fodder, fuel, medicines, gums, resins and material for shelter. Apart from consumption, NTFPs are also important traded commodities in local and national (and in some cases even international) markets. Proceeds from sale of NTFPs at the local markets contribute to the ability of the people, especially the poor, to obtain basic needs and to obtain employment. A study carried out within the framework of the EC-FAO partnership programme 'Information and Analysis for Sustainable Forest Management: Linking National and International Efforts in South Asia and South East Asia' compiled information on NTFPs for 15 Asian countries at the national level. It revealed that while cereals and food-grains form the staple food for the majority of the population, there is a growing importance of NTFPs and their role not only in increasing the availability of food in the lean period, but also in diversifying the food basket of the poor at normal times. (Vantomme et al., 2002)

Non-timber forest produce contributes in two ways to enhancing the food security of poor households. First, it increases the available quantity of food in line with the use of CPRs to increase food availability, as discussed previously. Second, economic access to food, otherwise not available, is increased by sale of NTFPs.[32] Food from NTFPs increases the nutritive value of the and adds palate to the diet of the poor, apart from increasing the total quantum of food available to them. Accessing food from NTFPs is, therefore, akin to increasing the total food availability through augmenting the usufruct entitlement of the poor or through exchange entitlement with nature.

As an example, the reader is referred to Table 5.7, which shows the main categories of forest foods recorded in field surveys in Lao PDR, depicting the rich biodiversity and diversified food base of NTFP. A total of 708 NTFPs from plants and animal sources were identified, including 238 varieties of plant-based NTFP food, and 470 sources of food from animals of different kinds.

Food Storage and Protection

Indigenous Methods of Storage

Food storage is critical in the food consumption chain. However, loss in storage of food in the Asia-Pacific region is significant. For example, in some countries like Sri Lanka, almost 40 per cent of the food produced is lost in storage and transmission,[33] and in

---

[32] Non-timber forest products such as aromatic oils and medicinal plants are also traded nationally and internationally and can achieve high prices in comparison with NTFP traded in local and even provincial markets. Since this chapter is concerned only with community-based responses to food insecurity, the issue is not explored any further here.

[33] Stated by Dr W. G. Somaratne, OXFAM (Australia), during the discussion at the South Asian Civil Society Forum on Responding to Food Insecurity in South Asia, South Asia Watch on Trade, Economics & Environment (SAWTEE), Kathmandu, 24 October 2008.

TABLE 5.7   Diversifying the Food Base through NTFPs in Lao PDR

Srl no.	Category	No. of products	Examples
1	Fruits, seeds	87	Sugar palm fruits, Baccaurea berries, Irvinga nuts
2	Leaves	86	Barringtonia, Lasia, Azadirachta, Centella
3	Shoots	23	Bamboo shoots, rattan shoots, palm hearts
4	Tuber, roots	22	Yam tubers (Dioscorea), Ganga roots
5	Mushrooms	16	Ear mushrooms, Shii take, Termite mushrooms
6	Flowers	4	Sesbania, Butea
	All plants	**238**	
1	Fish	300	Cyprinidae, Pangasiidae, Siluridae, Notopteridae
2	Birds	63	Dove, partridge, pheasant, bulbuls, estrildas
3	Mammals	54	Squirrels, wild boar, rats, civet rats, mouse deer
4	Reptiles, amphibians	41	Frogs, monitor lizards, snakes, turtles
5	Molluscus	7	Freshwater shrimps, crabs, snails, shells
6	Insects	5	Red ant eggs, bamboo grub, dung beetles
	All animals	**470**	
	Total	**708**	

Source: Foppes and Ketphanh (2004).

Bangladesh it is as high as 20 per cent.[34] Preventing wastage of food is tantamount to food production. Hence communities, especially farming communities in Asia and the Pacific, attach a lot of importance to devising and using good food storage systems. The advent of and progress in sophisticated post-harvest technologies are a manifestation of this phenomenon. Judged in that sense, food storage is a means of keeping intact the availability of food in the system and preventing its diminution.

Modern food storage techniques and methods are highly complex, involving a series of processes to prevent the loss of quantity, quality and nutritional value of the food stored. However, these facilities are available mostly in urban and peri-urban areas.

---

[34] Mr Atiur Rahman, Dhaka University, during the discussion at the South Asian Civil Society Forum on Responding to Food Insecurity in South Asia, South Asia Watch on Trade, Economics & Environment (SAWTEE), Kathmandu, 24 October 2008.

The poor and generally food-insecure communities across the region, therefore, do not often have access to these facilities, mainly due to cost implications and the location of such storage facilities. Communities facing food insecurity have in many cases developed their own systems and mechanisms to deal with food storage so that wastage is minimized and the quality and nutritional value of the food stored remain intact. These systems also include specialized techniques suitable to different agro-climatic zones and crops. The methods of handling, packaging, storing and transportation of foods and also their final consumption vary according to the nature of the crop for ensuring maximum utilization and least wastage. Examples of indigenous food storage structures in Himalayan regions of India (Verma et al., 1998) are instructive. Here, foodgrains like maize, wheat and paddy are stored in special containers made of bamboo called *peri* or *peru*. Prior to use, these containers are plastered on the inside with a mixture of cow dung and clay. The containers are placed in a separate room called *overi* on the ground floor of the house, and grain is loaded into them from a hole made in the roof of the ground floor, called *baurh*. To take out grains as needed, a special opening is provided near the bottom of each *peri*. Access to the *peri* is limited to very few people in the household. The use of bamboo containers allows the free flow of air inside the grain, and locating containers on the ground floor ensures a cooler temperature during the period of storage. Loading grains from the top and unloading them from the bottom makes the handling of stored material easy, and consequently losses in handling are kept at a minimum. Additionally, keeping the food storage structures away from the main living space protects the grain from potential fire hazards.

In some other parts of the Himalayan region, grain is also stored in wooden structures known as *darauntha* that are kept away from the main living spaces. *Daraunthas* made of deodar wood are preferred, as deodar wood inhibits the entry of insects and larvae. Wooden houses built away from the living quarters check the entry of rodents and other pests. Windows with wire mesh provide adequate ventilation. The size of the storehouse for each household is proportional to the size of landholdings. In some parts of Lahaul and Spiti in the northern Himalayas, special earthen rooms are constructed for storing cereal crops immediately after harvest. Storage in such rooms also helps keep the temperature cool, so essential for the storage of grains. Poor farmers in some parts of the temperate Himalayas have grain storage chambers made of bamboo.

These are built outside, but within safe vicinity of the farm house. The logic behind constructing these storage houses and rooms at some distance from the family units (also made of wood) is again to save the food-grains from common fire hazards.

As stated earlier, food saved is food grown. Efforts invested in setting up and maintaining a good storage system, thus, increase the availability of food in the system. In that sense, a good storage system has the effect of increasing different kinds of entitlements of the people, irrespective of the entitlement through which food was accessed in the first place.

Food Protection Responses

The success of a food storage system in many communities depends on how well they can protect their food. An important dimension, as a coping mechanism, is the use of indigenous and locally available material as pesticides and insecticides to protect food-grains from wastage. Communities who suffer from chronic food insecurity also invest resources and time to protect their food-grains from insects, mildew and dampness, irrespective of how the food was acquired. The extensive use, by communities in Asia, of leaves from trees having antimicrobial and pesticidal properties for the protection of stored food-grains has been well documented. The neem tree (*Azadirachta indica*) is a case in point. Neem has been known for centuries for its natural and comprehensive pest control properties. The neem is a tropical tree and grows in semi-arid climates in Asian countries. It is almost completely free from insects and pests due to the presence of the active compound Azadirachtin (*tetranortriterpenoids*) that acts as a growth regulator or anti-feedant. Use of neem to protect the stored grains by mixing the grains with dried neem leaves has been a common practice in rural Bangladesh, India and Pakistan. In 1994, the European Patent Office granted the US corporation W. R. Grace a patent on neem as a 'method for controlling fungi on plants by aid of a hydrophobic extracted Neem Oil' (Downes, 2003).

Like a good storage system that prevents wastage of food, food protection is also tantamount to producing food. Food saved from insects during storage is food grown. Efforts invested in protecting food from mildew, insects, rodents and the like, thus increase the availability of food. In that sense, as in the case of a good storage system, steps to protect food have the effect of increasing

---

BOX 5.3 The Ubiquitous Use of the UN 'Tree of the 21st Century' in Tropical Agriculture in Asia

Neem is planted in many parts of Asia: Bangladesh, Burma, Cambodia, India, Indonesia, Iran, Malaysia, Nepal, Pakistan, Sri Lanka, Thailand and Vietnam. It has recently been introduced in China (Hainan Island).

Neem cake has been applied to rice and sugarcane fields against stem borers and white ants since 1930. The mixing of neem leaves (2–5 per cent) with rice, wheat and other grains is still practised in some parts of India and Pakistan, and farmers still put green twigs and leaves in rice nursery beds to produce robust seedlings and simultaneously ward off attacks by early pests—leafhoppers, plant-hoppers and whorl maggots. It has been a common practice in rural India to improve the storage of grains by mixing the grain with dried neem leaves.

In the 1960s, Indian scientists reported the feeding-deterrent property of neem seed kernel suspension against the desert locust. Subsequently, several bioactive ingredients (particularly Meliantriol and Azadirachtin) were isolated from various parts of the tree. These findings aroused worldwide interest in the bioactivity of the neem tree. People all over the world are concerned about the safety of food that is treated with pesticides. Using extracts of the neem tree instead is a promising option, which is why the United Nations declared the neem tree as the 'Tree of the 21st Century'.

Source: Agro Extracts Limited (n.d.).

---

different kinds of entitlements of the people, depending on how they accessed the food in the first place. As a matter of fact, good storage and protection of food are often conjoined activities in the farming chain.

Community-based Provisioning of Basic Services: Community-based Information Systems

Community-based information systems are used to reduce crop loss and hence increase food availability: food saved is food grown. An example of a community-based information system that reduces the risk of loss of crops (read food) is the cyclone warning system in Bangladesh, a country visited by devastating floods every year during the monsoonal rains. The community-based information system in Bangladesh is based on a network of 33,000 village-based volunteers who are alerted of any impending floods via radio stations linked to Dhaka. The volunteers relay the warning through megaphones to villages at risk in the coastal areas. The villages take preventive action. This reduces the loss of crops and stored food (Miyan, 2008).

Self-help Groups and Micro-credit

Self-help groups (SHGs) and micro-credit initiatives are also important community responses to food insecurity. An SHG is a village-based financial intermediary, which could be either a registered or an unregistered group of micro-entrepreneurs (typically 10–15 local residents, especially women), having homogeneous social and economic backgrounds. These micro-entrepreneurs come together voluntarily to save regular, small sums of money, mutually agreeing to contribute to a common fund and to meet their emergency needs on the basis of mutual help (Wilson, 2002). The group members use collective wisdom and peer pressure to ensure the proper end use of credit and timely repayment. This system eliminates the need for collateral and is closely related to the practice of solidarity lending, widely used by microfinance institutions. Funds may be lent back to the members or to others in the village for any purpose at agreed, flat interest rates. Self-help groups are thus seen as instruments for a variety of goals, including empowering women, developing leadership abilities among poor people, increasing school enrolments and improving nutrition and food security, as shown in Figure 5.2. As discussed earlier in this chapter, SHGs have been used for the alternative public distribution system in Medak district, India; for grain, seed and fodder banks in Bangladesh and Gujarat, India; and even managing community-based irrigation systems in Bali (Indonesia), the Philippines and Thailand.

The link between microfinance through SHGs and food security is clear. There are three channels through which microfinance affects food security (Zeller et al., 1997). The first is through the familiar poverty-reducing path of improved income generation. Loans temporarily enhance a household's productive human and physical capital, and then savings and credit services increase a household's risk-bearing potential, leading to the adoption of potentially more profitable, though more risky, income-generating

FIGURE 5.2 Socio-economic Activities of a Group

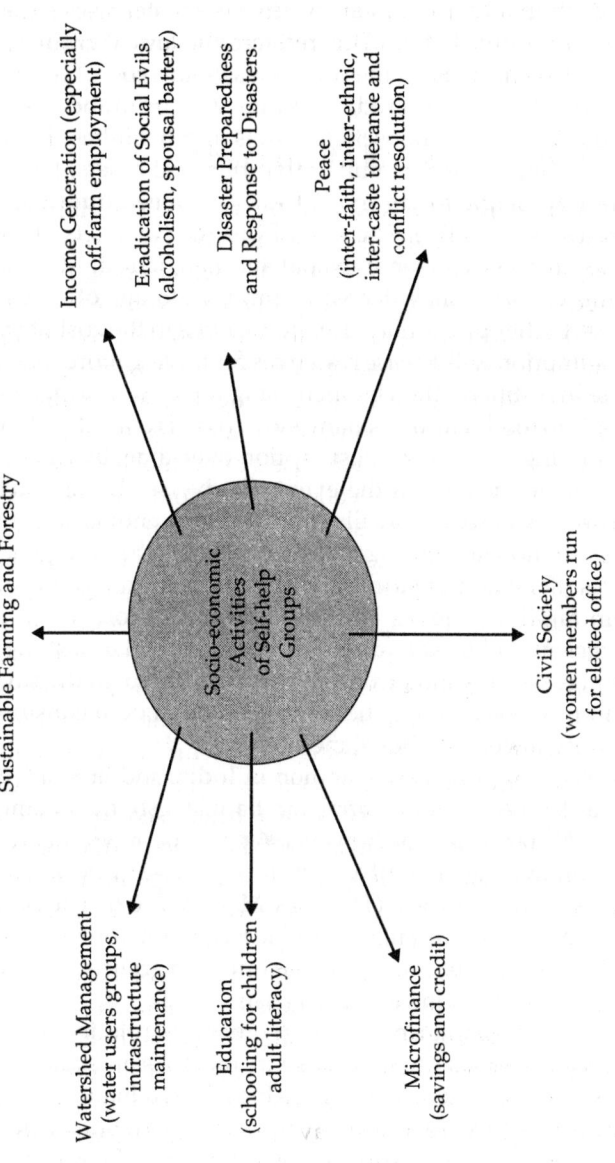

Sustainable Farming and Forestry

Income Generation (especially off-farm employment)

Eradication of Social Evils (alcoholism, spousal battery)

Disaster Preparedness and Response to Disasters.

Peace (inter-faith, inter-ethnic, inter-caste tolerance and conflict resolution)

Socio-economic Activities of Self-help Groups

Civil Society (women members run for elected office)

Watershed Management (water users groups, infrastructure maintenance)

Education (schooling for children adult literacy)

Microfinance (savings and credit)

Source: Author.

activities. The profitability and mix of productive activities may change, leading to increased income that contributes to virtuous production and investment cycles, which increases the economic access of people to food. Second, microfinance contributes to food security by decreasing the rural household's cost of self-insurance. Improved access to credit, savings and insurance services can induce changes in household assets and liabilities. For example, the holding of non-remunerative physical assets, such as cash, jewellery, staple foods and livestock, to meet unforeseen contingencies may decline. The sale of productive assets at low prices under distress, as seen in Nepal and Bangladesh (discussed later in this chapter), may decrease, and the storage of crops for later sale at higher prices may rise. Reductions in the cost of stabilizing consumption will release resources for buying more food and promote investment, thereby increasing the economic access of poor people to food. Third, consumption credit is critical for households attempting to smooth consumption over time, by adjusting their disposable income. In the event of adverse shocks, such as bad weather, accidents and illness, rural households use traditional consumption smoothing measures such as the emergency sale of assets, depletion of stocks and inventories, and grants and loans from family, relatives and the informal sector, which could be inefficient and bind households into social relationships that discourage savings and income generation. Micro-credit, savings and insurance services may help households smooth consumption so they use fewer traditional methods.

Self-help groups are common in India, and in South Asia and South-East Asia. For example, in Bangladesh, the Grameen Bank has 6,707,000 active borrowers (96 per cent of whom are women) with an average loan of $79. It also has 7,411,229 savers with an average saving of $37 (The MIX Market, 2009). The SHG movement in India represents one of the largest microfinance initiatives in the world, with over 1,000,000 self-help groups with 17,000,000 members formed in early 2000 (Ashe, 2002).

Since microfinance favours women, its role in securing food security is substantial, because in most Asian societies women gather, harvest, store and process food. Microfinance, however, as a means to achieve food security has some limitations. Microfinance provided by large microfinance institutions (MFIs) are in the eye of a storm on account of a variety of factors (Kalpana, 2011), including reports of over-lending, the spiralling indebtedness of poor borrowers, coercive and abusive loan recovery tactics deployed by the cutting-edge-level staff of MFIs, high rates of interest vis-à-vis those charged by banks (though they are lower than the rates charged by money-lenders), and reported suicides by poor borrowers who cannot repay their debts. There is also the criticism that that MFIs do not reach the poorest of the poor, and hence those who are likely to be the most food-insecure are left out.

A Suite of Community-based Responses to Transitory Food Insecurity

Increasing Economic Access

Exchange Responses: Internal Migration for Increasing Economic Access to Food

Since it increases economic access, migration is categorized here under exchange responses, because those who migrate for food usually earn either money or goods (usually food-grains) in exchange for labour. The money so acquired is exchanged for food, among other things. Judged in this perspective, migration increases the *exchange entitlements* of the emigrants, but can easily be categorized under distress responses. Income earned through labour migration can also be an important means of acquiring land and agricultural machinery, in which case migration increases the endowment entitlements of the migrants.

Though migration in search of alternative livelihoods is a widespread coping mechanism adopted by food-insecure communities across the region,[35] the benefits of migration for food security may vary from country to country. For example, in the case of Tajikistan, one study found that 'labour migration has allowed a significant part of the Tajik population to escape from starvation stemming from economic stagnation and massive unemployment' (Olimova, 2005). Mongolia provides another example of food security–related migration. Migration out of Kerala, India, on the other hand, has

---

[35] See Chapter 2 for a discussion of migration vis-à-vis price rise.

enabled the migrants to improve their dwellings in their home-towns and villages.

Migration can be either seasonal or long-term. In almost all countries of the region, there has been an explosion in short-term migration for work, not only across countries but also within countries.

The substantial movement of women as part of this process is relatively new in the Asia-Pacific region, especially as women are increasingly moving on their own. The presence of women in migration streams, whether into towns or into other rural areas for employment, generally signals extreme distress (Hugo, 2005). Whatever may be the form of migration, it exposes households to new risks in periods of food insecurity, such as risks arising from social exclusion, discrimination linked to their places of origin, inadequate housing and poor living conditions, low wages, sexual exploitation in the case of women, and lack of access to various basic services.

It needs to be underscored that in many cases of migration, the returns that individuals get from migrating may not make up for

---

**BOX 5.4   Engaging in Seasonal or
Long-term Migration in Vietnam**

For people with few other alternatives, seasonal or long-term migration in search of wage labour is a strategy to cope with shocks and earn wages primarily for satisfying food, clothing and other minimum needs of life. Migration is both an individual and a household risk management strategy, since remittances may provide important security to rural households exposed to production risk in the agricultural economy. In the Red River and Mekong deltas, men and women from vulnerable households often migrate to urban areas (such as Hanoi, Hai Phong and HCMC) for short periods for work to earn wages.

Landless labourers in the Mekong delta occasionally move to towns with their entire family to find employment when seasonal agricultural jobs in their home area are exhausted. This allows them to earn wages to meet their basic needs like food and clothing.

While migration is evidently a crucial way for vulnerable households in rural areas to increase their earnings or cope with events such as landlessness, migration also exposes households to new risks.

Source: ESAF (2005).

their toil, and the costs are often quite high in terms of bad health, broken families and deepening of their debt burden. The losses in terms of destabilization in social capital may also at times be significant, in that when migrants emigrate new groups are formed and social equations change. Emigrants returning home may find that the aged, vulnerable and infirmed who were left behind have suffered deterioration in health and nutritional status.

Take the case of migration in China (Shaohua, 2005). In China, migration has developed as a major coping mechanism through which people seek non-agricultural work opportunities to escape poverty and hunger. Migrant labourers mostly come from China's underdeveloped western and central provinces such as Sichuan, Anhui, He'nan and Gansu, and head mainly for urban areas, southeastern coastal locations, metropolises like Beijing and Shanghai, nearby townships, surrounding counties, small cities and provincial capitals. According to the estimates available for 2003, there were 114 million rural migrants in China constituting 20 per cent of the total rural work force in the country. It is further estimated that by 2020, another 100 to 150 million 'surplus' rural labourers will join the rural labour migration work force. Women account for around one-third of the total rural migrant work force, and their numbers are also on the increase (Zissis, 2007).

Around 70 per cent of the migrants are in the age group of 16–35 and, on an average, do not achieve more than junior middle school education, spending around nine years in school. They earn a monthly income ranging from 300 Yuan (US$36) to 600 Yuan (US$72). And most migrant labourers are employed in jobs which are generally considered dirty, dangerous and difficult (what are called '3D jobs') and which local inhabitants may be unwilling to take.

Like in other parts of the Asia-Pacific region, migration also exposes the migrants to new risks in the towns and cities in terms of residency certificates, legal rights, health care, skill training, housing and social integration. As a result, many of the migrants fall deeper into the poverty trap after migrating. Women face even greater difficulties and adversity leading to increased vulnerability, though their experiences as migrants promote better social status and gender equality for them. Today, migrant labourers in China contribute significantly to the development of industries such as construction, commerce, food, services and sanitation. The total contribution of labour migration from rural to urban areas has been about 16 per cent of the total growth of China's GDP over the past 20 years (Harney, 2008).

Child Labour: Increasing Economic Access

Common mechanisms of meeting food insecurity involve cutting down on consumption expenditure, which includes withdrawing children from school and increasing household labour inputs through employing the children as child labourers, thus increasing the household's economic access to food. Children are seen as providing pairs of hands that at least, fend for themselves in terms of their own food and earn scarce resources for the family to survive. The perceived lack of relevance of education and the cost associated with education, even though in most countries education is free, frequently contribute to the increased incidence of child labour.

Geographically, the Asia-Pacific region contributes 64 per cent of all working children globally. Working children engage in various types of activity regardless of their hazardousness and suitability. It is reported that around 70 per cent of working children are in the agricultural sector, 22 per cent in the services sector and 9 per cent in industry, including mining, construction and manufacturing.

Saving money to access food in the short term by taking children out of school and placing them as labourers does increase economic access to food. However, on the downside, it exposes children to risks such as work-related accidents and exploitation,[36] and engagement in unskilled jobs, the last restricting their choice of future livelihood options to low-paying jobs, trapping them in long-term poverty cycles and food insecurity. Across the region, children, once forced into child labour, are often unable to extricate themselves from the drudgery, and more often than not live a life of bare subsistence. Figure 5.3 depicts the timeline of a child labourer from an Indian village, showing how the typical child spends his years and the negligible returns he gets in exchange for his childhood. Similarly, Box 5.5 narrates the story of the real-life

---

[36] The working hours of children are usually long, with very little or no cash payments as wages. The wages are generally paid in the form of food-grains and occasional clothing and, in some cases, shelter, all of questionable quality.

FIGURE 5.3   Work Schedule of a Child

Source: Kumar (2002).

BOX 5.5　The Life of Panchawak Baag

Panchawak Baag is a child labourer from Khairmal village in Bolangir district in Orissa, India. He dropped out of school while he was in Class III, at the age of 7–8 years, and has been working since then. He was forced to come out of school and start working due to extreme poverty in his family and acute food shortage.

Panchawak has been grazing livestock since he started working to ensure food for himself and generate some surplus for his family. This assured him of two meals a day, one in the morning and another in the evening. In addition, he also got some clothes and some paddy, which came as a handy support for his family. Until the age of 12, he used to get one *lungi* (loincloth) every year along with additional paddy and thereafter two *lungis* and additional paddy. Panchawak's life has been restricted to livestock grazing and, in turn, having two meals a day to stay alive. He does not have any other options for breaking out of the cycle of poverty, which trapped him at an early age, limiting his possibilities in life. Panchawak is not a unique case.

Source: Kumar (2002).

---

example of a child labourer, Panchawak Baag. Such responses clearly indicate the need for a dual policy, one that addresses the immediate familial needs for food, and provides for the long run, for instance through investment in children's education so that they can escape from poverty.

In the short run, child labour may improve the food security of the child and maybe even the family of the child, through increasing exchange entitlements. However, because all other basic needs like education and health care take the back seat, the long-term food security of the child is jeopardized, through damaging his/her long-term trade and exchange entitlements.

Increasing Economic Access: Private Transfers and Remittances

Transfers as a mechanism specifically of hunger insurance include many examples of traditional institutions in the developing world of Asia and the Pacific. In general, private transfers of cash, food and clothing are seen frequently across the developing world.[37] A recent analysis by Cox et al. (2006), using comparable data from 11 low- and lower-middle-income developing countries, including some from Asia, finds that in 8 of these countries, 30 to 50 per cent of households are involved in private transfers, either as donors or recipients. The study also finds that transfers form a significant portion of household incomes. The existence of reciprocal gift giving, which can help in reducing food insecurity, where 'gifts' are sensitive to shocks that lead to food insecurity or to the observed income or expenditure level of individuals, is also fairly widespread in Asia. For example, the importance of private transfers in Vietnam is well known in helping people in distress to access their basic needs including food. Private transfers in Vietnam partake of the nature of means-tested public transfers, flowing from better-off to worse-off households and providing old-age support in retirement.

Changes in private transfers appear responsive to changes in household pre-transfer income, demographic changes and life-course events. Transfer inflows rise upon retirement and widowhood, for example, and are associated positively with increases in health expenditures. In countries like Kyrgyzstan, private transfers flow predominantly from older to younger households. It also appears that the inflow of private transfers increases for households affected by natural calamities. For example, private transfers increased significantly to households affected by Typhoon Linda, which devastated Vietnam's southern-most provinces in late 1997. During the 2005 earthquake in Pakistan, remittances were initially disrupted by the earthquake, but they increased significantly in the year following it (State Bank of Pakistan, 2006).

Private transfers in the form of remittances emanating from migrants, both internal and international, constitute an important means of increasing the economic access of people to food, especially vulnerable groups like elderly people, women and children, and the sick who are left 'at home'.[38] Globally, flows

---

[37] Transfers may also be for purposes other than risk management to ensure food security.

[38] The quantum of remittances can be very large in some cases. For examples, remittances from the US to Asian countries in 2006–07 were approximately US$26 billion to India, approximately US$14 billion to the Philippines and about US$23 billion to China.

of funds from migrants are predominantly used for household consumption (not investment and saving), typically to satisfy the basic subsistence needs of the household, such as food, clothing, medicines, education and housing (Meyers, 1998). For instance, in Bangladesh, remittances, which accounted for more than half of the household income of recipient families, were used for these purposes. Increased expenditure on food and housing and rising levels of living, combined with other benefits, also often lead to the development of human capital, as has been found, for example, in the Pacific Islands and Pakistan (Ghosh, 2006). Indeed, in Pakistan, remittances are predominantly used to meet daily expenses such as food, clothing and health care; however, funds are also spent on building or improving housing, buying land, cattle or durable consumer goods, the repayment of loans for migration and to fund pilgrimages to Mecca (Shahbaz, 2007). Notably, given the stability and reliability of the flow of remittances, they play a significant role in 'consumption smoothing' for the poor. While other forms of income received by poor households may vary and be unpredictable, income from remittances is relatively constant and allows poor households to absorb shocks and unexpected expenses. Judged from this perspective, remittances from migrant workers partake of the nature of insurance for use at times of need.

Overall, remittances of different kinds do increase the economic access of recipients to food. They are in the nature of transfer entitlements.

Consumption Responses

Reducing Consumption

Like production-related community-based responses, consumption-related community-based responses take several forms. Reduction in consumption is at the core of consumption-related responses to food insecurity. This could take the form of reduced food intake through smaller portions in each meal or reductions in the number of meals. Consumption adjustments can also be made by increasing the number of people consuming a given basket of food. The many variations of the phenomenon include migration, 'eating away from home' during the hungry seasons, marrying off daughters early, giving up children for adoption or, in extreme cases, the 'sale' of children, and abandoning old people or other vulnerable individuals (Swift, 1993). For example, due to the widespread and severe financial crisis in Indonesia in the late 1990s, food consumption dropped substantially and households reduced investments in health and education, all of which had long-term implications for food insecurity.

Similarly, during the 2007–08 food–fuel crisis, poor communities' first response to the situation was in the form of a cut in food consumption, seen in several countries of the Asia-Pacific region. Box 5.6 from Sri Lanka is a good example of this phenomenon. A cut in consumption as a coping mechanism adopted at the community level violates the first necessary condition·of food security: food

---

BOX 5.6   Higher Food Prices, Fewer Meals in Sri Lanka

Food in Sri Lanka is widely available mainly because of high imports of essential items to meet local demand. Out of the total food consumption basket in 2004, about 5 per cent of paddy, 30 per cent of potatoes and 78 per cent of dried milk were imported from abroad. These figures indicate that there is sufficient food at the national level. However, prices are increasing. 'In recent months, food prices in Sri Lanka have more than doubled,' said Sarath Fernando during the BBC World Debate in February 2008. 'About half the population of Sri Lanka is living in a situation where prices are beyond their economic reach.'

Rising food prices in 2008 meant rising food insecurity in Sri Lanka at the household level, hitting the poorest Sri Lankan people most. Impact assessments conducted in a project supported by the International Fund for Agricultural Development revealed that poor rural people have adopted a range of consumption-based coping strategies to face food insecurity: reducing the number of times they eat; altering the amount and the type of food they eat; adopting seasonal migration; and mortgaging and selling properties and other assets, jeopardizing their future earning capacity and hence food security. The first two coping strategies can lead to malnutrition. The fourth coping strategy can push poor people further into an unending cycle of food insecurity.

Farmers represent about 70 per cent of the rural Sri Lankan population. The rising food prices of 2007–08 affected the farmers, as they were also consumers. Ironically enough, therefore, farmers (the primary food producers) are themselves curtailing consumption and running into food insecurity in the face of rising prices and falling real incomes.

Source: Herath (2008).

availability. Speaking at the opening session of a workshop in Nadi, Minister Pokotoa Sipeli from Niue said: 'Higher food prices force poor islanders to reduce consumption of foods or buy cheaper foods of poorer quality and low nutritional value' (Reliefweb, 2008).

Contemporary studies from Bangladesh also show that the percentage of households resorting to limiting food portion sizes as a coping strategy has increased over the past three years. In one study, 74 per cent of the households interviewed in 2008, as compared to only 51 per cent in 2006, said that they had limited the size of their food intake at mealtimes (among other coping mechanisms). More recently, in four upazilas (a sub-district-level administrative unit) and two urban slums in Dhaka, it was found that reduction in the consumption of nutritious food was a coping strategy to deal with the increased prices of food in 2008 (Raihan, 2008). In about 87 per cent of rural households, children faced health-related problems due to lack of nutritious or quality food, while the corresponding figure was 75 per cent for urban areas.

There has been a cut in consumption even by farmers in relatively prosperous communities. Box 5.7 provides a short description of how the farming community in Fujian in China responded to the sharp rise in food prices in 2007–08. Additionally, the smaller farmers' diet in Fujian underwent a change in terms of dietary content. Significantly, while the Sri Lanka example demonstrates that the coping mechanism adopted at the community level led to a loss in *food availability*, the response of the communities in China shows that they suffered from food insecurity as the *nutritional value* of the food consumed was reduced.

There are also instances of food-insecure households cutting down on all other expenditures in order to meet food requirements. Curtailing expenditures, including expenditure on health, clothing and hygiene, results in the further deterioration of the physical conditions of people and their capacity to maintain their livelihoods for accessing food in future.

Discrimination in Consumption

In coping with food insecurity, discrimination in the allocation of food to different members within the household, or what is called intra-household prioritization in food allocation, is quite common. Women generally have last priority in the intra-household allocation of food in almost all food-insecure communities of the

---

**BOX 5.7 Poor Chinese Farmers Sell and Buy
Less in Response to Rising Food Prices**

For consumers (buyers) in China, food prices increased by 23.3 per cent between February 2007 and February 2008. Prices varied among different provinces, but generally meat and meat product prices increased by 45.3 per cent. Oil prices increased by 41 per cent and vegetable prices increased by 46 per cent. Soybean and maize prices increased by 24 and 15 per cent, respectively. Livestock prices increased by 31.4 per cent. The price of pork increased by 45.9 per cent, eggs by 16 per cent and milk by 6 per cent.

While farmers in China overall benefited from increased food prices, poor rural farmers generally lacked the capacity to adjust their production to this trend and were vulnerable to price fluctuations. In Fujian province in 2007, poor farmers actually reduced their sale of agricultural products during the first half of 2007 as the price of food increased. As a result, their income from selling agricultural products dropped by 3.7 per cent during the first half of 2007.

Higher food prices forced poor rural people to reduce their consumption of foods like meat and oil. Poor farmers in Fujian province reduced their consumption of pork by 15 per cent and of eggs by 20 per cent. These farmers additionally bought food that was available at a lower price, usually of poorer quality and of lower nutritional value. The poor farmers in Fujian now tend to consume more of their own agricultural products, thus reverting to subsistence agriculture.

The relatively better-off farmers were expected to produce more agricultural products and raise more livestock to take advantage of the higher prices that China saw in 2007–08. However, for these farmers, a fair part of the additional income they earned from selling grain at a higher price was lost due to the increased cost of production. For example, the increased price of pork enabled a farmer to sell a pig for an additional 300 Yuan (US$43). At the same time, the increased price of piglets and feed almost matched the increased sale price of a pig.

In China, food continued to be freely available and the grain harvest had been average and above average during the previous few years. But the price rise of 2007–08 created problems for the poorest members of society, who were unable to afford the price of more expensive food.

Source: Sun Yinhong (2008).

Asia-Pacific region. 'Discrimination against women and girls is an important basic cause of malnutrition' (Manuliak, 1999). Drèze and Sen remark: 'There is indeed considerable direct evidence of neglect of female children in terms of health care, nutrition, and related needs, particularly in North India' (Drèze and Sen, 2002). Similar evidence was found in Bangladesh (L. Chen et al., 1981). In some societies, widows are worst affected (M. Chen, 1999; Drèze and Sen, 2002).

Quite often the practice of 'maternal buffering' is also observed, where mothers deliberately limit their own intake of food in order to ensure that men and children in their households get enough to eat.[39] A participatory poverty profile study in Bolangir district of Orissa in India reflects this practice.[40] According to the study, the male members of the family, who were considered to be the bread earners in most households, consumed proportionately more food than other members in the family. Gender discrimination was quite visible in intra-household allocation of food during the study, in terms of the quantity of food. Among men *inter se*, those in the age group of 18–45 consumed the maximum quantity of food, as both men and women claimed that men's work involved a greater amount of physical labour. However, men belonging to the age group 45–65 consumed less food than their children in the former age category, as the work done by them was usually less arduous than that performed by their children. Similarly, women in the age group of 18–45 consumed more food than elderly women in the family (46–65 years) as most of them had to breastfeed their children.[41] Girls in the age group of 15–18 years ate less than their male counterparts, but more than girls of similar age if they worked as wage labourers.

These coping mechanisms did not lead to any increase in food availability or the entitlements of the community, but allowed the households in the community to manage the allocation of food among their members from the available food basket.

---

[39] See Chapter 3 for more details on intra-household gender disparities in food distribution.

[40] Bolangir may be a limiting case. Nevertheless, it has important lessons to offer.

[41] It must be noted, however, that the need for additional calories for pregnant and lactating women may not have been taken into account, and the work they did may not have been considered 'work' for purposes of food entitlement.

Multiple Coping Responses

Communities facing the prospect of food insecurity may adopt multiple coping responses. Community-based responses to food insecurity vary depending upon the seasonality, severity of food insecurity and failure of access to primary sources of food. A study of the impact of food price rise during 2007–08 in selected areas of Bangladesh (Raihan, 2008), covering four upazilas and two slums in Dhaka city, found seven major coping strategies of the households, namely: (a) reducing savings; (b) selling assets; (c) mortgage of assets and land; (d) taking loans; (e) reducing non-food expenditure; (f) reducing food intake; and (g) early marriage of daughters. Among these coping strategies, 'taking loans' was the technique most used to fight the price hike. In one upazila, as many as 85 per cent of the households surveyed resorted to taking loans to cope with the adverse food security situation obtaining in 2008. Almost 9 per cent and 6 per cent of households in rural and urban areas, respectively, reported that they had to cut their consumption of food; 14 per cent and 8 per cent of rural and urban households, respectively, sold their assets to cope with the food–fuel crisis. A few households were even forced to arrange the early marriage of their daughters.

Food prices also saw a sharp rise in Nepal, as in many other countries in the Asia-Pacific region, during 2007 and 2008. Among other reasons, the hike in food and commodity prices in 2007–08 and the closure of markets due to disputes over transport tariffs increased the severity of food insecurity in the country. In 2008, the food security monitoring and analysis system of the WFP, using survey data from the previous 12 months, undertook a comparative analysis for the Emergency Food Security Assessment in Nepal, conducted in nine districts of the country which had previously been identified as severely affected by drought. The analysis found that most responses that cause long-term damage to the food security of the community have increased, including borrowing money (presumably from unscrupulous money-lenders), purchasing food on credit, and migration (see Figure 5.4). Alarmingly, a high proportion of the population was increasingly selling land, household and agricultural assets, and resorting to increased out-migration (UNFWP, 2008). Figure 5.4 also shows a significant increase, over a 12-month period, in responses such as taking children out of

FIGURE 5.4 Changes in Coping Behaviour in Nepal

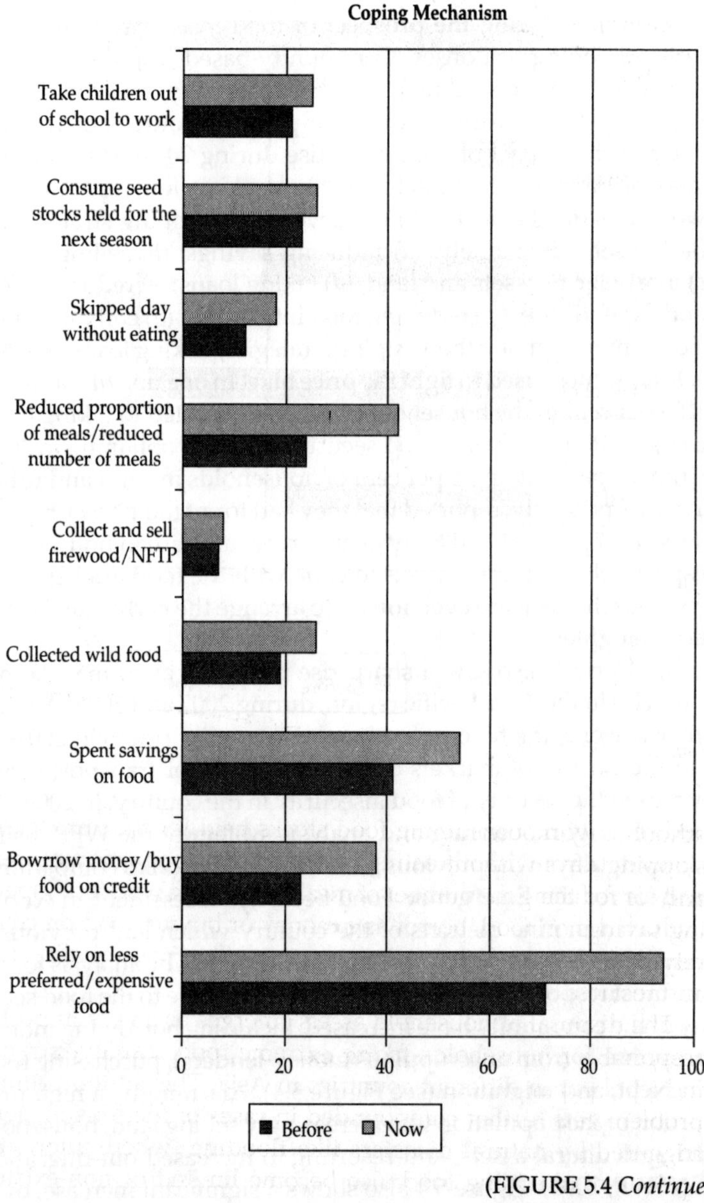

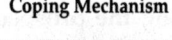

Coping Mechanism

(FIGURE 5.4 *Continued*)

(FIGURE 5.4 *Continued*)

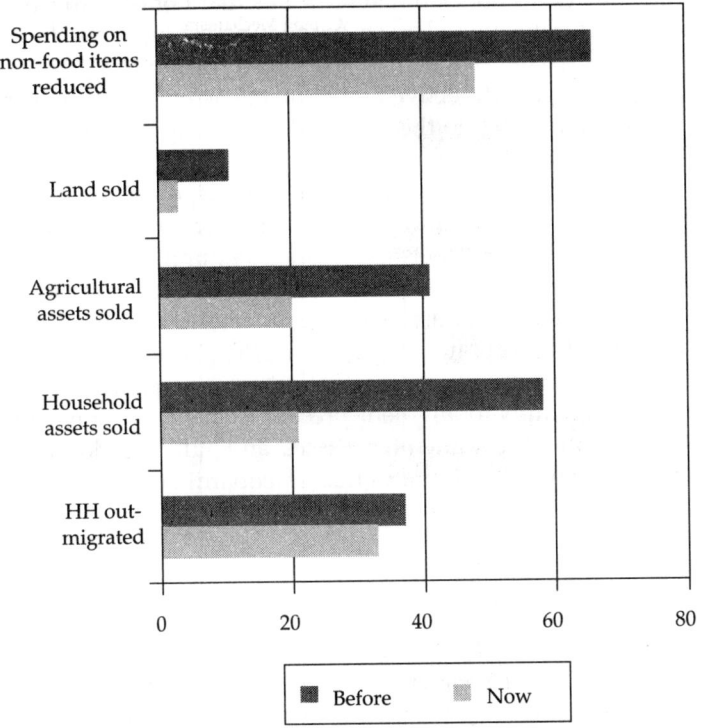

Source: UNWFP (2008).

school to work, consuming food stocks held for the next season, skipping days without eating, reducing the proportion of meals and/or reducing the number of meals, collecting wild food, spending savings on food, borrowing money or buying food on credit, relying on less preferred or less expensive (but poor-quality) food and reduced spending on non-food items.

The disposal of household assets emerges as a very important response for households facing extreme food scarcity, not only in Nepal but in different countries in Asia. This all-too-common problem gets further compounded in cases of food insufficiency triggered by natural disasters like flooding, where alternative options of obtaining food also become limited or non-existent.

A study in Bangladesh conducted over the flood and non-flood periods in 1997 and 1998 demonstrates that not only did poorer households have limited assets to sell to access food, but they also had to sell whatever they had. The study also found that people held on to their productive and valuable assets till there was no other alternative left to them (Carl del Ninno et al., 2001). Sale of assets was a measure of last resort.

People embarking on multiple coping responses, as in Nepal, have a resonance of entitlements in their approaches as well. For example, selling assets, land or agricultural assets or pulling children out of schools is an attempt to access more food through greater exchange entitlements, though admittedly at a cost. The collection and sale of fuel-wood from CPRs are also an attempt to access more food through greater exchange entitlements. Unfortunately, attempts to augment present exchange entitlements to food through sale of agricultural assets and pulling children out of school may have the ultimate effect of jeopardizing future endowment and/or exchange entitlements.

Distress Measures

Under conditions of food insecurity, many communities get down to adopting drastic measures like selling assets, land, etc., as illustrated earlier, but they also resort to distress measures like the sale of body organs and letting themselves be trafficked, with some resorting even to suicide. Distress measures are, strictly speaking, not community responses, but are in most cases individual responses to desperate intra-household poverty and food insecurity situations. Nevertheless, they are mentioned here in passing as they have serious social implications and warrant policy interventions in more than one way.

Being Induced into Commercial Sex

Being induced or trafficked into commercial sex is one of the extreme responses adopted to cope with food insecurity and poverty, a phenomenon visible in almost all countries in Asia. In some countries, the process is relatively easier as it has evolved over time, and because social and cultural norms do not explicitly militate against such activity (Dalrymple, 2008). These are in addition to cases where uneducated and unsuspecting girls who do not know how to protect themselves are trafficked by people they trust, including relatives or neighbours, on the promise of finding them lucrative employment in the city or abroad. These hapless women end up working in brothels and are forced to pay off the large sums of money which the trafficker paid their parents (IRIN, 2008a, 2008b).

Suicides

As a measure of last resort to cope with food insecurity (among others), suicide is not a very uncommon phenomenon in the region. Suicide as a distress measure, however, often entraps the affected family further in conditions of food insecurity of an even more pernicious variety. With often the main bread earner of the family gone, the family has little to fall back upon. In extreme cases, farmers' indebtedness and failure to discharge their liabilities have resulted in suicides, as seen in some parts of India (Government of India, 2007; Sharma, 2004) and to a less extent in China (*Strait Times*, 2004), South Korea (*USA TODAY*, 2005)[42] and Sri Lanka (Sunil, 2005).[43]

Advantages and Limitations

Community-based responses to improve food security have a number of advantages. First, community participation often results in improved targeting outcomes (Coady et al., 2004). A cross-country survey of safety net programmes across many developing countries finds that those that involve beneficiary communities, local groups and local NGOs achieve better targeting outcomes (Subbarao et al., 1997). A comparison of PDS in India with a system like the APDS discussed earlier in this chapter illustrates the point. Second,

---

[42] According to the Korean Peasants' League, at least six Korean farmers committed suicide in November 2005. (*USA TODAY*, 2005).

[43] The overall suicide rate per 100,000 population in paddy rice-growing districts was 34.33 in 2000, 34.83 in 2001, 31.41 in 2002, 29.04 in 2003 and 35.91 in 2004. See Sunil (2005).

community-based responses have the advantage of low information and enforcement costs. In addition, in community-based responses, both unwritten and informal contracts can be self-enforcing, as the short-term benefits from reneging on contracts are much smaller than the long-term costs to those who renege (Coate and Ravallion, 1993).[44] Thus, community-based responses can overcome the problems of moral hazard and contract enforcement that often plague state-administered responses and insurance contracts.

Community-based responses have their problems as well. There may be holes in the community safety nets. Certain subgroups—commonly including the poorest households or individuals and women, persons with disabilities and the voiceless—are often excluded from groups that benefit from responses, and enjoy limited, if any, advantages from community-based responses with others in their community. Group formation for coping with food insecurity, or for that matter any other purpose, is voluntary and endogenous, and may potentially exclude important subgroups, such as women, religious and ethnic minorities and those at the bottom of the pyramid. The exclusion of women is especially remarkable because they have such a key role in the production, harvesting, storing and processing of food. Therefore, some commentators have pointed out that 'community' typically includes the community members with the most power (Bollier, 2006) and, by implication, those who corner most of the benefits.

Then there is the potential inability of community-based responses to manage food insecurity arising from covariate risks. Community-based responses may fail in the wake of natural or man-made disasters (such as the Sichuan earthquake in China in

---

[44] This is not to deny the possible role of a 'moral economy' in tribal or peasant societies, which define solidarity as a moral obligation and subsistence as a right. In fact, as discussed by Fafchamps et al. (1998), social norms of mutual assistance can serve to increase the costs of reneging on (unwritten) contracts, thereby lowering enforcement costs, making informal insurance and moral economy explanations mutually reinforcing and not necessarily strict substitute explanations. On a related note, Platteau (1997) argues that traditional mutual insurance systems are based on 'balanced reciprocity' rather than the 'conditional reciprocity' that characterizes modern insurance systems, implying that members of these societies are not necessarily always motivated by altruistic reasons.

2008 or the tsunami that hit Banda Aceh, Indonesia, and other parts of Asia on the Indian Ocean in 2005). These are situations in which poor households have limited resources for self-food-insurance and often cannot avail of local risk-sharing systems. In addition, the efficacy of informal community-based responses to deal with food insecurity varies inversely with the severity of covariate shock, because risk sharing may break down in the face of more severe shocks such as those seen during the 2008 food crisis in Asia. An important limitation of community-based responses is their relatively small size and paucity of resources, both human and financial.

Conclusions

Poor populations are vulnerable to several shocks that could lead to food insecurity. Where food insecurity may have already existed and is compounded by the threat posed by sudden adverse shocks (such as the Asian tsunami in 2005), considerable returns may be had through improved *ex ante* risk management and providing improved coping strategies through insurance and insurance-like responses. Communities throughout the region (as in other parts of the world) are commonly seen to take these tasks upon themselves, bonding together to reduce the risk to food insecurity and to provide informal mutual insurance within the group, through various means. It is important to recognize such behaviours, and they merit reinforcement.

Most people who are food-insecure are the ones who suffer from idiosyncratic shocks. There is a need, therefore, to hear the voices of the poor. Though community-based responses respond to localized risks, nevertheless their wider propagation has substantial potential. As discussed earlier, much of the agricultural reform in China that began in the 1980s was started by 18 farmers in a village in Anhui province. This is both a lesson in history and a tribute to the Chinese leadership of the time which recognized the great potential in the small move of the 18 farmers.

In many cases, the food security situation would be worse but for community-based responses. The successful examples of community-managed responses to cope with food insecurity need to be disseminated for their wider replication and adoption. These may be second-best options in many situations, but several such responses taken together also form a critical element in the whole scheme of managing food security in the Asia-Pacific region.

Because the cost involved in community-based responses is low while their sustainability is higher, these responses have a lot to commend themselves. In resource-poor economies where large-scale macro interventions for food security, like the ICDS programme in India or the Food Loan Programme in China, are not possible, low-cost, community-based interventions could, at least in the interregnum, hold great promise.

Despite a range of threats, the system of community management, and what comes under the more general term of common-pool resource management, still offers a convincing and appealing option for management, as opposed to more commonplace emphases on state- or market-driven modes of regulation (Ostrom, 1994). However, as mentioned earlier, because there may be gaps in coverage in community-based responses to food insecurity, special safety net programmes for the potentially excluded groups should be in place. Further, population increases in many countries of Asia, the greater economic and geographic mobility of many Asians and increased exposure to covariate shocks associated with natural disasters, war and macroeconomic crisis (such as the one in 2008) place growing strains on community-based responses as a guarantee of food security.

In the case of covariate shocks, the state must act decisively or else people will take recourse to damaging coping responses. Additionally, mitigation as also prevention of food insecurity that is not born out of lack of food availability, like gender discrimination, need wider and deeper social policy interventions.

References

Agro Extracts Limited. n.d. 'Neem has been declared as the "Tree of the 21st Century" by the United Nations'. http://agroextracts.com/default.htm (accessed on 6 March 2012).

Ashe, Jeffrey. 2002. *Self-Help Groups and Integral Human Development*. Waltham, MA: Brandeis University and Catholic Relief Services.

Ashraf, Nava, Dean Karlan and Wesley Yin. 2006. 'Tying Odysseus to the Mast: Evidence from Commitment Savings Product in the Philippines'. *Quarterly Journal of Economics*, vol. 121, no. 2, May.

Bansil, P. C. 2003. 'Demand for Food Grains by 2020 AD', in S. Mahendra Dev et al. (eds), *Towards a Food Secure India*. New Delhi: Manohar.

Bhattamishra, Ruchira, and Christopher B. Barrett. 2007. 'Community-based Risk Management Arrangements: An Overview and Implications for Social Fund Program Design', Working Draft, Cornell University, Ithaca, 1 October.

Bollier, J. 2006. 'Environmental Wealth and Gender Justice', 22 May. www.onthecommons.org.

Brocklesby, Mary Ann, and Emily Hinshelwood. 2001. *Poverty and the Environment: What the Poor Say—An Assessment of Poverty*. Environment Linkages in Participatory Poverty Assessment, Centre for Development Studies, University of Wales Swansea.

Chambers, Robert. 1992. *Micro-environment Unobserved*, Gate Keeper Series Publication. London: International Institute for Environment and Development.

Chen, Lincoln, Emdadul Haq and Stan D'Souza. 1981: 'Sex Bias in the Family Allocation of Food and Health Care in Rural Bangladesh'. *Population and Development Review*, vol. 7, no. 1, pp. 55–70.

Chen, Martha Alter. 1999: *Widows in India*. New Delhi: SAGE.

China Daily. 2008. 'Bumper Harvests Lift Farm Income'. 29 December, p. 1.

———. 2009. 'Enlighten Policies Spread the Wealth'. 13 January, p. 18.

Chotiboriboon, Sinee, Sopa Tamachotipong, Solar Sirisai, Sakron Dhanamitra, Suttilak Smitasiri, Charana Sappasuwan, Praiwan Tantivatnasathien and Pasamail Eg.kontrong. 2009. 'Thailand: Food System and Nutritional Status of Indigenous Children in a Karen Community', in Harriet V. Kuhnlein, Bill Erasmus and Dina Spigelski, *Indigenous People's Food System*. Rome: Food and Agriculture Organization and Centre for Indigenous Peoples' Nutrition and Environment.

Coady, David, Margaret Grosh and John Hoddinott. 2004. *Targeting of Transfer in Developing Countries: Review of Lessons and Experience*. Washington, D.C.: World Bank.

Coate, Stephen, and Martin Ravallion. 1993 'Reciprocity without Commitment: Characterization and Performance of Informal Insurance Arrangements'. *Journal of Development Economics*, vol. 40, pp. 1–24.

Cox, Donald, Emanuela Galasso and Emmanuel Jimenez. 2006. 'Private Transfers in a Cross Section of Developing Countries'. Centre for Retirement Research Working Papers, Centre for Retirement Research, Boston College, Boston.

Cruz, Ma. Conception J. 1989. 'Water as Common Property: The Case of Irrigation Water Rights in the Philippines', in Fikret Berkes and Carl Folke (eds), *Linking Social and Ecological System*. Cambridge: Cambridge University Press.

Dalrymple, William. 2008: 'The Daughters of Yellamma', in Negar Akhavi (ed.), AIDS Sutra, *Untold Stories from India*. London: Vintage.

del Ninno, C., et al. 2001. *The 1998 Floods in Bangladesh—Disaster Impacts, Household Coping Mechanisms and Responses*. Washington, D.C.: International Food Policy Research Institute.

Downes, Gerard. 2003. 'Implications of Trips for Food Security in the Majority World'. Prepared for Comhlámh Action Network, Dublin, October 2003.

Drèze, Jean, and Amartya Sen. 2002: *India: Development and Participation*. New Delhi: Oxford University Press.

Economist. 2008. 'Briefing China's Reforms: The Second Long March'. 13 December, p. 30.

ESAF. 2005. 'Food Insecurity and Vulnerability in Viet Nam: Profiles of Four Vulnerable Groups'. Food Security and Agricultural Projects Analysis Service (ESAF) ESA Working Paper No. 04-11, ESAF, Vietnam.

ESCAP, UNDP and ADB. 2005. *A Future within Reach: Reshaping Institutions in a Region of Disparities to meet the Millennium Development Goals*. New York: United Nations.

Fafchamps, Marcel, Christopher Udry and Katherine Czukas. 1998. 'Drought and Saving in West Africa: Are Livestock a Buffer Stock?' *Journal of Development Economics*, vol. 55, no. 2, pp. 273–305.

FAO (Food and Agriculture Organization). 1984. 'Food and Fruit Bearing Forest Species 2: Examples from South Eastern Asia'. Forestry Paper No. 44/2, FAO, Rome.

———. 2002 *Non-Wood Forest Products in 15 Countries Of Tropical Asia: An Overview*. Bangkok: FAO.

FATA PPA Team. 2003. 'Between Hope and Despair'. Pakistan Participatory Poverty Assessment: FATA Report, Pakistan.

Foppes, J., and S. Ketphanh. 2004. 'NTFP Use and Household Food Security in Lao PDR'. Paper prepared for the NAFRI/FAO EM-1093 Symposium on 'Biodiversity for Food Security', Vientiane, 14 October.

Ghosh, Bimal. 2006: *Migrants Remittance: Myths, Rhetoric and Realities*. Geneva: International Organization for Migration.

Goodman, Masami Iwasaki, Satomi Ishi and Taichi Kaizawa. 2009. 'Traditional Food Systems of Indigenous Peoples: The Ainu in the Saru River Region, Japan', in Harriet V. Kuhnlein, Bill Erasmus and Dina Spigelski, *Indigenous People's Food System*. Rome: FAO and Centre for Indigenous Peoples' Nutrition and Environment.

Government of India. 2007. 'Report of the Expert Group on Agricultural Indebtedness'. Banking Division, New Delhi, July.

Hardin, Garrett. 1968. 'The Tragedy of the Commons'. *Science*, vol. 162, no. 3859, 13 December, pp. 1243–48.

Harney, Alexandra. 2008. 'Migrants Are Like China's Factory without Smoke'. CNN, 3 February. http://www.cnn.com/2008/world/asiapcf/02/01/china.migrants/?iref=newssearch (accessed on 28 December 2008).

Herath, Anura. 2008. 'Higher Food Prices, Fewer Meals in Sri Lanka', in IFAD, *Making a Difference in Asia and the Pacific*, No. 21, June–July.

Hugo, G. 2005. 'Migration in the Asia-Pacific Region'. Paper prepared for the Policy Analysis and Research Programme of the Global Commission on International Migration, USA.

IRIN (Integrated Regional Information Networks) (Pakistan). 2008a. 'Women Suicide Cases on the Rise—But Why?' *IRIN News*, June. http://www.irinnews.org/Report.aspx?ReportId=78948

———. 2008b. 'Indonesia: Poverty at Root of Commercial Sex'. *IRIN News*, July. http://www.irinnews.org/report.aspx?ReportID=79441

Jodha, N. S. 1986. 'Common Property Resources and Rural Poor in Dry Regions of India'. *Economic and Political Weekly*, vol. 21, no. 27.

Kalpana, K. 2011. 'An Anthropological Critique of Microfinance in Bangladesh'. *Economic and Political Weekly*, 27 August, pp. 33–34.

Kumar, S. 2002. *Methods for Community Participation: A Complete Guide for Practitioners.* New Delhi: SAGE.

Lansing, J. S. 1991. *Priests and Programmers: Technologies of Power in the Engineered Landscape of Bali.* New Jersey: Princeton University Press.

Lansing, J. S., and J. N. Kremer. 1993. 'Emergent Properties of Balinese Water Temple Networks: Co-adaptation on a Rugged Fitness Landscape'. *American Anthropologist*, vol. 95, no. 1, pp. 97–114.

Li Honggu. 2009. 'Land of Opportunity'. *China Daily*, 13 January, p. 18.

Mae-Wan Ho et al. 2008. *Food Futures Now.* Beijing: Institute of Science in Society & Third World Network.

Manuliak, Kristy. 1999: 'Women and Children Last?' in Bread for the World Institute, *The Changing Politics of Hunger.* Maryland: Bread for the World Institute.

Mazhar, Farahad, Daniel Bickles, P. V. Satheesh and Fairida Akhter. 2007. *Food Sovereignty and Uncultivated Bio-diversity in South Asia.* New Delhi: Academic Foundations, and Ottawa: International Development Research Centre.

Meyers, D. W. 1998. 'Migrant Remittances to Latin America: Reviewing the Literature'. Inter-American Dialogue/Tomas Rivera Institute Working Paper.

Miyan, M. Alimullah. 2008. *Cyclone Disaster Mitigation in Bangladesh.* University of Dhaka. http://www.fao.org/forestry/media/11285/1/0/ (accessed on 30 November 2008).

Mukherjee, Amitava. 1994. *Structural Adjustment Programme and Food Security: Hunger and Poverty in India.* Avebury, Aldershot, Brookfield, Hong Kong, Singapore and Sydney.

Mukherjee, Amitava, and Neela Mukherjee. 2008a. 'Use of Common Property Resources for Averting Food Insecurity in a Tribal Village in India'. Mimeograph.

———. 2008b. 'Rural Women and Food Insecurity: A Longitudinal Study', in Neela Mukherjee, Meera Jaiswal and Bratindi Jena (eds), *Learning to Share Experiences and Reflections on PRA and Other Participatory Approaches.* New Delhi: Concept Publishing Company.

Murgai, Rinku, Paul Winters, Elisabeth Sadoulet and Alain De Janvry. 2002. 'Localized and Incomplete Mutual Insurance'. *Journal of Development Economics*, vol. 67, no. 2, pp. 245–74.

Nanavaty, Reema. n.d. *Sustainable Livelihood and Agriculture: Observations and Experiences.* Ahmedabad: SEWA.

Olimova, S. 2005. 'Impact of External Migration on Development of Mountainous Regions'. Paper prepared for a workshop on Strategies for Development and Food Security in Mountainous Areas of Central Asia, Dushanbe, Tajikistan, 6–10 June.

Ostrom, E. 1994. *Neither Market Nor State: Governance of Common-pool Resources in the Twenty-first Century.* IFPRI Lecture Series. Washington, D.C.: International Food Policy Research Institute.

Platteau, Jean-Philippe. 1991. 'Traditional Systems of Social Security and Hunger Insurance: Past Achievements and Modern Challenges', in Etisham Ahmad, Jean Drèze, John Hills and Amartya K. Sen (eds), *Social Security in Developing Countries.* Clarendon: Oxford University Press.

———. 1997. 'Mutual Insurance as an Elusive Concept in Traditional Rural Communities'. *Journal of Development Studies*, vol. 33, pp. 764–96.

Praxis. 1998. *Participatory Poverty Profiling Study, Bolangir District, Orissa.* Patna: Praxis.

Raihan, Selim. 2008. 'Impact of Food Price Rise on School Enrolment and Dropout in the Poor and Vulnerable Households in Selected Areas of Bangladesh'. *Bangladesh Economic Outlook*, vol. 2, no.1, September. Dhaka: SHAMUNNAY.

Reliefweb. 2008. 'Urgent Support Needed for Pacific Farmers and Food Consumers', 22 September. http://www.reliefweb.int/rw/rwb.nsf/db900SID/KKAA-7JR87 (accessed on 16 December 2008).

Shahbaz, B. 2007. 'Analysis of Institutional Changes in Forest Management and Their Impact on Rural Livelihood Strategies in NWFP Pakistan'. PhD thesis, University of Agriculture, Faisalabad, Pakistan.

Shaohua, Zhan. 2005. 'Rural Labour Migration in China: Challenges for Policies'. Policy Paper IL0, UNESCO, Paris.

Sharma, Devinder. 2004. 'Farmers Suicide', 24 January. http://www.zmag.org/content/showarticle.cfm?ItemID=4871.

Shivakoti, Ganesh P. 2008. 'Participatory Interventions in Farmer Managed Irrigation Systems in Northern Thailand: Dynamism in Resource Mobilization'. Asian Institute of Technology, Bangkok. http://www.wca-infonet.org/servlet/binarydownloaderservlet?filename=1066126267718_shivakotig042400.pdf (accessed on 26 December 2008).

Sim Chi Yin. 2008. 'The Village Where It All Began'. *Straits Times*, Singapore, 6 December.

State Bank of Pakistan. 2006. 'The State of Pakistan's Economy: Third Quarterly Report for the Year 2005–2006'. Karachi: State Bank of Pakistan.

Strait Times. 2004. 'Plight of Chinese Farmer', *Strait Times*, 5 September. http://www.hartford-hwp.com/archives/55/333/html (accessed on 19 January 2012).

Subbarao, Kalanidhi, Aniruddha Bonnerjee, Jeanine Braithwaite, Soniya Carvalho, Carol Graham and Alan Thompson. 1997. *Safety Net Programs and Poverty Reduction: Lessons from Cross-country Experience.* Washington, D.C.: World Bank.

Sunil, W. A. 2005. 'Suicides Highlight Desperate Conditions Facing Sri Lankan Farmers'. WSWS News and Analysis, 10 June. http://www.wsws.org/articles/2005/jun2005/sril-j10.shtml.

Swift, J. 1993. 'Understanding and Preventing Famine and Famine Mortality'. *IDS Bulletin*, vol. 24, no. 4. Institute of Development Studies, Sussex.

The MIX Market. 2009. http://www.mixmarket.org/en/demand/demand.show.profile.asp?ett=1658 (accessed on 13 January 2008).

UCANEWS. 2008 'South Asian Farmers Exchange Ideas on Organic Farming', 27 August. http://www.ucanews.com/story-archive/?post_name=/2008/08/27/south-asian-farmers-exchange-ideas-on-organic-farming&post_id=49086 (accessed on 24 January 2012).

UNWFP. 2008. *Food Security Bulletin–20.* United Nations World Food Programme, Rome.

USA TODAY. 2005. 'South Korean Farmers Doggedly Target Trade Talks'. com/money/world/2005-12-15-wto-fri-usat_x.htm.

Vantomme, N. Paul, Annu Markkula and Robin N. Leslie (eds). 2002. *On-wood Forest Products in 15 Countries of Tropical Asia: An Overview.* Bangkok: FAO, under the EC-FAO Partnership Programme on Information and Analysis for Sustainable Forest Management: Linking National and International Efforts in South Asia and Southeast Asia.

Verma, L. R., et al. 1998. *Indigenous Technology Knowledge for Watershed Management in Upper North-west Himalayas of India*. Participatory Watershed Management Training in Asia (PWMTA) Programme, Kathmandu, Nepal.

Wilson, Kim. 2002. 'The New Microfinance: An Essay on the Self-help Group Movement in India'. *Journal of Microfinance*, vol. 4, no. 2.

Xinhua. 2008. 'Kazakhstan Gears Up to Tackle Food Crisis', 12 July. On line Edition.

Yinhong, Sun. 2008. 'Poor Chinese Farmers Sell and Buy Less in Response to Rising Food Prices', in IFAD, *Making a Difference in Asia and the Pacific*, No. 21, June–July.

Zeller, M., G. Schrieder, J. Von Braun and F. Heidhues. 1997. *Rural Finance for Food Security for the Poor: Implications for Research and Policy*. Washington, D.C.: International Food Policy Research Institute.

Zissis, Carin. 2007 'China's Internal Migrants'. Council on Foreign Relations, New York and Washington (accessed on 28 December 2008).

Further Readings

Barrett, Christopher B., Christine M. Moser, Oloro V. McHugh and Joeli Barison. 2004. 'Better Technology, Better Plots or Better Farmers? Identifying Changes in Productivity and Risk among Malagasy Rice Farmers'. *American Journal of Agricultural Economics*, vol. 86, no. 4, pp. 869–88.

Bhattamishra, Ruchira. 2007. 'An Institutional and Impact Analysis of Village Grain Banks: Evidence from Tribal Orissa'. PhD dissertation, Department of Economics, Cornell University, Ithaca.

Dev, S. Mahendra. 2003. *Right to Food in India*. Hyderabad: Centre for Economic and Social Studies.

Dev, S. Mahendra, et al. 2002. *Towards a Food Secure India: Issues and Policies*. New Delhi: Institute for Human Development, Hyderabad: and Centre for Economic and Social Studies.

Jalan, Jyotsna, and Martin Ravallion. 1999. 'Are the Poor Less Well Insured? Evidence on Vulnerability to Income Risk in Rural China'. Policy Research Working Paper 1863, World Bank, Washington, D.C.

Meyer, Richard L. 2002 'Microfinance, Poverty Alleviation, and Improving Food Security: Implications for India', in Rattan Lal (ed.), *Food Security and Environmental Quality*. Boca Raton, FL: CRC Press LLC.

Ostrom, E. 1992. *Crafting Institutions for Self-Governing Irrigation Systems*. San Francisco: Institute for Contemporary Studies.

Popkin, Samuel L. 1979. *The Rational Peasant: The Political Economy of Rural Society in Vietnam*. Berkeley: University of California Press.

Praxis. 1999. 'Consultations with the Poor, India 1999—Site Reports from Andhra Pradesh'. Patna: Praxis.

SAARC. 1992. 'Meeting the Challenge'. Report of the Independent Commission on Poverty Alleviation, SAARC Secretariat.

SAARC. 2003. 'Our Future Responsibility: Road Map towards a Poverty Free South Asia'. Report of the Independent Commission on Poverty Alleviation, SAARC Secretariat.

Shapouri, Shahla, and Stacey Rosen. 2001. *Food Security Assessment: Regional Overview.* Agriculture Information/Bulletin No. 765-1, United States Department of Agriculture-Economic Research Service, USA.

Smith, Justin. 2009. 'Self-help Groups: Economic Capacity Building through Cooperative Associations'. http://www.publicsphereproject.org/patterns/pattern.pl/public?pattern_id=624 (accessed on 13 January 2009).

Tang, S. Y. 1992. *Institutions and Collective Action: Self-governance in Irrigation.* San Francisco: Institute for Contemporary Studies Press.

6

Social Access and Social Protection for Food Security*

Famines could thrive even without a general decline in food availability.

—Amartya Sen

Access to and utilization of food are as important for food security as producing food itself. Social protection helps provide access to food, and furthers its utilization,[1] among subpopulations that would otherwise not be able to access and utilize food. It should be clarified that the social protection regime encompasses three interrelated areas of support for the poor and vulnerable, and can be viewed as the concept of protecting and promoting social policies. Social security, social assistance (transfers) and social insurance are the three operational components that address the promotion and protection of human development. Given the larger role of social security schemes and programmes in developing countries, the terms social protection and social security are often used interchangeably.

This chapter discusses social protection, its role in food security, and the social policy challenges of ensuring access to food for all.

*The author is grateful to Sri P. Upendranath, Mr Marco Ronacarti and Ms Beverly Jones.

[1] The great tragedy of the Bengal famine of the early 1940s, where people died of starvation despite granaries being full of food because they had no purchasing power to buy the food (Sen, 1986), could have been avoided had there been social protection measures in place.

It starts with a discussion of the significance of social protection in human development, followed by an examination of demographic trends in the context of pressures on food security related to migration and changing dependency ratios. Migrants, both international and internal, ethnic minorities, tribal people and 'other' discriminated groups are vulnerable to food insecurity. The chapter discusses the need for social policies for these food-insecure groups. It also addresses the question of reducing gender-based food insecurities and discrimination based on other attributes towards making access to resources more equitable for all. It explores the issue of social protection for people who are exposed to shocks, with special reference to agriculture and employment. This is followed by a design of social protection measures for small farmers who are subject to shocks, seasonality and venerability. Finally, a brief description of social protection measures for tackling climate change has been included, bearing in mind that the final chapter provides a fuller discussion of climate change and food security. This chapter also considers institutional arrangements in relation to food security and shows how good governance is fundamental in improving the life of those at greatest risk of food insecurity.

BOX 6.1   Social Protection, Social Assistance and Social Insurance Etc.

Social Protection: An umbrella term which has evolved considerably in the past few decades. In general, it is a society's set of policies and programmes designed to reduce poverty, vulnerability and inequality. Thus, social protection programmes can prevent individuals from slipping into poverty, and promote opportunities, while also transforming communities and societies through investment in human capital and health. This can include social insurance, social assistance, social services and labour market policies.

Social Insurance: This provides benefits, such as unemployment pay or pensions. The schemes are usually funded by contributions from up to three sources: individuals, employers and governments. The underlying principle is that of shared risk across the whole society. Certain events trigger payments—such as unemployment, sickness, maternity or reaching pensionable age. The benefits generally vary according to the level of contribution and are, therefore, not always progressive or

---

(BOX 6.1 *Continued*)

(BOX 6.1 *Continued*)

redistributive: those who contribute more or work longer, for example, typically receive more.

Social Assistance: This refers to those parts of social protection systems that are funded from general government revenues and by some NGOs and international donors. Benefits are paid according to need and bear no relation to what people have paid in taxes. Indeed, the beneficiaries are very often those who have paid the least tax. This would include, for example, food subsidies and emergency food distribution. Insofar as they combine interventions aimed at strengthening the productive capacity of the poor, they play an essential role in integrating strategies for both poverty reduction and development.

Social Services: These are essential and basic services and are included in social protection because they often substitute for missing markets. The provision of free or subsidized health services for children and pregnant women, safe water and basic electricity fall under this category.

Social Pensions: These are non-contributory benefits, for older persons for example, or persons with disabilities, paid to everyone out of general taxation.

Social Safety Nets: These are measures that cover a small fraction of the population in need and aim at compensating for the impact of crises and shocks on basic living conditions. They often involve limited coverage, financial constraints and significant targeting.

Social Security: This term is used in different ways. Sometimes it refers to social insurance, sometimes it corresponds to social protection. To avoid confusion, this report only uses the umbrella term 'social protection'.

Source: ESCAP (2011).

Contextualizing Social Protection

While historically, welfare states have given priority to well-being through systematic social security measures to meet the adversities of all populations, the social policies of developing countries have not yet been able to fulfil the basic needs of millions of poor and vulnerable people. Meeting calamities, shocks, adversities and deprivations through public policy and strengthening the supply side has continued to be the focus of many developing countries of Asia.

The systematic evolution of social protection policies can be traced back to the structural adjustments of the 1990s, wherein 'adjustment with a human face' and 'safety nets' guided nations in meeting the adverse impact of macro policies.

East Asian economies experienced rapid economic growth in the 1980s, which culminated in a financial crisis in the late 1990s, demonstrating the inability of states to protect the basic livelihoods of citizens. Limits to the notion that economic growth would deliver poverty reduction were exposed. The crisis has in fact exposed the limits of the then existing social protection mechanisms, and a systematic approach to social protection has been found to be the *sine qua non* for addressing shocks and vulnerabilities. Dovetailing social protection policies with mainstream growth and development policies has also been advocated, in contrast to the 'residual' nature of social protection policies hitherto followed. Pro-poor growth and development policies would necessarily mean integrating social protection policies into the scheme of macroeconomic development.

Impressive growth was registered across the region at the turn of the century. Once again, the limits of growth can be seen in the rise in income inequalities across countries in the region. The subprime-triggered financial crisis of the late 2000s has yet again demonstrated the need for coherent and systematic social protection policy responses in order to further social cohesion. Chronic poverty and social exclusion have been identified as important obstacles to translating growth into food security, and these two obstacles can be addressed only through systematic social protection mechanisms.

The promotional and protectional roles of various social protection policies have increasingly been recognized as a response to the dynamic nature of poverty and food deprivation and the changing character of society and polity across countries of the region. Programmes related to the enhancement of basic human capabilities, mechanisms of targeted social transfers and social assistance demonstrate the scope and the unfinished agenda of universalization of social protection measures for meeting adversities exemplified by food insecurity.

Given labour market segmentation and the exclusion mechanisms that operate owing to differential resource endowments, mechanisms to address employment, decent work and workplace social security are important elements of social protection policies for sustained food security. Market-based mechanisms and state-led responses have been acknowledged in such contexts to address shocks and vulnerabilities. However, employment-related risks in the formal sector are only one part of the overall spectrum of vulnerabilities that the poor face. Given the changes in work—informalization of work, the rise in the informal sector and rise in employment in the unorganized sector—all these by themselves create vulnerabilities among the poor that impinge on their food security. Existing social protection mechanisms respond to these challenges in inadequate measure.

Further forms of vulnerability that reflect relational disadvantage—for example, disadvantages arising from gender, caste or other forms of identity—and that lead to chronic forms of poverty and food insecurity are also not sufficiently addressed by the existing social protection regimes in many countries of Asia. It is also important to acknowledge the implementation failures of programmes and schemes in order to build robust mechanisms of social protection.

Food Price Inflation: Risks to Human Development, Issues of Food Accessibility and Social Protection

There are four dimensions along which rising food prices may have a negative impact on human development, all of which are relevant to the issue of food security in Asia and the Pacific. Inflation can increase poverty and inequality, exacerbating an already difficult food security situation; it can worsen nutrition; reduce the utilization of education and health services, having a bearing on food security in the long run; and deplete the productive assets of the poor (World Bank, 2008a), with a deleterious impact on access to food and/or food availability. If there is a deterioration in any of these four dimensions, it is hard to reverse the damage caused to food security, and its implications may be felt for years, and in some cases for generations to come (ibid.). Indeed, the increases in food prices in 2007–08 and again in 2010–11 have been harmful to the poor and people with special needs, in that they cut into their economic access to food, reducing their consumption (noting that women in particular have reported cutting food consumption)

(IFAD, 2008a). Food price inflation in 2007–08 worsened the nutrition situation of the people. For example, in Indonesia, for 76 per cent of the poor who are net rice buyers, including 72 per cent of the rural poor, each 10 per cent increase in rice prices reduced the real value of the expenditure of the poorest tenth of the population by some 2 per cent (World Bank, 2006). In the Philippines, it is estimated that 73 per cent of rural and all urban households were adversely affected by high rice prices (Brahmbhatt and Christiaensen, 2008) as their economic access was impaired. And then, as seen in Chapter 5 on community responses, in many cases, for instance in India, Bangladesh and Nepal, the utilization of education services has been reduced and productive assets depleted.

The Asian and Pacific region over time has seen dramatic aggregate economic growth and a fall in the number of people living in poverty and, by implication, in food insecurity. However, food insecurity affects different people differently, and so do increases in food prices. It is, therefore, important to examine how specific population subgroups have been and may continue to be affected by food insecurity, and plan appropriate responses. Food security differs among countries, and within countries by age, sex and rural versus urban residence. It also varies by the form of livelihood and whether one is a member of a particularly disadvantaged social group.

Take the case of under-5 children. In the Asia-Pacific region, 28 per cent of children under 5 years of age are underweight. On a global scale, the region accounts for about two-thirds of children who are underweight; the percentage of underweight children is almost 50 in certain countries. The relatively high prevalence of underweight children in rural areas in selected countries in Asia is revealed in Figure 6.1. Related to this is the issue of many infants and children receiving inadequate nutrition, including that provided by breastfeeding. The extent to which infants are breastfed is influenced by the socio-economic status of mothers as well as by factors such as the role of institutions and social frameworks; when the latter function well, they can enhance food security including that related to improved nutrition of infants and children.

Then there are children at risk of undernutrition and micronutrient deficiencies. Deficiencies of some micronutrients (essential vitamins and trace minerals that are essential for chemical processes

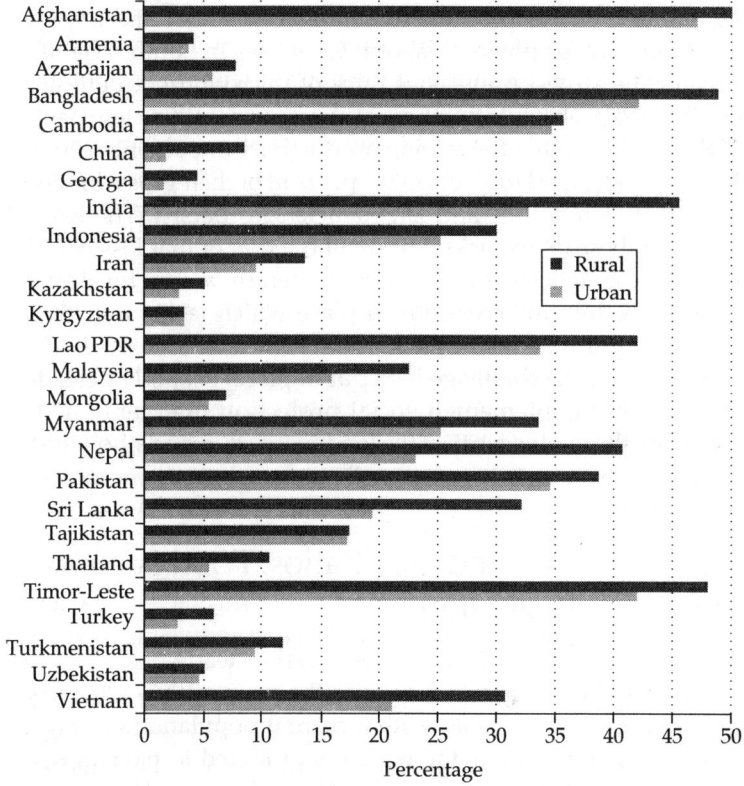

FIGURE 6.1   Underweight Prevalence (%) by Background Information: Residence, 1998–2006

Source: UNICEF (2008a).

Note: All data are for children of 0 to 59 months unless otherwise indicated. A year span followed by an asterisk indicates that the data are from the most recent year in that period for which data are available. For Afghanistan, data are for the age group 6–59 months, and for Azerbaijan and Sri Lanka data are for the age group 3–59 months.

that ensure the survival, growth and functioning of vital human systems) are prevalent in low- and middle-income countries in Asia-Pacific. Such deficiencies increase the risk of illness or death from infectious diseases by reducing immune and non-immune defences and by compromising normal physiology and/or mental development. Deficiencies in three key micronutrients—iron, vitamin A and zinc—can contribute to maternal and child mortality and morbidity. Sufficient micronutrients help prevent disabilities such as neural tube defects and child blindness, and protect learning abilities, enable progress to be made in school and improve adult capacity for physical labour. Countries in South Asia have some of the largest prevalence rates of micronutrient deficiency and the largest absolute numbers of micronutrient-deficient people (UNU, 2007). Studies measuring productivity impacts have found that countries stand to lose about 1 per cent of their gross domestic product if iron, iodine and zinc deficiencies persist. The cost of reversing these deficiencies is a small fraction of that loss of GDP due to these deficiencies. There is, therefore, a case for putting social protection interventions in place which tackle nutritional and micronutritional deficiencies.

In all cases, the challenge is to put in place, properly designed and properly implemented social protection interventions for enhancing the food security of the food-insecure, especially children and infants who by themselves will not be able to do this.

Population Dependency Ratios, Food Security and Social Protection

As a region, Asia-Pacific has experienced a steady decline in the annual population growth rate over the period 1997–2008, from 1.3 per cent to 1.0 per cent. With a current population of slightly over four billion people, the region is projected to have close to 5.2 billion people by the year 2050. With this population growth, the number of older persons (60+) is also growing rapidly in the region (see Table 6.1). It is projected that the region will experience an increase in older persons from 410 million in 2007 to about 1.3 billion in 2050. The proportion of older persons relative to the total population is also increasing, with a possible rise from about 10 per cent at present to 25 per cent in 2050. In Japan, about 28 per cent of the population is already over 60 years of age. Population ageing has been occurring at a rapid pace in certain countries, such as China, Georgia, Japan, the Republic of Korea, the Russian Federation and Singapore, and at a relatively fast pace in others such as Malaysia and Thailand. While the absolute and relative numbers of older persons is increasing and many of them are in the

TABLE 6.1 Percentage of Population 60 Years and Older to Total Population, 2007, 2025 and 2050, by Fertility Level

Fertility level	Country	Total fertility rate 2000–05	Percentage of population 60 years and older 2007	2025	2050
	ESCAP		10.3	15.4	24.3
Lowest–low	Macao, China	0.8	11.5	29.2	42.8
TFR<=1.5	Hong Kong, China	0.9	16.4	30.2	39.4
	Republic of Korea	1.2	14.6	27.4	42.2
	Japan	1.3	27.9	35.8	44.0
	Russian Federation	1.3	17.4	23.8	32.4
	Armenia	1.3	14.6	22.6	33.9
	Singapore	1.4	13.8	31.6	39.8
	Georgia	1.5	18.3	25.4	34.9
Low	Azerbaijan	1.7	9.2	16.7	27.6
TFR 1.6 to 2.1	China	1.7	11.6	20.0	31.1
	Australia	1.8	18.7	25.8	30.2
	Thailand	1.8	12.0	21.5	29.8
	Democratic People's Republic of Korea	1.9	13.7	16.6	24.6
	New Zealand	2.0	17.3	24.9	30.2
	Kazakhstan	2.0	10.4	15.3	24.1
	Sri Lanka	2.0	10.6	19.7	29.0
	Mongolia	2.1	5.9	10.8	25.1
	Iran (Islamic Republic of)	2.1	6.6	10.9	25.6
Near-replacement	Turkey	2.2	8.5	13.8	24.5
TFR 2.2 to 2.9	New Caledonia	2.2	10.6	16.9	26.0
	Myanmar	2.2	8.2	13.9	25.6
	Vietnam	2.3	7.7	13.4	26.1
	Indonesia	2.4	8.6	13.7	24.8

(TABLE 6.1 Continued)

(TABLE 6.1 Continued)

Fertility level	Country	Total fertility rate 2000–05	Percentage of population 60 years and older		
			2007	2025	2050
	French Polynesia	2.4	8.5	15.0	24.7
	Kyrgyzstan	2.5	7.3	11.6	21.5
	Brunei Darussalam	2.5	5.1	11.2	20.1
	Uzbekistan	2.7	6.2	10.6	20.9
	Guam	2.7	10.1	16.6	22.3
	Turkmenistan	2.8	6.2	10.8	21.2
	Maldives	2.8	5.6	8.5	20.0
	Malaysia	2.9	7.1	13.2	22.2
	Bhutan	2.9	7.1	10.1	23.3
Transitional	Fiji	3.0	7.4	12.7	19.6
TFR 3.0 to 4.9	India	3.1	7.7	11.5	20.2
	Bangladesh	3.2	5.9	9.2	17.0
	Philippines	3.5	6.2	9.8	18.2
	Lao People's Democratic Republic	3.6	5.1	7.6	16.3
	Cambodia	3.6	5.4	7.9	15.2
	Nepal	3.7	5.9	7.8	14.0
	Tonga	3.7	9.1	9.1	15.6
	Tajikistan	3.8	5.1	8.1	16.1
	Pakistan	4.0	6.0	8.6	16.5
	Vanuatu	4.2	5.2	7.7	15.9
	Micronesia (Federated States of)	4.2	5.6	9.2	15.3
	Papua New Guinea	4.3	4.0	6.3	11.2
	Solomon Islands	4.4	4.8	6.3	13.0
	Samoa	4.4	6.6	10.3	17.4
High	Timor-Leste	7.0	4.6	5.3	7.6
TFR =>5	Afghanistan	7.5	3.7	3.8	5.6

Source: United Nations (2007).

Note: TFR is the total fertility rate.

category of 'young old' (60–79), a growing number are classed as 'oldest old' (80+). In these countries, with a decreasing proportion of young people and young adults and an increasing proportion of older workers and pensioners, there are growing strains on the ability to ensure care for older persons. The category of older persons disproportionately comprises women, who are generally more dependent on others for food and other basic necessities than men due to relatively low levels of property ownership and pension entitlements. Social protection for the elderly does not always exist, and in many cases where it exists, it needs reform. In several countries, there will not be enough money to pay pensions for the elderly in the future.

In contrast, several countries in South and South-West Asia, South-East Asia and the Pacific have very young populations. In South Asia, total fertility rates range from 1.9 in Sri Lanka to 7.2 in Afghanistan (UNICEF, 2008b). Total fertility rates in South and South-West Asia are at 2.8. With current population growth rates, both Afghanistan and Timor-Leste are expected to double their populations in the coming 25 years. The Pacific Island countries tend to have high total fertility rates and a very high proportion of their populations under 25 years of age. For example, the total fertility rate of Papua New Guinea and Solomon Islands was 4.0 in 2006, while for Samoa, it was 4.1. Many countries with high fertility and population growth are especially vulnerable to constraints on resources, including food; and, as is the case for countries with a high proportion of older persons, countries with young populations will also have high dependency ratios. This implies that young people as well as older persons in Asia-Pacific are, or will be in the short run, largely dependent upon others for their food security. Of these countries, those experiencing slower social and economic development and less capable of ensuring food security for their people will face greater population pressure than others (United Nations, 2007).

Social protection for the elderly, the young and women is therefore critical. For example, a price-indexed old-age pension scheme for the elderly and supplementary nutrition like the midday meal scheme in India for the young hold promise. In some countries, the need may be to reform the social protection system, finding innovative ways to finance social protection and target social protection measures.[2] The coverage of social protection by government and market instruments remains very low in the Asia-Pacific region, and is not always focused on the food-insecure, with roots in fiscal and implementation constraints and in the political economy of the countries of the region. The 'business-as-usual model' cannot be allowed to proceed unhindered, and governance has to improve. By thinking 'outside the box', there are opportunities for social protection to engage and improve risk management using innovative approaches for greater food security. For example, the new financing being made available for climate change is an opportunity for scaling up social protection instruments to counter threats to climate change–induced food insecurity.

Migration and Food Security

International Migration

Asia has witnessed a dramatic increase in international migration in the past few decades, with 58 million migrants in 2005 (United Nations, 2006). There is a high level of migration from China, the Philippines and many of the South Asian and Pacific Island countries, and some Central Asian countries. Increased disparities between countries with regard to income and opportunities, and demographic imbalances among countries, encourage people to move so as to improve their lives. In many cases, the financial remittances sent by migrants enable families to purchase more and better-quality food, and are a very important contributor to poverty alleviation and overall socio-economic development. In 2007, over

---

[2] There are, however, countries with a large proportion of people of working age (15–64), such as the Islamic Republic of Iran, Mongolia and Myanmar, which are at the stage of the demographic transition, called the *demographic window of opportunity*. This implies that they are optimally placed to benefit from economically productive investment, the 'demographic dividend', as a result of low economic dependency levels and a relatively large number of potential workers to support those in the non-productive age groups. In the context of food security, such a situation allows for the diversion of a significant proportion of national output from consumption to investment without sacrificing existing living standards.

$121 billion worth of remittances were received in the Asia-Pacific region, more than double the figure in 2000 (World Bank, 2008b).

Remittances from migrant workers enable many families to purchase food. India, Bangladesh, Philippines and Sri Lanka receive large amounts of remittances. The Philippines, for example, receives over $7 billion in remittances each year from the United States alone, while India receives about $12 billion annually from the Middle East (Asian Migrant Centre, 2003). While it is difficult to predict the impact of the global financial crisis on migration and remittances, judging from other crises, migrant worker flows may slow down, but stocks should remain close to what they are. Remittances may be sent by fewer migrant workers.

Although remittances can have a positive effect on food security, various negative consequences of migration remain worrisome. Migrant workers frequently constitute vulnerable populations in the host country or in the urban areas where they move to, because they are often not entitled to government assistance, may be very dependent on their employers for food (such is the case for domestic workers), may not be entitled to health and education services and may lack the social networks which they had in their communities of origin. Box 6.2 gives an illustration of a limiting risk faced by migrants. As a consequence of such vulnerability, they face heightened risks of food insecurity. Social protection for migrants is critical.

Internal Migration

Migration (or internal displacement) also takes place within countries and territories, often as a response to livelihood or food insecurity, landlessness, loss of land tenure or crises. Conflicts such as those in Afghanistan and Timor-Leste have displaced millions of people and contributed to greater food insecurity. On the other hand, migration is often seasonal and is used as a coping mechanism to improve food security, such as in the case of some Central Asian countries, China and India as well as other countries in Asia and the Pacific.

Migration trends within countries are generally from rural to urban areas, again by those seeking greater opportunities, including the landless poor. For example, China's economic development and rapid urbanization have increased the number of people moving

BOX 6.2   Myanmar Migrants Cast Adrift at Sea by Thai Security Forces

About 200 Myanmar migrants found adrift off Sumatra told rescuers they were beaten, towed out to sea and left to their fate by Thai security forces, reported an Indonesian navy officer. The migrants, all men from Myanmar's minority Muslim Rohingya community, said they had been cast adrift by Thai security forces as they tried to flee persecution in their military-ruled homeland. They showed scars from beatings they said they had received at the hands of the Thais, matching similar allegations from another group of 174 Rohingya who were found off Sumatra on 7 January 2009.

'They were drifting for about 21 days. Most of them are in critical condition and are receiving treatment at a local state hospital in East Aceh district.' Using limited Malay, the Bengali-speaking migrants said they had left their homes in Myanmar's western Arakan state because they were being forced to embrace Buddhism, according to the navy officer.

Myanmar's military rulers effectively deny citizenship rights to the Rohingya, leading to discrimination and abuse and contributing to a regional humanitarian crisis as hundreds try to flee the country by boat every year.

Source: Hasan (2009).

BOX 6.3   Migration and Food Security

Garhwa, India, is a district characterized by a high percentage of poor and landless families. Recurrent drought and famine have resulted in large-scale migration to the neighbouring states of Bihar, Uttar Pradesh and Chattisgarh. Ultimately, it is food security that lies at the heart of it all. The whole family migrates to different states during the course of a year. It is women who have to bear the greatest burden. They comprise the majority of the population below the poverty line and are very often in situations of extreme poverty, given the harsh realities of intra-household and social discrimination. Migration also results in a breakdown of social life: this is true both in the case of men migrating alone and when entire families migrate. Women and young girls are especially vulnerable to sexual exploitation. Migrant workers have no access to subsidized grain at their destinations and spend a sizeable proportion of their wages on basic food supplies.

Source: Alternative for India Development, 'Food Security and Migration'. http://www.aidindia.org.uk/nl_mig_nov06/mig_oct01.htm (accessed on 15 January 2009).

within the country, largely from rural areas to towns and cities in search of job opportunities. Although exact figures vary by tens of millions, official statistics place the number of internal migrants in China at over 130 million (Tan and Xin, 2009) despite the *hukuo* system.[3] The implications for food security are significant, with more people becoming dependent on market transactions for the purchase of food, and thus susceptible to price fluctuations. These migrants become even more vulnerable during recessions and down-turns. To continue with our China example, 20 million of the 130 million workers lost their jobs in 2009 (ibid.) due to the recession and would have been further at risk of food insecurity.

Mongolia provides an example of migration to urban areas driven by food insecurity. Summer droughts, extremely harsh winters, overgrazing and locusts as well as unaffordable, privatized veterinary services have at times caused the death of livestock and pushed herders off the land and into urban poverty, even though undernourishment in urban areas is more severe than in rural areas. A similar situation exists in many parts of South Asia, where high levels of urban poverty and related food insecurity often prevail.

Thus, due to the phenomenon of internal migration, countries of Asia and the Pacific have experienced, *inter alia*, high rates of urbanization, and are expected to continue becoming increasingly urbanized in the years to come. For instance, from having an urban population of 42 per cent in 2008, the region is projected to have 51 per cent of its people living in urban settings by 2025 (ESCAP, 2008a). The expansion of urban areas often involves encroachment on arable land that could otherwise be used for food production, hence threatening food security. In addition, within urban areas there is increasing pressure on public services, including those related to hygiene. Millions of people live in slums where sanitation and water are inadequate, increasing the burden of disease and exacerbating food insecurity. Furthermore, in Asia and the Pacific, demand for food is especially high in poorly planned and managed cities, which pushes up the price of food, given limited supply, and makes the poor and other vulnerable groups especially susceptible to food insecurity due to lack of access to food.

---

[3] A system by which people were tied to where they lived by making government services contingent upon their occupation and place of residence.

Growing urban populations spur changing food habits and, frequently, increasing dependence on imported food at the expense of locally produced foods (United Nations, 2008). So, special social protection for those living in slums in cities and urban areas can no longer be ignored. Due to urban economies being highly monetized, it is especially crucial to create employment leading to income security as a social protection measure for people living in cities and urban areas for improving food security.

As we stated in Chapter 2, in the Pacific, though high levels of internal migration and poverty in towns and cities are seen, these phenomena generally do not imply food insecurity or destitution seen in, say, South Asia. They rather imply a continuous urge to meet essential living expenses especially in urban areas and make choices between the competing demands for expenditure on food and other basic needs, with limited cash income, requiring trade-offs, say, between food or school fees Migrant households may borrow regularly from 'loan-sharks' at extortionist rates of interest to meet basic family commitments and obligations. Thus, many families are frequently, and some families are constantly, in debt and in a more monetized economy, they are at greater risk of food insecurity. Improvement in employment and social protection, including access to affordable credit, are thus, fundamental in enhancing food security for the migrants in the Asia Pacific region.

Ethnic Minorities, Tribal People and Discriminated Groups of People: Vulnerabilities, Food Security and Social Protection

With the great diversity of the Asian and Pacific region, groups who are vulnerable to food insecurity often vary given local socio-economic contexts such as traditional, religious and cultural practices. The Asia-Pacific region is home to 60 per cent of the world's population and 70 per cent of the world's indigenous populations (UNFPA, 2007). The indigenous peoples and tribal people of Bangladesh, India, Pakistan and Philippines, as also the minorities in China, India and Nepal, suffer from food insecurity. In India and Nepal, for example, caste-based discrimination increases the vulnerability to food insecurity of people belonging to lower castes, also called dalits, who make up over 13 per cent (over three million persons) of Nepal's total population (Shrestha, 2003), and over 16 per cent of India's population, numbering 167 million people (Navsarjan, 2008–09). People in countries like Timor-Leste who have been displaced by conflicts and those in Myanmar who have been displaced by natural disasters are especially vulnerable to food insecurity. Essentially, whoever has power, access to and control over resources (financial, natural, human and other) generally has better access to food. Inequality and misuse of power have always caused food insecurity, among other maladies, both within countries in the Asian and Pacific region as well as within each community and household. Thus, the indigenous people who are made vulnerable due to discrimination and displacement by conflicts or natural disasters, because they lack power and control over resources, confront the risk of food insecurity. The same applies to people with disabilities, numbering more than 400 million in the Asia-Pacific region (ESCAP, 2009), and people living with HIV/AIDS, numbering six million or so in the region (ESCAP, 2007).

Furthermore, numerous indigenous groups and people belonging to tribal groups in many countries in Asia and the Pacific have, over many years, lost access to their traditional land and forests and other common property resources on which they depended for food and livelihoods. The consequences have been disproportionately higher levels of food insecurity suffered by such groups, many of which have also not benefited equitably from socio-economic development that caused them to lose access to their land, forests and common property resources. These socially marginalized people face greater barriers in gaining access to food, due to disadvantages with regard to knowledge, income-earning capacity and a variety of other factors.

Guaranteeing land rights and rights to access forest produce, and even the restoration of land to indigenous peoples where politically feasible, are very special social protection measures. This is often a messy process, as the powerful may stand to lose from it, but it has to be accomplished (Meinzen-Dick, 2011). The right to access forest produce is often overlooked as a social security measure in the context of food insecurity despite the fact that forest produce provides income, vital nutrition and palate to the diet of indigenous peoples. In fact, in the semi-arid regions of India, common property resources and forests provide up to 35 per cent of the food consumed by poor people.

Financial inclusion and universal obligations to create financial access need to be an important strategy of social protection for these groups of people to bolster their food security. Evidence indicates that in countries where formal financial infrastructure exists, a robust downscaling of the formal sector to reach out to meet the needs of these groups would be a compelling strategy as a means of providing social security in the context of globalization. The ideas of 'self-help groups' and micro-finance as a means of reaching out to these socially excluded groups are important innovations, as discussed in Chapter 5. Besides credit, microfinance includes the provision of services of savings, pension, insurance and money transfers. From this perspective, contextualizing microfinance in the realm of social protection becomes crucial to reaching out to the poor, and has several policy implications. The Grameen Bank II experience in this regard emerges as one of the best practices in social security through microfinance, with potential replicability for the entire region.

BOX 6.4   Bangladesh Programme on Food Security

Challenging the Frontiers of Poverty Reduction/Targeting the Ultra Poor (CFPR/TUP) extends the 'laddered strategic linkage' approach of Vulnerable Group Development (VGD) to the very poorest. It builds up the asset base of the poorest, beginning with the transfer of income-generating assets, health and education support, training, social development and later integrating with micro-credit programmes, to reduce poverty among the poorest and support income-generating activities. In a 2004 mid-term assessment study on the 2002 entrants and a comparison group, it was found that the programme participants fared significantly better in nutrients and in overall calorie intake, with a calorie gap from RDA (Recommended Dietary Allowance) at 8 percentage points lower for participants; 97 per cent of participants reported to be in 'food deficit' at the baseline, but this was reduced to only 27 per cent two years later. Severe malnourishment among children under 5 years was reduced by 27 percentage points for participants but only 3 percentage points for the comparison group.

Source: World Food Programme (2007).

Social Protection and the Challenge of Gender-Based Food Insecurities

The multifaceted inequalities that women face in Asia-Pacific are matters of public history. Such inequalities increase the vulnerability of women and girls to food insecurity, more so if they are elderly and/or disabled, or belong to a lower caste or ethnic minority. It is, therefore, necessary to understand inequality within households and how intra-household distribution of food and other resources affects individuals' food security within the household unit, in particular for women (Cromwell and Slater, 2004).

Women face seven inequalities, viz., mortality inequality, natality inequality, inequality in access to basic facilities, inequality in access to special services, professional inequality, ownership inequality and, finally, household inequality (Sen, 2001).[4] Based on these seven inequalities faced by women, it is easy to identify seven kinds of food insecurity that women face:

1. *Food insecurity based on mortality inequality*: In some Asian countries, there are unusually high mortality rates among women and a consequent preponderance of men in the total population, as opposed to the preponderance of women found in societies with little or no gender bias in health care and nutrition. Women and girls may be denied adequate nutritious food, resulting in higher infant and child mortality rates among them than among boys.

2. *Food insecurity based on natality inequality*: In many male-dominated societies, male children are preferred and female children are often aborted, or seen as a burden. This preference can lead to girls and women being in a weaker position to be food-secure, by suffering from limitations to physical and social access to food.

3. *Food insecurity based on basic facility inequality*: Females have less access to education and learning or fewer opportunities to develop their talents and skills. This limits their productivity in producing food and their opportunities for employment, jeopardizing their long-term food security.

---

[4] This has been more fully discussed in Chapter 3.

4. *Food insecurity based on special facilities inequality*: Females may have access to basic facilities such as primary education, but their opportunities for higher education and professional training may be fewer than for young men because, *inter alia*, 'the culture does not see this as "feminine"'. Girls may be discouraged from studying subjects that are deemed to be 'the province of men'. These include agricultural sciences and training in techniques for improving agricultural productivity. Such inequality prevents women from achieving improved access to better food.

5. *Food insecurity based on professional inequality*: In terms of employment as well as promotion in work and occupation, women often face greater handicaps than men. Women's income-earning potential is therefore hindered, which in turn reduces their ability to purchase food.

6. *Food insecurity based on ownership inequality*: Women do not have the same rights as men to inheritance or ownership of productive resources such as land[5] and capital. The absence of claims to property can not only reduce the voice of women, but can also make it harder for women to enter and flourish in commercial, economic and even some social activities. With less access to land and other resources, women often become dependent on others for food, especially as widows, when divorced, or when abandoned. These factors together contribute to the food insecurity of women. Moreover, ownership inequality reduces women's ability and incentives to invest in agricultural land, having a bearing on their food security.

7. *Food insecurity based on household inequality*: Even in cases where there are no overt signs of anti-female bias in, say, survival or son preference or education, family arrangements can be quite unequal in terms of sharing the burden of housework and childcare, limiting women's opportunities

---

[5] Some studies in China also show that increasingly the landless tend to be women. Figures derived from a survey undertaken by the All-China Women's Federation and the State Statistics Bureau of 2000 showed that 70 per cent of people without their own land were women and, among these women, 20 per cent had never held land, while the rest had lost their land upon marriage, divorce or reallocation (Li, 2003).

for earning income. Such inequality may take the form of girls being fed less food and food of lower nutritional value than boys. Intra-familial distribution of resources including food within households, such as women being expected to eat the least and after all others in the family, makes women vulnerable to food insecurity.

There are other aspects of gender-based food insecurities as well. The labour force participation of women remains very low in some parts of Asia, and gender-based discrimination in conditions of employment and wage gaps remain problems for the region as a whole. All this has an adverse impact on the food security of poor households. Women's wages are consistently lower than those of men for the same work. This has serious implications for the food security of households that depend on female earners. In the case of Bangladesh, for example, the female wage rate is so low that a day's wage cannot maintain a family of three, even if the female worker is employed full-time. Rural women face the bleakest situation. With the poor tending to spend such a high proportion of their earnings on food, and women usually being expected to put food on the table, the low wages and limited employment opportunities of poor women affect the quantity and nutritional quality of food that can be purchased and consumed.

Women play a critical role in rural economies, particularly so in food production. In most parts of the developing countries of the Asia-Pacific region, they participate in crop production and livestock care, including foraging for fodder; provide food, water and fuel for their families; and engage in off-farm activities to diversify the family income. In South Asia, 61 per cent of working women are employed in agriculture (ILO, 2008). According to the International Fund for Agricultural Development (IFAD), in Asia, women contribute about 65 per cent of the total food production (Inter-Press Service, 2008). In the case of Nepal, 40 per cent of women are economically active, with most employed in the agricultural sector and the majority working as unpaid family labourers in subsistence agriculture. A similar picture emerges from other parts of the Asia-Pacific region. Yet rural women are discriminated against in matters of access to a range of resources such as credit, land, agricultural inputs, extension services and employment, within both the community and the household. Such discrimination has an obvious bearing on both availability and access to food for women and children.

With men increasingly moving out of farming and livestock rearing due to migration, among other factors, or shifting to only cash crop production, agriculture and food production in particular are becoming more and more 'feminized' in much of the region. For example, in Bangladesh and in the hills of northern India, in village after village only children and women are to be seen: the male adults have all migrated. The women left behind often have few skills outside of agriculture and have little direct access to land, credit, inputs or information. Special measures are required so that the existing institutional mechanisms for delivering credit, technology, agricultural extension and so on, which are typically oriented toward male farmers, are remoulded to be fair to women. If women are increasingly the farmers, guaranteeing security of land tenure for women as a social security measure rather than a land reforms intervention can prove key to both agricultural revival and providing food security to rural families as well as tackling poverty (Agarwal, 2008). In this context, it is also important to look at access to homestead plots for women and not only agricultural land.

Discrimination faced by women in matters of health also needs to be mentioned. The high percentages of underweight children, particularly girls, in the region should be a matter of concern,[6] since some of these could be related to the health of women and girls, which in turn is influenced by cultural practices that discriminate against them. For example, women are denied certain foods as widows or when pregnant, based on often incorrect beliefs. Such beliefs include pregnant women avoiding certain types of bananas and other foods for fear of having difficulty during delivery and producing children with disabilities (Roncarati and Taoprasert, 2000). Many babies are born underweight, largely due to maternal undernutrition or malnutrition. South Asia has especially high

---

[6] UNICEF notes that approximately 47 per cent of India's under-4 population is underweight, with Bangladesh and Nepal having a higher child undernourishment rate at 48 per cent. Over half of the world's underweight children live in just these three South Asian countries, with South Asia being the only region in which girls are more likely to be underweight than boys (UNICEF, 2006).

levels of maternal undernutrition and anaemia, which in turn are related to high maternal mortality. In a continuous cycle of undernutrition, underweight and undernourishment, girls grow up to become undernourished mothers. Such mothers often have limited access to appropriate food and good-quality health care during pregnancy and, therefore, are in turn at risk of giving birth to children who are underweight (ESCAP et al., 2008), who will have lower cognitive skills and hence be vulnerable to food insecurity. In India, for example, one out of every three adult women is underweight and therefore at risk of delivering babies with low birth weight (ibid.: 12).

The challenge is to develop social protection measures that deal with food insecurities and discrimination faced by women in a systematic and sustainable manner. Social protection measures for women have to be long-term and systemic in order to be effective, as they have to circumvent deeply entrenched socio-economic trends. Some of the social protection measures could be micro health insurance specifically targeted to women, stricter laws and their diligent implementation against wage discrimination, creating employment opportunities targeted specifically towards women, nutritional education on a mission mode to prevent pregnant women being fed on a diet which leads to their undernutrition, and greater access to educational opportunities at elementary, secondary and tertiary levels.

Social Protection to Deal with Shocks

As discussed in Chapter 2, there are two kinds of shocks, idiosyncratic and covariate, that people face.[7] These shocks can be of every description, ranging from food–fuel inflation, epidemics and pandemics, earthquakes, floods, droughts, hailstorms, pest attack, recession and economic down-turn, civil strife and conflicts, to fire, loss of income due to illness, loss of limbs, sudden loss of jobs, breakdown of assets for livelihood and on and on, on a growing list. While various preventive socio-economic measures can be put in

---

[7] See World Bank (2000/2001) for a discussion of these kinds of shocks.

place to allow greater access to food by all—from macroeconomic changes to specific taxes, subsidies and incentives—certain social protection interventions will also be needed to help those affected by sudden shocks. Social protection in the form of cash transfers, food aid, school feeding or public works programmes helps people cope with these shocks, whether idiosyncratic or covariate.

Devising social protection measures to deal with such a wide range of shocks is a serious challenge. The challenge for governments to help low-income groups attain food security through social protection is likely to become more difficult because of their increasing numbers, due in part to the deepening financial crisis and rising income disparities. In addition, along with demographic changes, traditional forms of social capital such as kinship systems have been weakened, raising the importance of social protection for food security all the more. Three kinds of social protection measures to deal with shocks commend themselves for their reach and implementability: improvements in agriculture, employment generation programmes and *ex ante* management of shocks.

Agriculture

Improved agriculture and much greater investment in agriculture are some of the best forms of social protection. Agriculture is a critical source of livelihoods and employment for a huge majority of people (up to 70 per cent in some cases), especially women, in most developing countries in the Asia-Pacific region, and a key pathway out of poverty (World Bank, 2007) and food insecurity. During economic down-turn, when industrial and service-sector jobs shrink, agriculture provides enormous respite. Investment in agriculture is, therefore, good not only for increasing food availability and economic access to food, but for providing good social protection. Since, in the Asia-Pacific region, women perform much of the agricultural work and produce most of the food crops (about 65 per cent of the total food production in Asia), expansion in agriculture will expand more income opportunities for women.[8] Furthermore, when women have access to income they tend to

---

[8] Notwithstanding the fact that wages earned by women are lower vis-à-vis those earned by men.

spend a higher percentage on food for the family, so that food security at the household level, especially of children and the elderly, stands a better chance of improving.

Making agriculture more socially and economically viable is important for reducing poverty and addressing food insecurity in terms of both improved availability and access. Social protection strategies that seek to revitalize agriculture could include measures that help connect the rural poor to cities and markets; improve service delivery to boost the health and education of the rural poor; diversify agriculture to tap new markets and opportunities; introduce crop insurance to mitigate crop failures and price declines; revamp land policy for socially inclusive growth; and promote social mobilization to influence agricultural policy. In some situations, strategies which seek to facilitate people moving out of agriculture, for example by empowering people to enter labour markets through better training and promoting the rural non-farm sector, hold great promise (ESCAP, 2008c).

Employment

Employment as a social protection measure to deal with shocks is crucial for food security, because it creates income directly and hence increases economic access to food directly. Under normal circumstances, managing the interdependence between consumption and income is the key to ensuring food security (Sen, 1997). Hence, normally, while food needs to be available, people need to have access to income and other resources to access food, to ensure food insecurity. The World Commission on the Social Dimension of Globalization points out that employment to enable (economic) access involves positive potential for food security (ILO, 2004).

Providing employment to people during shocks becomes even more important for food security. That is why the Comprehensive Framework for Action, presented in July 2008 (United Nations, 2008), emphasizes that employment is crucial to enable people to purchase food. In this connection, guaranteed employment for a minimum number of months or weeks during disasters acts as *de facto* insurance, for example, in the form of public employment guarantee schemes exemplified by food-for-work or cash-for-work projects. Employment guarantee schemes and food-for-work additionally offer considerable insurance against idiosyncratic shocks, and serve as a means of protecting the valuable productive assets of vulnerable groups who suffer unexpected income loss, unemployment or other adversities (Barrett et al., 2004; Jalan and Ravallion, 1999). It may be underscored here that the food security outcome of employment for women is better because as stated earlier, women have a higher propensity to allocate incremental income to food, especially for their children, as we have mentioned earlier.

It ought also to be added that employment will generally improve people's ability to access food, but it needs to be accompanied by measures to reduce and ultimately eliminate wage discrimination. Otherwise there will always be groups who do not benefit equally from employment opportunities and will be at greater risk of food insecurity. That is probably one reason why Sen (1997) argued at length that 'expansion of employment and decent rewards for work' is one of the eight imperatives for a food-secure society. Implementing measures for fair and equitable wages is, therefore, a good social protection measure. Elimination of wage discrimination as a social protection measure will additionally help the already disadvantaged subpopulations who are often relegated to the most insecure '3D jobs' (dirty, dangerous or degrading), usually in the informal sector, which is characterized by non-enforcement of labour laws and lack of social protection and workmen's compensation in case of occupational accidents causing loss of life or limbs.

Ex Ante Management of Shocks

As discussed in Chapters 1 and 2, large sections of the population in the Asia-Pacific region are vulnerable to several shocks that could lead to food insecurity, be they from natural disasters or climate change. Where food insecurity may have already existed and is compounded by the threat posed by sudden adverse shocks (such as the Asian tsunami in 2005 or the Sichuan earthquake in China during 2008), there is considerable scope for improved ex ante risk management and providing improved coping strategies through insurance and insurance-like responses. The key challenge is to have counter-cyclical risk management programmes in place before the onset of natural disasters, with flexible targeting, flexible financing and flexible implementation arrangements (Alderman and Haque, 2007; de Janvry et al., 2006).

Most people who are food-insecure are the ones who suffer from idiosyncratic shocks caused by water-borne diseases due to lack of access to potable drinking water and other health shocks; by transitory shortfalls in access to and utilization of food; and shocks to assets and income. These are disproportionately specific to individuals and households in the Asia-Pacific region. It is, thus, important to tackle idiosyncratic shocks to prevent an overwhelming majority of food-insecure individuals or households existing outside of 'humanitarian emergencies' that are associated with covariate shocks. This underscores the need for effective, ubiquitous and continuing *de jure* insurance programmes, whether through financial innovations such as micro insurance, index insurance schemes or community-based insurance programmes.

Measures to revitalize and make the food distribution system more effective and responsive to poor, especially vulnerable groups also need to be looked at. Public food distribution systems need to be universalized for the old, infirm, disabled, widows and ethnic minorities, and community-managed alternative public distribution systems[9] should be upscaled for food security, environmental management and biodiversity conservation. Governments should create an appropriate regulatory environment that facilitates the spread of APDS for all those who are vulnerable to food insecurity.

Social Protection, Small Farmers and Food Production

Small farmers play a crucial role in providing food security in the Asia-Pacific region. Social protection to small farmers is of special interest, as such protection plays a pivotal role in protecting the vulnerable livelihoods of small farmers, and has beneficial effects on agricultural production, leading to improvements in food security. How various instruments can alleviate liquidity constraints for smallholders, create demand for farm products, create multiplier effects throughout the local economy, help deal with seasonality and help risk taking by smallholders is especially important. These challenges are by no means exhaustive.

---

[9] See Chapter 5 on community-based responses for details of the alternative public distribution system (APDS).

The Challenge of Alleviating the Liquidity Constraints of Small Farmers: Cash and Food Transfers and Their Multiplier Effects

One of the major barriers to agricultural production is lack of access to liquidity during agricultural operations to buy agriculture inputs (Ravallion, 2003). Smallholders with limited or no access to credit, or access to credit at predatory rates of interest, are often able to purchase only a fraction of the required inputs, which leads to lower output, lower net incomes and lower returns to labour and capital (World Bank, 2007), with all the attendant consequences for the food security of the most vulnerable population groups. Evidence suggests that cash transfer programmes prevent smallholder households at risk from adopting damaging coping strategies like asset sales, indebtedness or removing children from schools (some of which were discussed in Chapter 5 on community responses), and relieve, at least partially, the liquidity constraints faced by smallholder farmers, allowing small-holder farmers to accumulate productive assets (Coady, 2004; Harvey, 2007). Cash transfers can unleash untapped productive and income-generating potential by boosting household investments in farming as well as non-farm micro-enterprises, all contributing to betterment of food security (Martinez, 2004). It has been seen that a small percentage of transfers (even when the amount is small) received by some poor households is invested in fertilizer and seeds, leading to higher food production. The challenge is to make cash available to small farmers in time.

Like cash transfers, food transfers also have a bearing on small farmers' food security. Food transfers to small farmers increase labour supply to agriculture, wage work and business activities (Abdulai et al., 2004). It must be noted, however, that bringing food from outside in the form of food transfers can harm local farmers, due to falling prices and commercial displacement (Barrett, 2006), while, on the other hand, if food is procured locally, the net food purchasers can be harmed by driving food prices up. However, well-targeted and well-timed food transfers have minimal negative price effects, because they reach households who are already priced out of the market (Barrett and Maxwell, 2005). Besides, food transfers have very positive effects by helping local production, labour markets and consumption patterns when the food for the transfer is locally sourced or derived from elsewhere within the region. For example, school feeding schemes or food-for-education programmes have been found to have positive impacts on local agricultural production. Local purchases of food for school meals can stimulate production by augmenting demand, not only for staple crops but also for vegetables, meat, eggs and dairy products (Ahmed and Sharma, 2004). During Indonesia's economic crisis in the 1990s, the government initiated a countrywide school feeding scheme, stipulating that the local staple should not be included in school meals, to avoid meal substitution at home, and that only locally grown commodities should be used. Meals were prepared by local women, organized through local women's associations. A survey found that 72 per cent of farmers interviewed said that the school feeding scheme gave them more opportunities to sell produce from their fields and vegetable gardens (Studdert et al., 2004). As a social protection measure, well-targeted and well-timed food transfers to small farmers can be beneficial for food production because they reach small farmers who, being already priced out of the market, would otherwise resort to depletion of productive assets, jeopardizing their long-term food security.

While local sourcing of food can generate demand for local production, cash transfers are likely to have more positive secondary and multiplier effects than food transfers. This is because cash is spent on purchasing goods and services, which in turn create employment and income for the providers of these goods and services. These multipliers apply equally to transfers given to economically inactive groups (like social pensions or child support grants) as to transfers given to small farmers. However, the synergies with agriculture are likely to be higher were the recipients to be the farmers, who spend at least a part of their incremental income on agricultural operations. The magnitude and distributional impacts of economic multipliers depend on a number of factors, including the openness and structure of the local economy, its linkages with urban centres and other large markets (Taylor and Yunez-Naude, 2002), as well as the expenditure patterns of different groups receiving cash transfers. It is true that the macroeconomic benefits claimed for cash transfers are based on limited empirical findings, and the evidence to date is ambivalent (Devereux and Coll-Black, 2007), but there is sound evidence in support of the localized multiplier effects of social transfers. In sum, then, both cash transfers and food transfers as social protection measures for small farmers deserve serious consideration in Asian countries.

Small Farmers, Timing and Seasonality

The detrimental effects of seasonality in terms of smallholder food insecurity and vulnerability are well known. Given the seasonal character of agricultural production, the importance of facilitating access to inputs for smallholders who face seasonal cash constraints cannot be overemphasized. While subsidies to inputs, like fertilizer subsidies or free seed distribution, are at times controversial due to their adverse market and distributional effects, they have successfully increased food production in many countries in the Asia-Pacific region, such as China, India, South Korea, Thailand and Vietnam, with positive impacts on food production and consequently upon household and national food security. There is thus a case for providing resources to small farms in different ways, including through microfinance provisioning, to smooth out cash deficits due to seasonality.

The seasonality of fluctuations in food and asset prices undermines the household food security of small farmers by raising the cost of accessing food while reducing the market value of assets sold at 'distress prices' to buy food. In the longer term, uncertainty in commodity markets hinders the efficient allocation of productive resources by farmers, and may cause producers, consumers and traders to engage in risk-reducing strategies such as diversification into lower-value but more stable products, not using purchased inputs, and not trading in remote locations. Prior to the era of globalization and liberalization, many governments typically intervened in grain markets in an attempt to ensure price stability throughout the year for both consumers and producers, through providing farmers with minimum support prices for selected crops and state procurement of their produce through parastatals like the Food Corporation of India. There are those who disfavour 'interventionist' measures in favour of market-based solutions. Nonetheless, large countries like China and India still intervene in grain markets to ensure price stabilization for the benefit of small farmers,[10] and to good effect. There is a case, therefore,

---

[10] Market-based tools such as futures markets may also able to insulate producers from short-term price volatility, but they are typically not accessible in low-income countries in the Asia-Pacific region. Commodity exchanges and futures markets have been established in China, India and

for arguing in favour of minimum support prices to farmers for selected crops, and state procurement of their produce.

When it comes to seasonality in the labour market, well-timed public works programmes partly address the seasonal underemployment that is typical of the rain-fed agriculture systems found in many parts of Asia. As an 'employment-based safety net', food-for-work or cash-for-work programmes offer smallholders a supplementary source of food or income for consumption-smoothing purposes during non-agricultural seasons (also called lean seasons), because small farmers often fail to produce enough to feed their households round the year. A known employment-based safety net is the one provided by the National Rural Employment Guarantee Act of 2005 in India, which entitles every rural household to 100 days of employment at the local average agricultural wage. Apart from smoothing consumption in farming households during food-insecure seasons or bad years, the assets constructed by the public works activities are intended to boost agricultural production by facilitating transportation and enhancing market access and soil fertility. One risk with public works is that participation may force smallholders and landless agricultural labourers to divert their labour away from vital on-farm activities such as de-weeding, towards public works initiatives, especially if the employment (which is primarily off-farm) is offered during periods of high agricultural activity, which also happens to be the 'food-insecure season'. This creates a trade-off between social protection for the immediate consumption needs of farmers and returns to agriculture (McCord, 2005) that needs to be balanced properly.

The challenge of reducing the vulnerability of small farmers to seasonal variations in agricultural production, food availability, prices of food and assets, and demand for labour through timely and appropriate social protection interventions to mitigate such stresses is significant for the Asia-Pacific region. And given the structural composition of the economies of the region, different

---

Thailand, but the establishment of such instruments is dependent on good financial and legal institutions (World Bank, 2007). It is argued that governments should facilitate the private sector's adoption of measures such as warehouse receipts and the purchasing of futures and options, but such market instruments are themselves dependent on integrated markets and may not be accessible to small-scale farmers in the Asia-Pacific region.

countries have to adopt their unique sets of social protection measures to tackle the vagaries of seasonality.

Predictability and Risk Taking by Small Farmers

The synergies between social protection and agricultural policies are more powerful than in any other areas of risk reduction. Social protection mechanisms (like cash transfers, seasonal public works and insurance schemes) can influence productivity by stimulating risk-taking behaviour. Where cash transfers are predictable and are perceived as a secure source of income, risk-averse farming households and other rural small scale entrepreneurs will be more than willing to increase investment in productive activities, despite perceptible risks because predictable cash transfers acts as a 'safety net' against future shocks (Gertler et al., 2005 as quoted in Sabates-Wheeler et al., 2009).

The 'Employment Guarantee Scheme' in Maharashtra State, India, (which is the precursor of the current Mahatma Gandhi National Rural Employment Gurantee Programme in India) provided manual employment to unskilled manual labourers, who are willing to work in rural areas. The guarantee of paid work served as an insurance, which reduced precautionary savings to make more resources available for more productive purposes. Consequently, the farmers in the State started planting more of higher-yielding (as against drought tolerant) crop varieties than farmers in other states (Ravallion, 2003 as quoted in Sabates-Wheeler et al. 2009) after the introduction of the Employment Guarantee Scheme. However, more understanding about the magnitude of such insurance effects on risk-taking behaviour is required (Dorward et al., 2006 as quoted in ibid.).

In this connection, it is also important to recognize the positive effects of insurance on ensuring predictability and encouraging risk-taking. Granted that most smallholders do not have access to crop insurance, crop failure due to vagaries of nature or crop loss due to pest attack etc., results in loss of agricultural output and productive assets. This is preventable if accessible insurance markets or social insurance mechanisms exist. Crop insurance for small-holders has generally failed due to high transaction costs, moral hazard, adverse selection, covariate risk and delayed payouts (Alderman and Haque, 2007; Hellmuth et al., 2007 as quoted in Sabates-Wheeler et al. 2009), all of which make private crop insurance economically unviable for insurers and inaccessible or unresponsive to client needs (IISD, 2006 as quoted in ibid.).

There has been a move towards 'weather-indexed' insurance, where the insurance is against a local index, say, rainfall shortage or days of hailstorm or snow or frost, which is correlated with harvest outcomes (see Box 6.5). Under the 'weather-indexed insurance', farmers get paid compensation if the index reaches a 'trigger' level, without reference to crop losses. Index-based weather insurance can perform both a protective and a productive function. Because payments are disbursed rapidly, farmers are able to smooth their

---

BOX 6.5   Insuring Against Adverse Weather

Weather-based index insurance can sometimes substitute for traditional crop insurance. Weather-based index insurance uses objectively defined trigger events (e.g., rainfall, soil moisture) in an area to set contingent damage payments according to an index. Contracts and indemnity payments are the same for all buyers per unit of insurance; there is no use of field- or household-specific damage and loss data. This model discourages moral hazard and cheating, avoids adverse selection problems, and lowers transaction costs. It also makes the insurance instrument accessible to the broader rural population (Skees et al., 2002). However, index insurance weakens the correlation between losses and payouts. This is known as 'basis risk'—an insured party may suffer a loss yet not receive a payout.

Weather insurance through index-based insurance for farmers and for local and national governments is used increasingly, but is not a panacea: it may not be appropriate for slow-onset climate impacts; and preventing losses is sometimes more cost-effective than loss-based insurance (Alderman and Haque, 2007). Additionally, many countries in the Asia-Pacific region lack insurance markets and may not find insurance easily affordable. It may also not be desirable for some developing countries to take out insurance if indemnities are likely to crowd out concessional emergency funding. This being said, it may be noted that many humanitarian crises are caused by factors other than climatic variability (e.g., conflict, poor governance, lack of infrastructure, political and macroeconomic crises). Safety net and emergency responses thus should not be tied exclusively to index instruments.

Source: Presentation by Joanna Syroka (2007) *Experiences in Index-Based Weather Insurance for Farmers: Lessons Learnt from India & Malawi,* World Bank, Innovative Finance Meeting, New York, 18th October 2007.

consumption following a poor harvest, while avoiding damaging coping strategies such as selling productive assets. Since households and farms covered by weather-indexed insurance are creditworthier, investment in productive assets and higher-yielding crops is also promoted (Mechler et al., 2006). However, the main challenge to the widespread adoption of 'weather-indexed insurance' is its high cost. On a commercial basis, premiums may be too high for smallholders, which is why their financing should partake of the nature of a social protection measure. Crop insurance throughout the world is, therefore, highly state-subsidized.

Social Protection, Climate Change and Catastrophe Safety Nets

Chapters 1 and 7 explore how climate change has affected and will affect a large number of people in the Asia-Pacific region. For instance, in Bangladesh, an increase in sea level of 1 metre would lead to the inundation of 15–18 per cent of the country's landmass; and by the year 2050 some 30 million people could become environmental refugees (UNDP, 2007). Pacific Island countries and the Maldives are at particular risk from rising sea levels. In fact, a rise in water stress is also likely to affect 185 million to one billion people in South and South-East Asia (IPCC, 2007). The region will face significant socio-economic impacts because of geographic exposure, reliance on climate-sensitive sectors, low incomes and weak adaptive capacity (IPCC, 2001, 2007; Stern, 2006; UNDP, 2007).

The United Nations Convention to Combat Desertification (UNCCD, 2009) notes that Asia-Pacific contains some 1.7 billion hectares of arid, semi-arid and dry sub-humid land, with degraded areas including the expanding deserts in China, India, Iran, Mongolia and Pakistan, the steeply eroded mountain slopes of Nepal and the deforested and overgrazed highlands of the Lao People's Democratic Republic. Since dry-land agriculture constitutes a large proportion of food production in the Asia-Pacific region, the projected decline in agricultural productivity and other natural resource–based livelihoods will trigger, *inter alia*, relatively large income losses and food insecurity. If business as usual continues, climate change will alter the pattern of shocks faced by households in the Asia-Pacific region toward more and more frequent covariate shocks (like floods, droughts and hot spells), greater uncertainty about the shocks, and interactive shocks, all of which pose significant challenges to the coping mechanisms of people at all levels.

As mentioned earlier, managing climatic shocks, especially of the idiosyncratic variety, has traditionally been the responsibility of households with little external support for managing common climatic shocks such as localized floods, pest attack or crop failures due to natural conditions. This may have to change. Large and repetitive covariate climatic events could overwhelm the risk-coping capacity of many community institutions. Furthermore, as seen in Chapter 5, household and community adaptations are not always equitable, sustainable or desirable and, without external support, many poor households and communities will choose adaptation strategies that are inequitable, unproductive or asset-degrading. It is true that there are unresolved issues surrounding the design of external support for adaptation, relating to both governance and design, to ensure support for pro-poor local climate action. But they have to be resolved sooner rather than later. Efforts to enhance community resilience need to feature more prominently in the response to climate change–induced food insecurity. Social protection interventions hold promise for this purpose (Heltberg et al., 2008; IDS, 2007). Adaptation responses have to span a range of sectors and draw on a variety of approaches and disciplines. And while some responses must be at the national or international level (like the development of new crop technologies, such as those being attempted by the International Rice Research Institute, Manila), other responses need to be very local (like the adoption of new crops or water-saving techniques being practised in China). Thus, a key challenge is to identify the most effective instruments of social protection to support adaptation and deploy them at the right level. Arguably, community-based adaptation could become the most important element of adaptation response, where social protection could contribute to national-level responses. Catastrophe safety nets should be put in place. The benefits and challenges of catastrophe safety nets are uncontested, and it is now agreed that even subsidized insurance systems are preferred to post-disaster aid, particularly because the reinsurance market in the region is not yet prepared to commit sufficient and affordable capital to markets serving the poor. Governments must ensure that catastrophe

---

BOX 6.6   The Munich Climate Insurance Initiative

The Munich Climate Insurance Initiative (MCII) has proposed a two-pillar, international risk management programme (financed fully by Annex 1 countries) as part of a comprehensive adaptation strategy that enables risk management and insurance through the funding of a global adaptation strategy. A risk prevention pillar would directly support risk reduction measures. A two-tiered insurance pillar would address high and medium layers of risk. The first tier takes the form of a climate insurance pool (CIP) that indemnifies victims of extreme catastrophes in non–Annex 1 countries by a percentage of their losses. A second tier takes the form of a climate insurance assistance facility and provides support to enable micro and national insurance systems to offer cover for middle-layer risks in vulnerable developing countries. The support includes providing technical assistance, capacity building and possibly absorbing a portion of the insurance costs. Low-level risks would continue to be absorbed fully by the respective governments and the private sector.

The MCII two-pillar proposal meets the challenge of providing support to promote sustainable, affordable and incentive-compatible insurance programmes for vulnerable households, small and medium businesses and governments in the developing world, at the same time enabling private-sector involvement. Because of the substantial economies of pooling public- and private-sector risks, there are strong arguments for creating facilities like the CIP at the global or regional scale.

Source: MCII (2008).

---

safety nets are closely coupled with risk management programmes, including a vulnerability assessment.

Governance and the Provisioning of Food

Overall, the effectiveness of social protection interventions in enhancing food security is dependent on a wide spectrum of variables which are country-specific. As with many issues in the social sector, political will holds the key to protecting the vulnerable and establishing social protection interventions which fit within the general rubric of sustainable and equitable socio-economic development. Social protection programmes adopted to deal with food insecurity should largely be cognizant of local and country contexts. Whatever system is chosen, it is vital to have effective administration, institutional setup and governance.

Indeed, food insecurity is exacerbated by poor governance.[11] Many actors are involved in governance, though in the context of ensuring food security governments should play the main role, given that without government interventions vulnerable groups suffer disproportionately more from food insecurity. Good governance has eight major characteristics, in that it is: participatory, consensus-oriented, accountable, transparent, responsive, effective and efficient, equitable and inclusive, and subject to rule of law (ESCAP, 2008b). One of the most important starting points for judging good governance at the national level is a government's performance in providing basic public goods to all of its citizens, including those in rural areas (Paarlberg, 2002). In Asia, the eight major characteristics of good governance are hardest to fulfil so as to bring about food security. The time has come for the countries of the Asia-Pacific region to be cognizant of these characteristics of good governance and take steps towards achieving them.

For the purpose of food security, another key challenge in the Asia-Pacific region is that the numerous vulnerable households are in communities and countries that have the weakest institutional capacity and resource base to respond. Addressing the challenge of food insecurity effectively additionally requires responsive and accountable institutions. Interventions are unlikely to meet with success unless institutions can demonstrate accountability and responsiveness. These are the areas that are weak in many countries of the continent. The challenge is to establish effective institutions—from legal ones to those providing credit, and from training establishments, including those instilling behavioural change, to the norms and practices of communities—to enable the participation of all in development processes, including processes encompassing issues relating to ensuring food security.

---

[11] The concept of 'governance' relates to the process of decision making and the process by which decisions are (or are not) implemented.

One weakness in governments' ability to respond to food insecurity relates to poor horizontal coordination within government structures. As food security touches on a range of issues—land, agriculture, social welfare, water, environment and trade, to name a few—if ministries that deal with issues such as food and agriculture, fisheries, rural development, land, health, social welfare, disaster management, trade and others are not well coordinated, inefficiencies and inequities will result. Jean Ziegler, the former Special Rapporteur of the Human Rights Council on the Right to Food, in his last report (2008), emphasized the lack of coherence in policies implemented by some specific governments in Asia and the Pacific as well as other regions.[12] Though policies on issues relating to food security have been in place, there appears to be no comprehensive strategy fully addressing issues of food security in terms of access to food.[13]

Good governance, a great virtue in itself, is also an important social protection measure.

Conclusions

Not enough has been done to ensure food security through interventions that improve food accessibility and promote development for all. The right to food is a basic human right; however, millions of people across Asia, especially the socially marginalized groups, suffer from food insecurity. This in turn puts them at risk of greater poverty and ill health, though progress has been achieved in reducing the number of food-insecure people in the aggregate. These particularly vulnerable groups also suffer from fewer and inferior employment opportunities, as well as greater discrimination along various dimensions. Women are often discriminated against, and

---

[12] In that report, Ziegler also states that refugees from hunger are among the most excluded and discriminated against people and are also among those who suffer most from the lack of coherence in states' policies. He argues that to be coherent, states must extend legal protection to people fleeing from hunger and facing other severe violations of their right to food.

[13] See Ziegler's report on his mission to Mongolia (Ziegler, 2005).

this can be exacerbated by belonging to a particular caste, being a widow, or having HIV/AIDS or a disability. Continued high population growth in some countries and large ageing populations in others, bringing about high dependency ratios, mean that food availability and access are likely to become increasingly important issues. Migration and urbanization also pose significant challenges to food security, as do gender inequalities within households.

Food insecurity is exacerbated by poor institutional setups and poor governance. Interventions are unlikely to meet with success unless institutions can demonstrate accountability and responsiveness. These are the areas that are weak in many parts of the region. Poor coordination within government structures, including ministries that deal with diverse issues, results in reduced ability to respond to food insecurity as well as inefficiencies and inequities. This is especially relevant when considering the role of social protection and how food security can be enhanced when there is good governance and other factors that promote inclusive development. Issues concerning how social policies and social protection can work to reduce food insecurity are covered in the recommendations in the next section of this chapter.

Social protection measures need to be counter-cyclical and not pro-cyclical, as is normally the case. Social policy–related issues need to be effectively integrated into all polices and interventions that seek to address food insecurity, especially those dealing with economic factors. The challenge is to persuade governments to adopt multi-sectoral approaches that entail horizontal coordination between different ministries, include various stakeholders, and accord with the principles of good governance, as well as taking into account subnational and national idiosyncrasies.

Cooperation at the regional level can be an effective means of preventing and reacting to crises. Regional action can also involve investments in the effective establishment of regional agricultural insurance, e.g., crop and cattle insurance (Deacon et al., 2007).

However, social protection programmes have intended and unintended gender implications. For example, conditional cash transfer programmes, which are based on the concept of 'co-responsibility', may impose unintended demands on overworked mothers (in poorer households especially), who more often than not have the responsibility of meeting conditionalities such as ensuring that children attend school and clinics. Apart from reinforcing 'traditional' gender roles, these conditions can displace women's labour from farming or income-generating activities. Similarly, efforts to target women in public works projects by setting gender quotas can lead to 'perverse effects', if women who are already 'time-poor' and overburdened are obliged to increase their workload in order to access social transfers (Devereux, 2002).

There is evidence that targeting transfers to women yields better results, because in many countries men have a higher propensity to spend incremental income on themselves, while women have a higher propensity to allocate incremental food or cash to their families, especially their children (Haddad et al., 1997). If, however, the objective of a safety net programme is to raise household productivity and incomes, the case for targeting individuals who own and work with productive assets may be stronger. For instance, if women have no access to land and men are responsible for ploughing, a programme that transfers draught bullocks for ploughing, if targeted to men rather than women, would maximize synergies between social protection and agricultural productivity, till such time as the institutional arrangements are changed so that ownership rights over productive resources are vested in women on par with men.

Recommendations

1. *Bring to bear the full force of the long arm of the state for the food security of special groups.* There are marginalized groups like people with disabilities, people living with HIV/AIDS, ethnic minorities, the elderly, widows, small farmers, migrants, internally displaced persons and women, who must have the protection of the state for food security. Towards that end, governments may on a priority basis:

   • Tackle the seven food insecurities faced by women, through a multi-sectoral programme of, *inter alia*, social protection, affirmative action, changing laws relating to inheritance and ownership of productive resources and making rights to food, education and health care for women constitutional rights.

- Undertake *ex ante* management of covariate shocks by boosting the coping strategies of those at risk of suffering from such shocks. This may be done through the setting up of insurance and insurance-like programmes with flexible targeting, flexible financing and flexible implementation arrangements, before the onset of natural disasters.
- Eliminate gender-based inequalities by: (*a*) adopting an agent-oriented approach to the women's agenda and regarding women as potentially active agents of major social change rather than as solicitors of social equity; (*b*) creating an enabling environment for 'cooperative conflict' between genders and devising ways and means for them to be resolved amicably; and (*c*) affirmative action, including reservation of seats in all legislatures and the parliament for women as a fair outcome and realization of the benefits of law.
- Provide social protection to migrants by establishing bilateral agreements for minimum standards for the migrants' protection, wages and access to health services, and explore ways and means to regularize migration across borders.

BOX 6.7   Singapore Offers Migrants Some Help

Singapore has set up a migrant workers' centre to help foreign workers who are involved in disputes with their employers and can be stranded without wages, food and shelter. The centre, which was to be opened by April 2009, is to be guided by the Migrant Worker Forum, an enterprise of the state-backed National Trade Union Congress and Singapore National Employers' Federation. Singapore employs 757,000 foreign workers. An increasing number of cases have surfaced of workers left unemployed without food, shelter and health care.

Source: Business News, 14 February 2009.

- Provide guaranteed employment for marginalized groups and people who face discrimination on the basis of sex, race, religion, caste, ethnicity, disability and communicable diseases, who are often among the poorest of the poor, commensurate with their needs, noting that employment guarantees are among the best forms of *de facto* insurance[14] for marginalized groups.

2. *Provide widespread protection against idiosyncratic shocks.* Given that food insecurity is often wrongly thought of as being caused only by covariate shocks, while much of food insecurity is caused by idiosyncratic shocks, governments in the region may take measure to reverse the situation through:

- Provisioning for *de jure* and *de facto* insurance for idiosyncratic shocks including: (a) more effective, ubiquitous and continuing insurance programmes, whether through financial innovations such as micro insurance or index insurance schemes and community-based health insurance programmes; and (b) *de facto* insurance via, for example, supporting community-based responses, public employment guarantee schemes like the National Rural Employment Guarantee programme in India, underpinned by food-for-work or cash-for-work projects as a means of protecting vulnerable people from idiosyncratic shocks (Barrett et al., 2004; Jalan and Ravallion, 1999).
- Protecting common property resources (CPRs), which constitute a secondary food system[15] and act as insurance for vulnerable groups and ethnic minorities, who need the support of the state, against temporary food insecurity by smoothing consumption. This may be achieved through: (a) establishing a robust system for management of common property resources, forests, water bodies and biodiversity, including joint management by all stakeholders, to ensure that while these resources continue

---

[14] Such as public employment guarantee schemes exemplified by food-for-work or cash-for-work projects, discussed earlier.

[15] Adding vital nutrients as also providing supplementary food to the common people.

to contribute to food security, they are not misused or overused but are properly conserved; (*b*) recognizing the usufruct and traditional property rights of the common people to CPRs in a new set of laws regulating the use of such resources, duly enforced; and (*c*) building and reinforcing local institutions, especially informal social safety nets and indigenous NGOs, as reliable channels for food distribution and as forms of social protection.

Appendix 1: Social Protection and Social Security: Illustrations from Some Asian Countries

Food Security in Bangladesh

Food Security Funded by External Aid

Bangladesh has a long tradition of safety nets funded by external food aid. The three biggest programmes are Food-for-Work, which provides wheat in exchange for work in rural infrastructure projects; Food-for-Education, which provides wheat and rice to poor children in return for regular primary school attendance; and Vulnerable Group Development (VGD), which provides food-grain and training to disadvantaged women.

The Food-for-Work programme provides wages in cash and kind (usually wheat) to rural labourers working in labour-intensive public works, for example, water, roads, forestry and fishery, during the dry season. Studies to evaluate the programme show that there has been a large increase in food consumption and calorie intake at the household level and improvements in the nutrition of the population in the areas covered by the programme.

The Food-for-Education programme, changed to Cash-for-Education in 2002, provided cash transfers to households with children in poor areas, on the condition that children were enrolled at school and had a minimum attendance level. The main welfare outcomes showed a 9–17 percentage point rise in the school enrolment rate (from a base of 55 per cent), and nearly full attendance among beneficiaries, with improvement in long-term opportunities for children.

The VGD programme provides in-kind wheat transfers to enable destitute rural women to improve their economic and social condition. A complementary package of development services was introduced in

1988, including health and nutrition education, literacy training, savings, and support in launching income-earning activities. Evaluations show that in-kind transfers of wheat increased wheat consumption dramatically—by 70 per cent for VGD households compared to 13.9 per cent for its cash-based equivalent transfers. This high wheat consumption stems from the heavy transaction costs faced by female-headed households, largely Muslim, in accessing local markets to sell their grain.

BOX 6.8   An Innovative Micro-Health Programme, Bangladesh

The Gonoshasthaya Kendra (GK) micro-health insurance is a scheme that is partly self-financed, that reaches out to sections excluded by formal public and private schemes. It provides preventive and pro-motional care including immunization, family planning, and mother and child care free of charge to 100 per cent of the population in its catchment area. It provides the cost of curative care on a sliding scale with non-insured families being charged a higher cost. The scheme is financed by members' contributions, voluntary contributions by non-members, state contributions and non-state subsidies from development agencies and donors.

Source: Cook et al., 2003.

**BOX 6.9   Old-Age Security and Assistance
for Destitute Women, Bangladesh**

The Old-Age Allowance Scheme (OAAS) and Assistance Programme for Widowed and Destitute Women (APWDW) provide a cash transfer to the elderly population and to destitute widows for reducing extreme poverty and destitution among them. Analysis of household data from the Bangladesh Demographic and Health Survey (2000) indicates that the percentages of beneficiary households in quintiles of wealth index are (from the poorest to the richest) 6.4, 6.0, 2.5, 0.8 and 0.2, respectively. There is a concentration of beneficiary households in the lowest wealth index quintiles.

Source: Global Extension of Social Security: *Old Age Allowance Scheme (OAAS) and Assistance Programme for Widowed and Destitute Women (APWDW).* http://www.socialsecurityextension.org/gimi/gess/showcompendi-umprogrammesbycountrys.do.pid=53

Food and Nutrition Security in India

Public Distribution System (PDS)

The public distribution system is the most important intervention by the Government of India (GOI) towards food security. The Food Corporation of India (FCI) buys food-grains from farmers at an administered price (minimum support price), and then arranges to sell the food-grains to consumers at a subsidized price, through allocating these to different states who in turn run their respective PDS. In the beginning, the coverage of PDS was universal, with no discrimination between poor and non-poor. The objectives of the PDS are: (a) maintaining price stability; (b) raising the welfare of the poor; (c) averting food security situations of scarcity; and (d) keeping a check on the private trade (GOI, 2002). Of all these objectives, the first two are of overarching importance.

Over the years, the policy related to PDS has been revised to make it more efficient and targeted. In 1992, a Revamped Public Distribution System (RPDS) was introduced in 1,700 blocks. The target was to reach the benefits of PDS to remote and backward areas. From June 1997, in a renewed attempt at better targeting to the poor in all regions, the Targeted Public Distribution System (TPDS) was introduced to adopt the principle of targeting the 'poor in all areas'. A differential price policy for the poor and the non-poor was adopted for the first time. Further, in 2000, two special schemes were launched—the Antyodaya Anna Yojana (AAY) and the Annapurna Scheme (APS)—whose special target groups were the 'poorest of the poor' and 'indigent senior citizens', respectively. The functioning of these two schemes was linked with the already existing network of the PDS.

The PDS as a whole has succeeded in preventing large-scale food shortages during periods of natural disaster. However, the system has certain limitations. Owing to increasing minimum support prices and a more tar-geted approach to reach the poor, the food subsidy of the government has increased in leaps and bounds over the years. The food subsidy bill, which was ₹24.5 billion in 1990–91, increased to ₹300 billion by 2003–04.

A major concern, however, has been the marked leakages in the channel of subsidy reaching to the target group. Drèze (2004) notes that in 1999–2000, the average Indian citizen received a subsidy of about ₹2 per month from the PDS. This suggests that the real subsidy reaching the households for one year may be in the range of ₹25 billion to ₹30 billion. Even some of the best estimates show that only 20 per cent of the total food subsidy reaches the household level (Dev et al., 2004). These estimates succinctly exhibit the poor effectiveness of the PDS in targeting the poor.

One basic social security measure that has an intergenerational effect relates to nutritional security. With over 40 per cent of its children below 5 years malnourished, India stands at the threshold of malnutrition. Except for the states of Tamil Nadu and Kerala, there has been no concerted effort to accomplish nutrition security among children in any other state. The following two schemes are the crux of nutritional security for children and women.

Integrated Child Development Services (ICDS): The ICDS is an ambitious scheme aimed at delivering a package of services that cater to the comprehensive development of children below the age of 6. It seeks to integrate efforts at improving child health and nutrition, non-formal education as well as maternal health and nutrition into a single service delivery window. At present, the scheme reaches out to 4.8 million expectant and nursing mothers and 22.9 million children under 6 years of age, through a network of 4,200 projects covering nearly 75 per cent of the development blocks and 273 urban slum pockets in the country. The ICDS is a centrally sponsored scheme implemented through state governments.

All ICDS services are provided through Anganwadis (childcare centres), where the beneficiaries, including infants, mothers and pre-schoolers, gather on a daily basis to receive the services provided to them. These include nutrition supplements, health and nutrition education to mothers, pre-primary teaching for the 3–6 age group and immunization services for all. Under the supplementary nutrition component, food supplements are provided to the daily diet of the beneficiaries and are meant for consumption at the Anganwadi centre itself. Severely malnourished children are given double the daily supplement provided to the other children. In addition to calories and proteins, specific micronutrients are also provided in accordance with the requirements of various age groups.

As regards the actual performance of the scheme, it is generally observed that the programme has not been doing well mainly due to administrative shortcomings. A concurrent evaluation of the ICDS reveals that about 67 per cent of the centres across the country report irregular supply of food. The worst affected are the central and north-eastern states, where more than 75 per cent of the centres report irregularity in supplies. Himachal Pradesh, Maharashtra, Tamil Nadu and Goa are comparatively better off in this respect. A large portion of the food supplement often does not reach the intended beneficiaries. In terms of beneficiaries' satisfaction, more than 80 per cent of the surveyed households were satisfied with the functioning of the nutrition component. However, state-wise variations do exist, ranging from about 90 per cent of households in some states to only 30 per cent of households in others.

The 11th Five-Year Plan envisages the strengthening of the nutrition component of ICDS with a focus on both micro- and macronutrient undernutrition. This will be achieved through strengthening the nutrition and health education components to ensure appropriate inter-familial distribution of food, targeting of children in the 6–36 month age group, and pregnant and lactating women, ensuring universal screening of all children, at least quarterly, to identify faltering growth, focusing on training and supervision of ICDS personnel to improve the quality of services provided, shifting the focus to take-home food supplements as against feeding at Anganwadis, and emphasizing inter-sectoral linkages, especially between the health and ICDS programmes.

Mid-Day Meal Scheme: Mid-Day Meal Scheme was introduced on a large scale, in 1960s in Tamilnadu, which was universalised in 1982 for all children up to class 10. Tamil Nadu's midday meal programme is among the best known in the country though the scheme is as good in Pondicherry. Several other states of India like Gujarat started it in late 1980s and Kerala started the same since 1995 and so did Madhya Pradesh and Orissa in small pockets. On 28 November 2001, the Supreme Court of India gave a landmark direction to government to provide cooked meals to all children in all government and government assisted primary schools. The programme has become almost universal by 2005.

The success was so spectacular that in 1995, the Government of India began the 'National Programme For Nutrition Support to Primary Education'. According to the programme, the Government of India will provide grains free of cost and the states will provide the costs of other ingredients, salaries and infrastructure. Since most state governments were unwilling to commit budgetary resources, they just passed on the grains from Government of India to the parents. This system was called provision of 'dry rations'. On 28 November 2001, the Supreme Court of India gave a famous direction that made it mandatory for the State Governments to provide cooked meals instead of 'dry rations'. The direction was to be implemented from June 2002, but was violated by most states. But with sustained pressure from the court, media and in particular, from the 'right to food campaign', more and more states started providing cooked meals.

In May 2004, a new coalition government was formed in the Centre, which promised universal provision of cooked meals fully funded by it. This promise in its Common Minimum Programme was followed by enhanced financial support to the states for cooking and building sufficient infrastructure. Given this additional support, the scheme has expanded its reach to cover most children in primary schools in India. In 2005, it was expected to cover 130 million children.

Food Security Measures in Pakistan

In Pakistan, the food security programme was designed in the form of the Food Support Programme, which is intended to mitigate the impact of increase in wheat prices. Initially, the coverage of the programme was extended to 1.2 million poorest households with monthly income up to ₹2,000. A cash support of ₹2.5 billion has been allocated for disbursement on a biannual basis. However, with the recent increase in the support price of wheat from ₹300 to ₹350 per 40 kg, the government has increased this amount to ₹3.5 billion to offset the impact of likely increase in the price of flour.

Social Security Programmes

Social Security and the Informal Sector

If we scan the literature on social security measures for workers and working age populations across countries, universalization has remained an unfulfilled promise. With over 90 per cent of the work force being in the unorganized sector in all the countries of the region, addressing social security for *all* remains, or should become, an urgent agenda. All the countries of the region have various schemes run by governments and by employers for formal-sector workers, who constitute not more than 10 per cent of the work force. Here, too, equity- and growth-oriented models of social security are needed in place of those that distort labour markets. The need for social security in the formal sector assumes importance in the context of globalization as well.

Universalization forms a priority in the region as far as social security of informal workers is concerned. The Sri Lankan experience of the three-pronged social security measures for workers (working populations), namely, employment protection and promotion, social security/insurance, and safety nets, would be an illustrative approach for other countries to follow.

Employment Security and Promotion

Poverty reduction and employment strategies and their coverage appear to be limited in many countries, although they cover large numbers of the poor. Though growth- and equity-oriented poverty reduction strategies have been enunciated by all countries, they still need to be prodded into operationalizing such strategies. There is evidence of improved targeting and coverage from Sri Lanka and Pakistan; however, the globalization processes that all countries of the region are undergoing call into question the basic tenets of welfare mechanisms for social security provisioning. Political will is needed for countries to mediate growth-oriented social security measures.

Towards a Comprehensive Social Security—The Case of India

NREGS: Two important milestones have been reached in India towards a comprehensive social security system for the poor. The first is the implementation of the National Rural Employment Guarantee Scheme (NREGS) in 200 poor districts of India from February 2006.

India's recently introduced NREGS provides an excellent 'rights-based' framework for social security. The salient features revolve around recognition of the right to work and the dignity of work. The entitlement approach enhances the accountability of the state. While it is currently being implemented in about 200 districts of the country, the initial assessment is one of guarded optimism, as several field-level inadequacies have been identified (IHD, 2006).

Nevertheless, specific features of NREGS in India are worth emulating in the rest of the South Asian region. The scheme has been envisaged from the perspective of the 'right to employment', and guarantees 100 days of employment for the poor at a minimum fixed rate. More importantly, NREGS has introduced the concept of a 'right to employment' for the the poor. The act has also identified the roles and responsibilities of the central and state governments, district and block administrations and the panchayats. The onus of guaranteeing 100 days of employment rests with the government, and the applicant can demand unemployment allowance in case he/she does not get work. Apart from the creation of work opportunities, the act also provides basic facilities at the work site, namely, crèches, safe drinking water and medical aid. The scheme focuses not only on providing employment, but also on building village assets through the employment. These assets would further help in the development of villages.

Various concurrent evaluations carried out so far show the varying levels of implementation and success of the scheme. Though the level of success is not uniform across states and districts, yet most states and districts have started in right earnest. The reason for the success of the NREGS is the inbuilt mechanism of auto-elimination of ineligible aspirants due to manual labour involved, and of contractors.

The reasons for the varying performance of NREGS across states and districts are generally specific to the state or the district. A few may be illustrated here. The state of Jharkhand faced problems in implementation because of the non-constitution of panchayati raj institutions (PRIs) in the state. Bihar faced problems related to shortage of staff, poor infrastructure, low administrative capacity of the state, etc. The success stories too have specific reasons. For example, Dungarpur district in Rajasthan was more successful in implementing NREGS because of the presence of effective grassroots NGOs and their ability to mobilize the poor. Andhra Pradesh has done equally well and has successfully used computerization and e-governance mechanisms in implementing the scheme. For, example, the fund in the state is being transferred electronically; every job seeker has got a bank account and wages are paid through the bank account; and the whole process from job application to registration is computerized.

Even in Bihar, despite the low level of administrative capacity, there was a high level of social awareness; benefits mostly reached the target groups; and the process was found to be non-discriminatory (IHD, 2006). The nature of the NREGS Act and its provisions, if properly implemented, have the potential to make it a very successful programme of social security. Moreover, the response of the society has been very encouraging. Members of Parliament, political parties, the intelligentsia and NGOs are actively involved in the planning and implementation of the scheme. All these factors are expected to make NREGS a successful social security programme.

As mentioned, implementation-level issues emerging from the field studies reflect concerns about the setting up of facilities, identification of jobs suitable for the poor, lack of monitoring systems and the inability of local-level officials to appreciate the provisions of the scheme, among other factors (IHD, 2006). Notwithstanding the initial setbacks, the scope for introducing such a comprehensive employment generation programme in other countries needs to be explored.

Social Security Legislation for the Informal Sector: The other important step is towards formulating comprehensive social security legislation for low-income workers in the informal sector. The National Commission on Enterprises in the Unorganized Sector (NCEUS) has prepared a report and draft bill for ensuring protective social security for all means-tested workers. According to the proposal, all workers below a ceiling level of income, whether paid or self-employed, will be eligible for the social security package upon payment of a small premium by themselves or (if they are poor) by the government. The objective is to institute, with legislative backing, a national minimum social security that will act as a floor level for the estimated 300 million unorganized workers with independent earnings. The minimum social security package includes: health insurance for the worker and family and maternity benefit for the worker/spouse; life insurance for the worker; and old-age security for the worker after reaching the age of 60, in the form of the provident fund, or pension (for below-poverty-line workers). Except the old-age security, all the other social security benefits are based on the insurance nodal. It will be the responsibility of the national board as well as state boards to ensure the best possible deal by selecting the appropriate service provider. Registration of workers will be encouraged and facilitated through 'facilitation centres'. The registration and ID cards issued to the workers will be 'portable', and it will be possible to change addresses upon request. The scheme could be co-financed by employers (where they are identifiable) or through specific taxes or cesses. The total cost to the exchequer of financing this social security package, when all informal-sector workers are covered, is estimated at ₹254,000 million, of which around three-fourths will be contributed by the central government and the rest by the state governments. As a percentage of GDP, this works out to 0.5. However, the whole scheme is envisaged to be implemented within a period of five years covering one-fifth of the eligible informal workers every year (NCEUS, 2005).

Both NREGS and the proposed draft bill for a minimum comprehensive social security for unorganized-sector workers are fundamentally different from earlier schemes. They are rights-based, i.e., in the form of legally enforceable entitlements, unlike the very many schemes floated by the central and state governments at different points of time. Also, they are universal in nature—all workers in rural areas in NREGS and all unorganized earning workers in the proposed NCEUS are eligible. Further, while the former guarantees a minimum income to rural workers, the latter proposes to provide a national floor-level social security to all informal workers throughout the country, to which state governments may add on their contributions or additional benefits, if they so choose.

The scope for designing similar schemes with universal applicability needs to be explored in the case of other countries in South Asia. The proposed bill may be taken up as a model bill for other countries to adapt to their specific conditions.

Financial Access/Security and Reducing Vulnerability

Microfinance

Microfinance (MF) in the region is emerging as a major form of social and economic security. Evidence shows that in countries where formal financial infrastructure exists, a robust downscaling of the formal sector to reach out to the needs of the poor would be a compelling strategy in the context of globalization, to meet the needs of a growing number of the poor. Financial inclusion and universal obligations to create financial access need to be important strategies. The idea of 'self-help groups' as a means of reaching out to the very poor and socially excluded is an important innovation in the Indian context (Kabeer, 2002). Such an approach needs to be expanded in terms of both depth as well as breadth so as to reach the extremely poor. In fact, experience demonstrates the need for innovation in reaching the extremely poor, and the group collateral approach may not be best suited for such circumstances. The experience of BRAC in this respect is worth noting.

The adaptation of the Grameen Bank to changing rural realities and to the social security needs of borrowers (women) needs to be appreciated as a model for replication as well as for expanding outreach. During the past few years, the bank and its clients addressed challenges from many quarters, including floods and natural disasters, resistance from male members of families, and the rules and regulations laid down by themselves for managing micro-credit activities. Grameen Bank, which has been implemented since 2002, has made several changes in its programme, in terms of eligibility criteria, products and services as well as rules and regulations. In the current Grameen programme, instead of a single product like 'credit', members have been provided with several choices to meet their financial needs, which are primarily focused on addressing the social security of the household. Changes in institutional arrangements have also been effected in order to meet the needs of the clients.

Grameen has developed diverse products that meet all the needs of the poor, the destitute and people and families in the unorganized sector. The products include basic loans, flexible loans, special loans for education, housing and health, loans for the destitute, various savings schemes, loans and life insurance, the Grameen pension scheme, shares and scholarships. Coupled with these, measures like Grameen telephony and the use of mobile phones and technology enable members to benefit from communication opportunities.

It is to be noted that loan schemes, such as loans for the destitute, are meant to service the most vulnerable groups, like beggars, and those who are not members of the Grameen groups. Similarly, some of the savings products, like fixed deposits, attract the general public who would like to save money with Grameen Bank. Scholarships for needy children are also a special feature especially designed to encourage education among the girls and boys of Grameen members (UNDP Case Studies, 2005).

A similar effort is SafeSave, which works with the extremely poor. SafeSave, a financial service provider in the city of Dhaka, Bangladesh, has attracted much attention for its innovative products and services addressing the needs of the extremely poor. Financial services for poor people are viewed by SafeSave as a matter of helping the poor to turn their capacity to save into usefully large sums. It provides individual banking services to the poor at their doorsteps. Without resorting to group-based collaterals, SafeSave addresses their financial needs and ensures low operational costs by having collectors from among the slum dwellers and measures like the computerization of services (Kabeer, 2002). Innovations in the promotion of various community and microfinance-led social security measures from the region are worth recording here, as they demonstrate potential for replicability across the region.

Microfinance in the Region: Challenges: Efforts towards restructuring the rural banking sector and cooperatives are imperative to ensure

---

BOX 6.10   Women's Development Federation, Hambantota

The Women's Development Federation (WDF) of Hambantota, or the Janashakti Banku Sangam, began in the late 1980s and currently has over 33,000 members in five divisions in Hambantota district. It is owned and managed solely by females and its modus operandi is founded on social mobilization efforts and the participation of women in social and economic activities. Its community-driven activities benefit a large number of women-headed households and their micro-businesses. Examples of such community improvement activities are harvesting rainwater for cultivation and consumption, and enabling access to safe water from both micro and minor irrigation systems.

Source: Author.

BOX 6.11   A Savings and Credit Scheme
for Rural Women in Pakistan

The Marvi Welfare Association in the village of Arab Solangi is the first all-female community-based organization (CBO) in the district of Khairpur, Sindh province, Pakistan. Created in 1993 by a few young women, with the support of the Aga Khan Foundation, its membership has grown to 128. The CBO adopted the savings and credit strategy for employment creation, introduced in 1994 by the ILO/Japan Inter-Country Project on Strategic Approaches toward Employment Promotion (ILO/PEP). This has resulted in savings of over US$1,550, qualifying the CBO for matching funds of US$4,150 from the project. After receiving training in savings and credit management and skills development support, elected CBO officials maintain accounts and administer loans. So far they have given out 86 loans totalling US$12,835. A survey of the first borrowers shows that after just one year, the total household income rose by more than 41 per cent. Dirt lanes between mud houses have been drained and paved, a sports ground created and a girls' school opened.

Source: International Labour Organization (2000).

access to financial services and, *inter alia*, income security for substantial numbers of the poor in the South Asian region. While current microfinance growth in the region is somewhat modest, it is expected that leveraging resources from the formal sector would be one of the paths for growth of MF. Also, governments' efforts to create a conducive policy environment for MFIs and the creation of apex-level funds are some initiatives that enable MFIs to expand their outreach rapidly.

In all the countries of South Asia, the presence of the public sector—formal banking structures—provides an opportunity for the integration of MF to scale it up. At the same time, this is a challenge, as most of the public sector needs revitalizing and refocusing on sustainable microfinance as a portfolio for poverty reduction. Nancy Barry, President of Women's World Banking, in 2004 identified the challenges as extending outreach to millions of people by increasing MFI capacity, involving mainstream banks and mobilizing the domestic capital market, building assets on the borrowers' part and building a culture among MFIs, bankers, policy makers and funders of MF based on trust, transparency, shared standards, generosity and mutual accountability for results (Microcredit Summit, 2004).

Reaching out to the extremely poor through financial services is a challenge for all countries of the region. While innovations in this respect are here to stay and are ever expanding, there is a need to view such inclusion from a sectoral perspective in order to sustain innovations. In India, models like 'banking correspondence' for expanding the outreach of services are one example. Similarly, as discussed previously, the experiences of NGOs from Bangladesh like Grameen II and BRAC in reaching out to the extremely poor require replication and adaptation in many other countries.

In most countries, NGO-MFIs, MFIs and other institutional forms are primary suppliers, albeit with formal financial institutions providing promotional support, and in some cases creating apex-level wholesale funds. In none of the countries, with the exception of Pakistan, is a comprehensive MF sector development policy available. However, the promotion of MF banks, licensing of NGO-MFIs, developing rating and prudential norms, developing appropriate supervisory mechanisms, etc., are some important elements of currently available sectoral policies in the region, with varying degrees of sophistication. Such measures would ensure financial security to millions of the poor who are a part of microfinance schemes and programmes. The flow of private commercial capital, venture capital and foreign capital and access to various forms of financial instruments for NGO-MFIs and MFIs are the need of the hour for enhancing the outreach of MF. National governments need to take steps in this direction through financial-sector reforms. Similarly, the franchising of successful models (and MFIs), as in the case of Grameen Bank and CASHPOR, may also be permitted for rapid outreach in several other countries of the region.

Social Security and Women

With over 96 per cent of the women workers in the unorganized sector, they face gender based segregation of work and are concentrated in the lower end of the spectrum with low paying and insecure jobs. While microfinance enables financial security for women, there are issues of employment and income security that need to be viewed through a gender perspective. Women in the region have been seen primarily in production and reproduction roles. Comprehensive social security covering health care and other social deficiencies—moving beyond maternity benefits during childcare, like early childhood care and support to families, etc.—would lead not only to higher productivity but also to income security. Social security measures for women workers may include childcare needs in the form of institutional delivery, maternity benefits, crèches, universalization of child development schemes and promotion of community-based approaches to childcare. A flexible, autonomous childcare fund could be also created. This fund can be drawn on to provide childcare facilities to all women regardless of income, number of children or other considerations. The fund would be at the state/national level, for administrative convenience, and could have multiple sources of funding.

Similarly, several countries have maternity benefits for the general population, but support to women (mothers) during the early years of their children's lives is missing, which in fact, creates a barrier for them to enter/re-enter the labour market. This is more pronounced in the urban informal sector, where the traditional support mechanisms of the extended family are often missed.

Apart from this, a statutory scheme that provides financial support for childbirth, childcare and breastfeeding in the first few months of the child's life would be beneficial for women in the unorganized sector. It could also be linked with the maternal and child health provisions of the public health system. Similarly, health insurance is a major need for women workers. It is found that health expenses are the largest major source of debt. At present, women workers, especially in the unorganized sector, have no access to any form of health insurance, except those promoted by NGOs and other community-based organizations. The same is the case with other countries as well. It is necessary to promote forms of insurance for these workers. There are a number of successful micro insurance schemes which could be upscaled across all countries of the region.

It should be noted at this juncture that India's newly enacted NREGS envisages the setting up of childcare centres at the workplace. Similarly, the proposed NCEUS comprehensive social security scheme for informal workers, again from India, also envisages addressing gender concerns related to women workers.

Organizing Women and Institution Building

Another important dimension of social security relates to institution building among informal and unorganized workers. Experiences across countries of women's institution building reveal the strength of women's collectives to address issues of empowerment, financial autonomy as well as gender relations. The poverty reduction strategies of all countries of the region identify women as primary agents for institution building. Experiences of both government-sponsored women's development programmes as well as those of NGOs reflect the need for institutions as precursor empowerment. India's experience of organizing women workers of the informal sector provides lessons for other countries as well.

India's Self-Employed Women's Association (SEWA) and the Working Women's Forum (India) (WWF[I]) demonstrate the benefits of organizing workers of the unorganized sector, especially women. These two organizations have demonstrated the efficacy of implementing various social security measures for women workers of the informal sector. Interventions like life insurance, reproductive health and childcare, unionization, ensuring minimum wages and workplace safety and security, informal banking, micro loans and savings, micro health insurance and micro-enterprise promotion have had a substantial impact on women workers in terms of income security and empowerment. These organizations have been able to create a social platform for women and have ensured social security through mobilization and organization. Similar mobilization of informal-sector workers is the need of the hour and the role of non-government organizations, which have enjoyed substantial autonomy and space in many countries of the region, in this regard is worth recognizing.

References

Abbott, D., and S. Pollard. 2004. *Hardship and Poverty in the Pacific*. Manila: Asian Development Bank.

Abdulai, A., C. Barrett and J. Hoddinott. 2004. 'Does Food Aid *Really* Have Disincentive Effects? New Evidence from Sub-Saharan Africa'. *World Development*, vol. 33, no. 10, pp. 1689–1704.

Agarwal, Bina. 2008. 'Rethinking Collectivities: Institutional Innovations in Group Farming, Community Forestry and Strategic Alliances'. B. N. Ganguli Memorial Lecture, 11 April 2007. Occasional Paper, Centre for Study of Developing Societies, Delhi.

Ahmed, A., and M. Sharma. 2004. 'Food-for-Education Programs with Locally Produced Food: Effects on Farmers and Consumers in Sub-Saharan Africa'. Washington, D.C.: International Food Policy Research Institute.

Alderman, H., and T. Haque. 2007. 'Insurance Against Covariate Shocks: The Role of Index Based Insurance in Low-income Countries of Africa'. Africa Human Development Series, Working Paper No. 95, World Bank, Washington D.C.

Asian Migrant Centre. 2003. *Asian Migrant Year Book 2002–2003*. Hong Kong: Asian Migrant Centre and the Migrant Forum in Asia.

Barrett, C. 2006. 'Food Aid's Intended and Unintended Consequences'. Background Paper for State of Food and Agriculture 2006, FAO, Rome.

Barrett, C., and D. Maxwell. 2005. *Food Aid after Fifty Years: Recasting Its Role*. London: Routledge.

Barrett, Christopher B., Christine M. Moser, Oloro V. McHugh and Joeli Barison. 2004. 'Better Technology, Better Plots or Better Farmers? Identifying Changes in Productivity and Risk among Malagasy Rice Farmers'. *American Journal of Agricultural Economics*, vol. 86, no. 4, pp. 869–88.

Brahmbhatt, M., and L. Christiaensen. 2008. *Rising Food Prices in East Asia: Challenges and Policy Options*. Washington, D.C.: World Bank.

Business News. 2009. 'Singapore Offers Migrants Some Help'. 14 February. http://news.monstersandcritics.com/business/news/article_1459494.php/Singapore to_set_up_migrant_workers_centre

CGAP (Consultative Group to Assist the Poor). 2001. 'Linking Microfinance and Safety Net Programs to Include the Poorest: The Case of IGVGD in Bangladesh'. Focus Note 21, CGAP, Washington, D.C., May. http://www.cgap.org/gm/document-1.9.2566/FN21.pdf (accessed 7 March 2012).

Coady, D. 2004. 'Designing and Evaluating Social Safety Nets: Theory, Evidence and Policy Conclusions'. FCND Discussion Paper, 172. Washington D.C.: International Food Policy Research Institute.

Cook, Sarah, Naila Kabeer, Gary Suwannarat and Ford Foundation. 2003. *Social Protection In Asia*. New Delhi: Har Anand Publications.

Cromwell, Elizabeth, and Rachel Slater. 2004. 'Food Security and Social Protection'. Paper produced for Department for International Development, London, September.

Deacon, Bob, Isabel Ortiz and Sergei Zelenev. 2007. *Regional Social Policy*. DESA Working Paper No. 37 (ST/ESA/2007/DWP/37), June.

De Janvry, A., F. Finan, E. Sadoulet and R. Vakis. 2006. 'Can Conditional Cash Transfer Programs Serve as Safety Nets in Keeping Children at School and from Working When Exposed to Shocks?' *Journal of Development Economics*, vol. 79, no. 2, pp. 349–73.

Devereux, S. 2002. 'Can Social Safety Nets Reduce Chronic Poverty?' *Development Policy Review*, vol. 20, no. 5, pp. 657–75.

Devereux, S., and S. Coll-Black. 2007. 'Review of Evidence and Evidence Gaps on the Effectiveness and Impacts of DFID-Supported Pilot Social Transfer Schemes'. *DFID Social Transfers Evaluation*. Brighton: Institute of Development Studies.

Dorward, A. 2006. 'Markets and Pro-poor Agricultural Growth: Insights from Livelihood and Informal Rural Economy Models in Malawi'. *Agricultural Economics*, vol. 35, no. 2, pp. 157–69.

Dreze, Jean. 2004. 'Financial Implications of an Employment Guarantee Act: Preliminary Estimates'. Discussion Paper, National Advisory Council, New Delhi.

ESCAP (Economic and Social Commission for the Asia-Pacific). 2007. *Statistical Yearbook for Asia and the Pacific 2007*. Bangkok: UNESCAP.

———. 2008a. *ESCAP Population Data Sheet 2008*. Bangkok: UNESCAP.

ESCAP (Economic and Social Commission for the Asia-Pacific). 2008b. 'What Is Good Governance?' http://www.unescap.org/pdd/prs/ProjectActivities/Ongoing/gg/governance.asp (accessed on 9 October 2008).
————. 2008c. *Economic and Social Survey of Asia and the Pacific 2008: Sustaining Growth and Sharing Prosperity*. New York: ESCAP.
————. 2009. Social Policy and Population Section, Disability Programme. http://www.unescap.org/esid/psis/disability/index.asp (accessed on 27 January 2012).
————. 2011. *The Promise of Protection*. Bangkok: United Nations.
ESCAP, ADB and UNDP. 2008. *A Future within Reach 2008: Regional Partnerships for the Millennium Development Goals in Asia and the Pacific*. New York: ESCAP, ADB and UNDP.
Gertler, P., S. Martinez and M. Rubio-Codina. 2005. 'Investing Cash Transfers to Raise Long Term Living Standards'. Mimeograph, World Bank, Washington D.C.
Government of India (GoI). 2002. 'Report of the High Level Committee on Long Term Food Grain Policy'. New Delhi: Department of Food and Public Distribution, Ministry of ConsumerAffairs and Food and Public Distribution.
Haddad, L. and J. Hoddinott. 1997. 'Incorporating Work Intensity into Household Models: A Primer for Non-economists'. Paper presented for the Workshop on 'Gender Differentials in Gender and Social Protection, 28. Work Intensity, Sustainability and Development', School of Development Studies, University of East Anglia, Norwich, 2–4 July.
Harvey, P. 2007. 'Cash-based responses in emergencies'. HPG Report, 24. London: ODI.
Hasan, Nurdin. 2009. 'Myanmar Migrants Say Cast Adrift by Thais: Indonesia Navy'. France 24, International News: AFP News Briefs List. http://www.france24.com/en/20090203-myanmar-migrants-say-cast-adrift-thais-indonesia-navy (accessed on 5 February 2009).
Hellmuth, M., A. Moorhead, M. Thomson and J. Williams (eds). 2007. *Climate Risk Management in Africa: Learning from Practice*. New York: International Research Institute for Climate and Society, Columbia University.
Heltberg, Rasmus, Steen Lau Jorgensen and Paul Bennett Siege. 2008. 'Climate Change, Human Vulnerability and Social Risk Management', Social Development Department, The World Bank, 21 February.
IFAD (International Fund for Agricultural Development). 2008a. *Making a Difference in Asia and the Pacific*. Newsletter No. 21, June–July. http://www.ifad.org/newsletter/pi/21.htm (accessed on 15 January 2009).
————. 2008b. 'Food Prices: Smallholder Farmers Can Be Part of the Solution'. Factsheet, IFAD, Rome.
ILO (International Labour Organization). 2000. *Informal Sector*, Bangkok, ILO. http://www.ilo.org/public/english/region/asro/bangkok/feature/inf_sect.htm (accessed on 7 March 2012).
————. 2004. *A Fair Globalization: Creating Opportunities for All*. Geneva: ILO.
————. 2008. *Global Employment Trends for Women 2008*. Geneva: ILO.
Institute for Development Studies. 2007. 'Connecting Social protection and Climate Change'. *IDS In Focus*, no. 2, Brighton, UK.
Institute for Human Development. 2006. 'A Study of the NREGA in Bihar', Field Report, New Delhi, July.

Inter-Press Service. 2008. 'Q&A: Women Do Most, with Least Assistance—Interview with Lennart Båge, President of the International Fund for Agricultural Development', 14 August 2008. http://ipsnews.net/news.asp?idnews=43554 (accessed on 12 November 2008).

IPCC (Inter-governmental Panel on Climate Change). 2007. 'Climate Change 2007: Mitigation'. Contribution of Working Group III to the Fourth Assessment Report of the Inter-governmental Panel on Climate Change. http://www.ipcc.ch/pdf/assessment-report/ar4/wg3/ar4-wg3-frontmatter.pdf (accessed on 15 July 2008).

Jalan, Jyotsna, and Martin Ravallion. 1999. *Are the Poor Less Well Insured? Evidence on Vulnerability to Income Risk in Rural China*. Policy Research Working Paper 1863, World Bank, Washington, D.C.

Li, Zongmin. 2003. 'Women's Land Tenure Rights in Rural China: A Synthesis'. Mimeograph, Ford Foundation, Beijing.

Mahendra Dev, S., C. Ravi and B. Viswanathan. 2004. 'Economic Liberalization, Targeted Programmes and Household Food Security: A Case Study of India'. Discussion Paper No. 68, IFPRI, Washington, D.C.

Martinez, S. 2004. 'Pensions, Poverty and Household Investments in Bolivia'. Mimeograph, University of California at Berkeley, Berkeley, CA.

MCII (Munich Climate Insurance Initiative). 2008. 'Insurance Instruments for Adapting to Climate Risks: A Proposal for the Bali Action, Plan, Version 2.0'. MCII Submission to the Fourth Session of the Ad Hoc Working Group on Long-Term Cooperative Action under the Convention (AWG-LCA 3), Poznan, 1–13 December.

McCord, A. 2005. 'Win-Win or Lose-Lose? An Examination of the Use of Public Works as a Social Protection Instrument in Situations of Chronic Poverty'. Paper for presentation at the Conference on Social Protection for Chronic Poverty, Institute for Development Policy and Management, University of Manchester, Manchester.

Mechler, R., J. Linnerooth-Bayer, D. Peppiatt. 2006. 'Disaster insurance for the poor? A Review of Microinsurance for Natural Disaster Risks in Developing Countries', ProVention Consortium, Geneva and International Institute for Applied Systems Analysis, Laxenburg, Austria.

Meinzen-Dick, Ruth. 2011. 'Property Rights for Poverty Reduction?' in Jomo Kwame Sundaram and Anis Chowdhury (eds), *Poor Poverty*. New York: Bloomsbury.

Microcredit Summit. 2004. 'State of the Microcredit Summit Campaign Report 2004'. Written by Sam Daley-Harris, Microcredit Summit Campaign Director, Washington D.C.

Mitra, S. N., Ahmed Al-Sabir, Tulshi Saha and Sushil Kumar. 2001. *Bangladesh Demographic And Health Survey (2000)*. National Institute of Population Research and Training (NIPORT), Dhaka, Bangladesh.

National Commission for Enterprises in the Unorganized Sector (NCEUS). 2008. 'Report on Conditions of Work and Promotion of Livelihoods in the Unorganised Sector'. New Delhi: Academic Foundation.

Navsarjan. 2008–09. *Who Are Dalits?* http://navsarjan.org/navsarjan/dalits/who aredalits (accessed on 5 February 2009).

Paarlberg, R. 2002. 'Governance and Food Security in an Age of Globalization'. 2020 Brief No. 72, International Food Policy Research Institute, Washington, D.C.

Ravallion, M. 2003. 'Targeted Transfers in Poor Countries: Revisiting the Tradeoffs and Policy Options'. Policy Research Working Paper 3048, World Bank, Washington, D.C.

Roncarati, M., and Y. Taoprasert. 2000. 'Challenges Facing Mor Muangs' Beliefs and Practices on Maternal and Child Health', in *Indigenous Knowledge and Practices on Mother and Child Care: Experiences from Southeast Asia and China*. Philippines: International Institute of Rural Reconstruction, PLAN International and Save the Children Federation, Inc.

Sabates-Wheeler, Rachel, Stephen Devereux and Bruce Guenther. 2009. 'Synergies between Social Protection and Smallholder Agricultural Policies', *IDS*. Sussex: Future Agricultures Consortium.

Sen, Amartya. 1986. *Poverty and Famines: An Essay in Entitlement and Deprivation*. Oxford: Clarendon Press.

———. 1997. *Hunger in the Contemporary World*. Discussion Paper DEDPS/8 (retroactively created in June 2002 from the paper originally published in November 1997 as DERP No. 8), The Suntroy Centre, London, and Toyota International Centre for Economics and Related Disciplines, London School of Economics and Political Science.

———. 2001. 'The Many Faces of Gender Inequality'. *Frontline*, vol. 18, no. 22, 27 October–9 November.

Shrestha, Anita. 2003. 'Dalits in Nepal: Story of Discrimination'. http://www.hurights.or.jp/asia-pacific/no_30/04.htm (accessed on 5 February 2009).

Stern, Nicholas. 2006. 'What is the Economics of Climate Change?' *World Economics*, vol. 7, no. 2 (April–June).

Studdert, L., K. Soekirman and J. Habicht. 2004. 'Community-based School Feeding during Indonesia's Economic Crisis: Implementation, Benefits, and Sustainability'. *Food and Nutrition Bulletin*, vol. 25, no. 2, pp. 156–65.

Syroka, Joanna. 2007. *Experiences in Index-based Weather Insurance for Farmers: Lessons Learnt from India & Malawi*. Paper presented at World Bank, Innovative Finance Meeting, New York, 18 October 2007.

Tan Yingzi and Xin Dingding. 2009. '20 Million Migrants Lost Jobs: Survey'. *China Daily*, 3 February, p. 1.

Taylor, J., and A. Yunez-Naude. 2002. 'Farm/Non-farm Linkages and Agricultural Supply Response in Mexico: A Village-wide Modelling Perspective', in B. Davis, T. Reardon, K. Stamoulis and P. Winters (eds), *Promoting Farm/Non-farm Linkages for Rural Development: Case Studies from Africa and Latin America*. Rome: FAO.

The Intergovernmental Panel on Climate Change (IPCC). 2007. 'Fourth Assessment Report: Climate Change 2007 (AR4)', IPCC, Geneva, Switzerland.

UNCCD (United Nations Convention to Combat Desertification). 2009. 'Combating Desertification in Asia', Fact Sheet 12. http://www.unccd.int/publicinfo/factsheets/showFS.php?number=12 (accessed on 27 January 2012).

UNDP (United Nations Development Programme). 2007. 'Climate Change and the MDGs in Asia Pacific: Challenges and Opportunities'. *Inside Asia Pacific*, vol. 2, no. 2, July 2007. http://www.undprcc.lk/rcc_web_bulletin/Issue2/country_Bangladesh.shtml (accessed on 22 October 2008).

UNFPA (United Nations Population Fund). 2007. *Upholding the Full Spectrum of Indigenous Peoples' Rights*. Ms Rena Doña, Assistant Representative, Conference on the International Day of the World's Indigenous Peoples, 9 August 2007.

UNICEF (United Nations Children's Fund). 2006. 'Progress for Children: A Report Card on Nutrition', No. 4, May.

———. 2008a. 'Childinfo: Monitoring the Situation of Children and Women'. Statistics by Area: Child Nutrition. http://www.childinfo.org/undernutrition_weightbackground.php (accessed on 9 October 2008).

———. 2008b. 'Progress for Children: A Report Card on Maternal Mortality'. New York: UNICEF.

United Nations. 2006. *Trends in Total Migrant Stock: The 2005 Revision* (CD-ROM). Sales No. 06.XIII.8, United Nations Publication.

———. 2007. *World Population Prospects: The 2006 Revision*. New York: United Nations Population Division.

———. 2008. *Comprehensive Framework for Action*. Secretary-General's High-Level Task Force on the Global Food Crisis, United Nations, July. www.un.org/issues/food/taskforce/docs.html (accessed on 3 September 2008).

UNU (United Nations University). 2007. 'Vitamin and Mineral Deficiencies Technical Situation Analysis: A Report for the Ten Year Strategy for the Reduction of Vitamin and Mineral Deficiencies'. *Food and Nutrition Bulletin*, vol. 28, no. 1 (supplement). Tokyo: United Nations University Press.

Varangis, P., J. R. Skees and B. J. Barnett. 2002. 'Weather Indexes for Developing Countries', in Dischel, R. S. (ed.), *Climate Risk and the Weather Market: Financial Risk Management with Weather Hedges*. London: Risk Books.

World Bank. 2000/2001. 'World Development Report 2000/2001: Attacking Poverty'. Washington, D.C.: World Bank and Oxford University Press.

———. 2006. *Making the New Indonesia Work for the Poor*. Washington, D.C.: World Bank.

———. 2007. 'World Development Report 2008: Agriculture for Development'. Washington, D.C.: World Bank.

———. 2008a. *Guidance for Responses from the Human Development Sectors to Rising Food Prices*. Human Development Network, World Bank, 21 June. http://siteresources.worldbank.org/EXTSAFETYNETSANDTRANSFERS/Resources/HD_Response_Food_Prices.pdf?resourceurlname=HD_Response_Food_Prices.pdf (accessed on 3 October 2008).

———. 2008b. 'Migration and Remittances'. www.worldbank.org/prospects/migrationandremittances (accessed on 6 November 2008).

World Food Programme. 2007. *Vulnerable Group Development (VGD): Making a Difference to the Extreme Poor Women in Bangladesh through a Social Safety Net Programme*. Dhaka.

Ziegler, J. 2005. 'Report of the Special Rapporteur on the Right to Food'. E/CN.4/2005/47/Add.2, United Nations Economic and Social Council, Geneva.

———. 2008. Report of the Special Rapporteur on the Right to Food. A/HRC/7/5, United Nations Economic and Social Council, Geneva.

Further Readings

Ahmed, A., and N. Caldes. 2004. 'Food-for-Education: A Review of Program Impacts'. Washington, D.C.: International Food Policy Research Institute.

Asia Times. 2008. *Gulf States Covet Asian Farms.* www.Khilafa.com (accessed on 4 October 2008).

Barahona, C., and E. Cromwell. 2005. 'Starter Pack and Sustainable Agriculture', in S. Levy (ed.), *Starter Packs: A Strategy to Fight Food Insecurity in Developing Countries? Lessons from the Malawi Experience 1998–2003.* Wallingford: CABI Publishing.

Bellin-Sesay, F. 2003. 'Overview on Current HIV/AIDS in Asia: Some Implications for Food and Nutrition Security'. *Malaysian Journal of Nutrition*, vol. 9, no. 2. pp. 75–84.

Bollier, David. 2006. 'Bina Agarwal: The Links Between Environmental Wealth and Gender Justice', On the Commons, 22 May. http://onthecommons.org/bina-agarwal-links-between-environmental-wealth-and-gender-justice (accessed on 27 January 2012).

Carter, M., and C. Barrett. 2007. 'Asset Thresholds and Social Protection: A "Think-Piece"'. *IDS Bulletin*, vol. 38, no. 3, pp. 34–38.

Chambers, R., R. Longhurst and A. Pacey (eds). 1981. *Seasonal Dimensions to Rural Poverty.* London: Frances Pinter.

Coulter, J., D. Walker and R. Hodges. 2007. 'Local and Regional Procurement of Food Aid in Africa: Impact and Policy Issues'. *Journal of Humanitarian Assistance*, 28 October. http://sites.tufts.edu/jha/archives/category/jonathan-coulter (accessed on 27 January 2012).

Dercon, S., and P. Krishnan, P. 2000. 'Vulnerability, Seasonality and Poverty in Ethiopia'. *Economic Mobility and Poverty in Developing Countries*, vol. 36, no. 6, pp. 25–51.

De Souza, Roger-Mark, John S. Williams and Frederick A. B. Meyerson. 2003. 'Critical Links: Population, Health and the Environment'. *Population Bulletin*, vol. 58, no. 3, September.

Economist. 2008. 'The Starvelings'. 24 January. http://www.economist.com/node/10566634 (accessed on 27 January 2012).

FAO (Food and Agriculture Organization). 2005. *Rural Women and Food Security in Asia and the Pacific: Prospects and Paradoxes.* Bangkok: FAO Regional Office for Asia and the Pacific.

———. 2008a. 'The Impact of Rising Food Prices on the Poor'. ESA Working Paper No. 08–07, FAO, Rome, August.

———. 2008b. *The State of Food Insecurity in the World 2008.* Rome: FAO.

Financial Times. 2008. 'UAE to Invest in Kazakh Agriculture'. 16 July.

Gillespie, S., and S. Kadiyala. 2005. *HIV/AIDS and Food and Nutrition Security: From Evidence to Action.* Washington, D.C.: International Food Policy Research Institute.

Grain. 2008. 'Seized: The 2008 Land Grab for Food and Financial Security'. http://www.grain.org/front/ (accessed on 28 October 2008).

Gulfnews. 2008. 'Dubai Equity Firm Buys Agricultural Land in Pakistan', 13 May 2008. http://www.gulfnews.com/BUSINESS/Investment/10212801.html (accessed on 27 August 2008).

Howell, F. 2001. 'Social Assistance: Project and Program Issues', in Isabel Ortiz (ed.), *Social Protection in Asia and the Pacific.* Manila: ADB.

IFAD (International Fund for Agricultural Development). 1999. 'Rural Poverty Assessment: Asia and the Pacific Region'. Draft for Discussion, IFAD, Rome.

ILO (International Labour Organization). 2005. *The Dynamics of the Labour Market and Employment in Bangladesh: A Focus on Gender Dimensions*. Geneva: ILO.

———. 2008. 'Global Wage Report 2008/09: Minimum Wages and Collective Bargaining: Towards Policy Coherence'. Geneva: ILO.

IUCN (World Conservation Union). 2008. *Gender Makes the Difference*. Fact Sheet on Agriculture. www.genderandenvironment.org/admin/admin_biblioteca/documentos/agriculture.pdf (accessed on 6 October 2008).

Kapsos, S. 2008. 'The Gender Wage Gap in Bangladesh'. ILO Asia-Pacific Working Paper Series, ILO Regional Office for Asia and the Pacific, Bangkok.

Kelkar, Govind. 2007. 'Feminization of Agriculture in Asia'. One World Net, 24 December. http://archive.oneworld.net/article/view/156365 (accessed on 15 January 2009).

Lao Women's Union and Gender Resource Information and Development Centre. 2004. *Gender, Forest Resources, and Rural Livelihood*. Vientiane: Lao Women's Union.

Loevinsohn, M., and S. R. Gillespie. 2003. 'HIV/AIDS, Food Security and Rural Livelihoods: Understanding and Responding'. Food Consumption and Nutrition Division Discussion Paper 157, International Food Policy Research Institute, Washington, D.C.

Olimova, S. 2005. 'Impact of External Migration on Development of Mountainous Regions, 2005'. Paper prepared for a workshop on Strategies for Development and Food Security in Mountainous Areas of Central Asia, Dushanbe, Tajikistan, 6–10 June.

Pitayanon S., S. Kongsin and W. J. Janjarean. 1997. 'The Economic Impact of HIV/AIDS Mortality on Households in Thailand', in D. Bloom and P. Godwin (eds), *The Economics of HIV and AIDS: The Cases of South Africa and South-East Asia*. New Delhi: Oxford University Press.

Piwoz, E., and E. Preble. 2000. 'HIV/AIDS and Nutrition. A Review of the Literature and Recommendations for Nutritional Care and Support in Sub-Saharan Africa'. SARA Project, US Agency for International Development, Washington, D.C.

UNAIDS and WHO (World Health Organization). 2007. *AIDS Epidemic Update: December 2007*. Geneva: UNAIDS and WHO.

UN-HABITAT (United Nations Human Settlements Programme). 2008. *Urban World*, No. 1, November. Nairobi: UN-HABITAT.

UNICEF, UNAIDS and WHO. 2008. 'Children and AIDS: Second Stocktaking Report'. Geneva: UNICEF, UNAIDS and WHO.

7

Towards an Agenda on Food Security for Asia

As the number of hungry people in the Asia-Pacific region continues to rise,[1] and as competition for land and water resources and energy grows concurrently, there is no gainsaying the fact that yields in food production per hectare will need to grow dramatically, and that too with relatively fewer resources in most countries in Asia. This is particularly so when, with rising prosperity in countries like China and India, the nature of the demand for food sees even greater transition to animal protein–dominated diets, requiring more agricultural produce to generate the same amount of food. Yet the challenge facing the countries of Asia is not just to increase yields very substantially, essential though that task is, but also to impart three characteristics to the food system in 21st-century Asia, namely, *resilience, sustainability and equity*, and *poverty reduction*.

First, *resilience*. The next few decades are likely to be a period of pronounced turbulence caused by a range of drivers. One set of drivers will be the increased prevalence of shocks: sudden onset of crises, such as extreme weather events driven by climate change, or environmental disasters, or even spikes in the price of energy as seen currently and in 2008–09 (see Box 7.1).

Another driver of turbulence will be stresses: slower onset impacts such as land degradation, deforestation and desertification or gradual price inflation, that risk being overlooked by short-term policy or investment planning. These have been discussed at length in Chapter 2 on the causes of food insecurity. Then there is the risk caused by human action through ignorance or accident: think of the positive feedback loop caused by one set of countries suspending

---

[1] See Chapter 1. This section draws on Evans (2009).

BOX 7.1   Climate and Food Production in China and Its Impact Beyond

The United Nations' food agency issued an alert in early February 2011 warning that a severe drought was threatening the wheat crop in China, the world's largest wheat producer (in 2010 China produced almost twice as much wheat as the United States or Russia and more than five times as much as Australia), and resulting in shortages of drinking-water for people and livestock. China has been essentially self-sufficient in grain for decades, for national security reasons. However, the state-run news media in China, Xinhua, warned on Monday that the country's major agricultural regions were facing their worst drought in 60 years. It said that Shandong province, a cornerstone of Chinese grain production, was bracing up for its worst drought in 200 years unless substantial precipitation came by the end of this month.

The United Nations Food and Agriculture Organization said that 12.75 million acres of China's 35 million acres of wheat fields had been affected by the drought. It said that 2.57 million people and 2.79 million heads of livestock faced shortages of drinking-water. 'Minimal rainfall or snow this winter has crippled China's major agricultural regions, leaving many of them parched,' Xinhua reported. 'Crop production has fallen sharply, as the worst drought in six decades shows no sign of letting up.' Reportedly Shandong province, in the heart of the Chinese wheat belt, had received only 1.2 centimetres, or about half an inch, of rain since September. Currently, the ground in the country is so dry from Beijing south through the provinces of Hebei, Henan and Shandong to Jiangsu province, just north of Shanghai, that trees and houses are coated with topsoil that has blown off parched fields.

However, China had about 55 million tons of wheat in stockpiles as of last summer. That was equal to about half the annual harvest. And with $2.85 trillion in foreign exchange reserves, China has ample buying power to prevent any serious food shortages. 'They can buy whatever they need to buy, and they can outbid anyone,' Mr Zeigler said. China's self-sufficiency in grain prevented world food prices from moving even higher when they spiked three years ago, he said. But any move by China to import large quantities of food in response to the drought could drive international prices even higher than the record levels recently reached. 'China's grain situation is critical to the rest of the world—if they are forced to go out on the market to procure adequate supplies for their population, it could send huge shock waves through the world's grain markets,' said Robert S. Zeigler, the

(BOX 7.1 *Continued*)

(BOX 7.1 *Continued*)

Director-General of the International Rice Research Institute in Los Baños, in the Philippines.

World wheat prices are already surging, and they have been widely cited as one reason for protests in Egypt and elsewhere in the Arab world. A separate United Nations report last week said global food export prices had reached record levels in January. The impact of China's drought on global food prices and supplies could create serious problems for less affluent countries that rely on imported food.

The heat wave in Russia in 2010, combined with floods in Australia later in the year, has drawn worldwide attention to the international wheat market, because Russia and Australia have historically been big exporters. But China's wheat industry has existed in almost total isolation from the rest of the world, with virtually no exports or imports, until last year, when modest imports began. Yet it is enormous, accounting for one-sixth of global wheat output. And trouble there, is trouble for the world.

Source: Bradsher (2011).

---

exports while another attempts to build up imports. For example, during the food crisis of 2007 and 2008 (UNESCAP, 2009), when international food prices were rising, rice exports were drawn down by India, Thailand and Vietnam, further exacerbating the volatility of the food prices (ESCAP et al., 2009: 78–79). Finally, the food system could be disrupted by malicious action—for example, during conflicts or through intentional systems disruptions by terrorists or insurgent groups. Such events are rampant in South Asia and in parts of South-East and East Asia, such as the terrorist attacks in Afghanistan, Bangladesh, India and Indonesia, or the civil strife in certain provinces in China, Nepal and Thailand.

While not all of these risks to food security can be prevented, a strong focus on resilience in food supply systems can at least help to mitigate their impact when they do occur. Resilience has to be built in throughout the food value chain: from evaluating crops for their resilience to droughts or pests to assessing vulnerability to disruption of trade relationships and domestic-level supply chains. The question of the resilience or vulnerability of poor people and poor countries—typically those most exposed to risks—needs to be a prime concern for policy makers.

The second objective for 21st-century food supply must be *sustainability*. It is well known that food supply is not only vulnerable to scarcity issues; it is also often a driver of them. Poor husbandry, such as overgrazing of grasslands and steppes or over-ploughing of agricultural land, can be a major contributor to land degradation. Inefficient and wasteful use of fertilizers or water contributes directly to demand for energy resources against a backdrop of tight supplies. Profligate use of water (at times caused by faulty policies like free electricity supply) for irrigation in agriculture depletes water tables and aquifers, apart from causing soil salinization, effects seen in parts of the state of Punjab in India and in Punjab province in Pakistan. And then, agriculture and food supply chains are significant emitters of greenhouse gases (see Box 7.2), noting, however, that agriculture can also be part of the solution.

BOX 7.2   Agricultural Greenhouse Gas Emissions

Agricultural greenhouse gas emissions come from several sources.

Agricultural Soil Management

Nitrous oxide emissions account for about 60 per cent of the total emissions from the agricultural sector. Nitrous oxide is produced naturally in soils through the microbial processes of nitrification and de-nitrification. During nitrification, ammonium (NH_4) produces nitrates (NO_3). During de-nitrification, nitrates (NO_3) are reduced to nitrogen gas (N_2). An intermediate step in both of these processes is the creation of nitrous oxide (N_2O). The large increase in the use of nitrogen fertilizers for the production of high-nitrogen-consuming crops like corn has increased the emissions of nitrous oxide.

Enteric Fermentation

Methane is produced as part of the normal digestive processes in animals. During digestion, microbes in the animal's digestive system ferment feed. This process, called enteric fermentation, produces methane as a by-product, which can be emitted by the exhaling and belching of the animal. Because of their unique digestive system, ruminant animals (e.g., cattle) are the major emitters of methane. Beef cattle account for

---

(BOX 7.2 Continued)

(BOX 7.2 *Continued*)

---

about 70 per cent and dairy cattle for about 25 per cent of these methane emissions. If beef and dairy cattle numbers increase, methane emissions will also increase. Feed quality and feed intake influence the level of methane emissions. In general, lower feed quality and higher feed intake lead to higher methane emissions.

Manure Management

Methane is produced by the anaerobic (without oxygen) decomposition of manure. When manure is handled as a solid or deposited naturally on grassland, it decomposes aerobically (with oxygen) and creates little methane emission. However, manure stored as a liquid or slurry in lagoons, ponds, tanks or pits, decomposes anaerobically and creates methane emissions. Dairy cattle and swine produce about 85 per cent of methane emissions. Methane emissions will increase as the number of large-scale livestock confinement systems increases.

Carbon Dioxide from Fossil Fuel Consumption

Use of fossil fuels in agricultural production accounts for 8 per cent of the emissions from agriculture. These emissions are primarily from combustion of gasoline and diesel fuel.

Others

A variety of other sources produce greenhouse gas emissions. For example, most rice is grown on flooded fields, which prevents atmospheric oxygen from entering the soil. When rice is grown with no oxygen, the soil organic matter decomposes under anaerobic conditions and produces methane that escapes into the atmosphere.

Source: Takle and Hofstrand (2008).

Minimizing the exposure of food systems to scarcity issues through enhanced resilience is only half of the story. Food and agricultural systems will also need to be made part of the solution, both through reducing their environmental impact and (wherever possible) through contributing actively to environmental restoration.

Finally, the experience of the Green Revolution of the 1960s in the countries of the region also shows that *equity* and *poverty reduction* should be a core objective in 21st-century food supply. As we noted in Chapter 1, one reason why nearly a billion people are undernourished in Asia-Pacific today is *not* that there is insufficient food to go around, but because there are bottlenecks to accessing and distributing the food that is available (see Box 7.3). If the region's total food production were added up and then divided equally between the region's populations, then each person would have enough calories per day, which would easily be sufficient to eradicate food insecurity in Asia and the Pacific. But that is not the case: there are many millions who are undernourished, as seen Chapter 1. On a global scale, 'the number of undernourished people is almost perfectly mirrored by the billion who are overweight or obese—primarily in developed countries, but also (increasingly) among new middle classes in emerging economies' (WHO, 2008).

Thus, the real problem is one of lack of *access*[2] and failure of *entitlement* to food, which result from a variety of causes: people

---

BOX 7.3   Defeating Hunger More Than a Question of Producing More Food

Food security exists when all people, at all times, have access to sufficient, safe and nutritious food to meet their dietary needs and food preferences for an active and healthy life. People can grow the food themselves (endowment entitlements) or earn money and buy their food (exchange entitlements). In rural areas where agriculture is the main economic activity, intensified crop production would obviously mean more jobs and, therefore, lower levels of food insecurity. But there is more to the story.

Even in countries with booming farm sectors, there is rural food insecurity: landless labourers, orphans, widows, the elderly and the poorest of the poor often don't get enough to eat. Even in rural families with regular incomes, ignorance about good nutrition may result in malnourished children and pregnant women. And then there are gender barriers to access to food for women even in non-poor families.

Source: FAO (2010).

---

[2] This can be economic, social or physical access.

may be unable to grow enough food on land that they own or can access through renting and in-leasing (failure of endowment entitlement); or they may be unable to buy enough, because their income is too low or they cannot get the money needed to buy food (failure of exchange entitlement); or they cannot acquire enough food as gifts or loans from relatives or neighbours, or through government rations or aid programmes (failure of transfer entitlements) (Sen, 1981). Yield increases on their own are not enough: resilience, sustainability and equitability are vital too. If, moreover, scarcity issues mean that agriculture struggles to deliver yield increases on the scale needed, then these three other policy objectives assume even greater importance.

The Framework for Food Security

Given this scenario, a successful strategy to deal with the present and future food security needs of the Asia-Pacific region would aim at:

- Ensuring sustainable supply of appropriate food in adequate quantity, especially to close the imminent gap between demand for, and supply of, food.
- Enhancing environmental quality and the natural resource base upon which food production depends.
- Protecting people against shocks.
- Meeting the challenges of water scarcity, energy security and climate change.
- Meeting the challenge of making trade and transportation work for food security.
- Providing people with economic access to food.
- Ensuring that people have physical and social access to food.
- Ensuring that people utilize and absorb the food that is consumed, for instance through public provisioning of potable water.

As stated in Chapter 3, in order to achieve these goals, governments of the Asia-Pacific region, will need to ensure, *inter alia*, increase and diversification of food production; enhancement of general economic growth, expansion of employment and guaranteeing decent rewards for work; providing protection against from *both* idiosyncratic and covariate shocks and social protection to vulnerable subpopulations; reduction in gender-based inequalities; enhancement of literacy and access to health care as also potable water; strengthening good governance and institutions such the news media and civil society organizations.

Increasing the Availability of Food: Change in the Underlying Approach to Agricultural Production

To move towards the framework just outlined, there is a need for a new system of agriculture with an emphasis on strengthening community support and self-help, for these would be the bedrock on which communities could be galvanized to increase food production. A brief outline of such a system follows.

First, in the current age when agriculture has acquired the characteristics of business, maximizing profits through agricultural production seems to have caught on as the main goal of agriculture. This is reflected in the movement of members of farming households towards cash crops, as also to seek employment elsewhere. There has to be a conscious effort at changing this mindset. Meeting human needs has to be restored to the centre stage as the goal of agriculture, with livelihood security being accorded a very high priority.

Second, the current thinking that decisions regarding agricultural operations should be made by experts, and that only experts can question each other on these issues, is fallacious. Common people need to have access to information that enables and empowers them to question the decisions of experts, participate in the planning process by making informed choices, and undertake evaluation of development interventions for the agricultural sector and even of paradigm. Thus, a majority of the natural resources management decisions pertaining to agriculture must be made at whatever level by the primary users of resources. Technology development priorities and decision processes must be established through an interactive participatory process involving all the primary stakeholders. The passive participation of people in the agricultural sector is insufficient. There must be interactive participation, the essence of which has been beautifully articulated in another context by mass educator Jimmy Yen of China:

> Go to the people,
> Live with them,
> Serve them,
> Respect them,
> Plan with them,
> Start with what they know,
> And build on what they have.

Third, the current philosophy in the era of liberalization, privatization and globalization (the LPG era), where self-centredness and competitiveness are great virtues, and profit the only basis of our progress, needs thorough re-examination in the context of food production. We need to ponder seriously whether the Keynesian edict (discovered and reinvented several times over in different incarnations) that the more we consume, the more we contribute to development by generating effective demand, has to be subjected to searching scrutiny. There is no alternative to sharing, caring and collaboration, and the degree of cooperation and harmony that exists in our socio-economic environment is a true barometer of the quality of life. Unbridled consumption is self-destructive, as has so eloquently been borne out by the precipice of environmental calamity to which the consumption of the rich has brought the world.

Fourth, the technical parameters of sustainable agriculture have also to be changed. The new parameters for sustainable agriculture can be best summarized in three words: 'poly-culture, integration and reduction'. Poly-culture, multi-storey farming and diversification of the agricultural production system to improve the stability of food output all through the year and over the years, must replace the current trend towards monoculture. It is essential to integrate seasonal, annual and perennial crops, trees and shrubs, birds and animals, aquatic plant and animals, insects, reptiles and microorganisms, to ensure energy efficiency and low-external-input, sustainable agriculture (LEISA), as against high-external-input agriculture (HEIA). Further, all agro-chemicals and agricultural practices that cause injury to the life of the soil must be reduced and eventually eliminated to free agriculture from high fossil-fuel intensity, together with adopting biological, physical techniques and methods of soil and water conservation to restore the health of the soil.[3]

Fifth, the design parameters of sustainable agriculture have to be rewritten. It has to be recognized that every element in the agricultural process is multipurpose. For instance, a fence to ward off predatory animals can also be a source of fuel-wood, of green leaves for manuring, of food and shelter for birds and animals, and a windbreaker, as well as helping prevent soil erosion. In the design parameters, all problems are also to be viewed as potential. Termites and white ants that create havoc in villages in India like Uncha Gaon, Kadhaoli and Sagarpur (Mukherjee, 2001) are also a rich source of poultry feed. Gullies created by erosion are ideal sites for afforestation and horticulture to provide fuel, fodder and food for the people. Farm design must change its orientation from use and exploitation to conservation. Soil, water, energy and genetic resources conservation has to be built into the farm design. Finally, the design parameters of sustainable agriculture must build the farming system on the foundation of indigenous plants and animals, local knowledge and traditional farming systems. The focus of the farming system should be to improve the entire system, rather than being geared to improving only yield or income by changing a single element.[4]

Increasing the Availability of Food: Food Production under Conditions of Scarcity[5]

In finding solutions on the supply side, agriculture and allied sectors like horticulture, dairy farming, poultry and aquaculture will all be of fundamental importance. These solutions may be classified into six broad areas, namely, finance and investment, research and development, promotion of ecologically integrated approaches to agriculture, agricultural innovations, changing the

---

[3] We shall return to this issue later in this chapter.

[4] This has been elaborated later in this chapter, in the discussion of the second Green Revolution.

[5] The next two sub-sections are adapted from Evans (2009), Chapter 4.

agricultural paradigm and supporting small farmers. These areas are now discussed, in that order.

Finance and Investment

For various reasons, investment in agriculture has fallen dramatically throughout the Asia-Pacific region. Globally, the proportion of official development assistance aid going to agriculture fell precipitously between 1980 and 2006 from 17 per cent to 3 per cent. In real terms, the total amount of aid spent on agriculture fell 58 per cent over the same period (Diouf, 2008). In the Asia-Pacific region, the share of agriculture in official development assistance (ODA) fell from a little below 20 per cent in 1980 to 5 per cent in 2010. This is bad news, given that the developing countries in the Asia-Pacific region need to double their food production by 2050 in order to feed their growing populations. 'A production increase of this magnitude will require the developing world alone to invest over $ 200 billion per year in agriculture till 2050, of which almost 120 billion U.S. dollars would have to be invested in the Asia-Pacific region alone.'[6] Asia and the Pacific, home to two-thirds of the world's one billion hungry people, need increased investment in agriculture of $120 billion per annum (up from the current level of $80 billion) till 2050 to contain hunger and future spikes in food prices, according to the United Nations and Asian Development Bank (Associated Press, 2010). The UN has called for increasing the percentage of official development assistance to food and agriculture from 3 per cent now to 10 per cent within five years (Diouf, 2008; World Bank, 2008). Many developing country governments had also scaled back public support to agricultural extension services over the past two decades: in Asia, for instance, only a small percentage of public spending on agriculture goes to agriculture, making the total for the whole of Asia just $13 billion. There is already widespread consensus on the need to reverse these trends, although estimates of how much money is required vary widely. However, while greater financial deployment is essential, it is also not sufficient.

---

[6] Statement by Dr Jacques Diouf, Director-General, Food and Agriculture Organization of the United Nations, read at the two-day Investment Forum for Food Security in Asia and the Pacific, 7 July 2010.

In the context of investment in agriculture, it needs to be underscored that three kinds of public investments are needed in Asia and the Pacific: (*a*) direct investment in agricultural research and development; (*b*) investments in sectors strongly linked to agricultural productivity growth, such as agricultural institutions including extension services, roads, ports, power, storage and irrigation systems; and (*c*) non-agricultural investments to bring about positive impacts on human well-being, like reduction in food insecurity and malnutrition, which would include investments in education, especially of women, sanitation, clean drinking-water and health care (FAO, 2009a).

Research and Development (R&D)

Research and development was a central element in the Green Revolution's success. The rates of return on investment in this area are well established: a study by Yale University, for example, found that crop yields in developing countries globally would have been 19.5 per cent to 23.5 per cent lower without investment by the Consultative Group on International Agricultural Research (CGIAR). Public investment in agricultural R&D is especially important for poor countries and poor small and marginal farmers, given that private-sector R&D tends to focus on the major high-value crops, on labour-saving technologies and on the needs of capital-intensive approaches to farming, whereas most farmers in the Asia-Pacific region are small and marginal farmers who are capital-poor, and the agricultural sector is labour-abundant. Hence, R&D in the private sector may not be the best bet for small and marginal farmers in Asia-Pacific. Additionally, R&D in the private sector will be protected under intellectual property regimes and is not easily accessible to these numerous smallholders. Thus, what is required in the Asia-Pacific region is R&D in the public sector that will attempt to cater to the needs of small and marginal farmers and poor people, involving long lead times and more marginal lands where outcomes are less assured, and which will, above all, provide benefits to people who have lower ability to pay for research undertaken in the private sector. Investment in agricultural research and development should focus first and foremost on improving crop yields, and bridging the gap between research and development in the main cereals and for staples (sorghum and millet) that are critical for small farmers in the Asia-Pacific region.

It may be noted that the net annual investment in R&D in agriculture globally, up to 2050, will be US$83 billion, of which US$29 billion will need to be spent in just two countries of the Asia-Pacific region, China and India; US$20 billion in South Asia; and US$24 billion in East Asia. Second, capital stock per worker is likely to double, approximately, in East Asia and South Asia. This has implications for the labour force employed: it will reduce the number of people absorbed by the agricultural sector.

It is true that there may be limits to how much further new seed varieties can take us in matters relating to increasing food availability, but there is no escape from the fact that yields must increase. Prior to the heavy emphasis on plant hybridization, domesticated cereals devoted only a small proportion of their energy from photosynthesis (i.e., from light) to seeds. In the case of wheat, for example, the figure was around 20 per cent (Brown, 2005, quoted in Evans, 2009). Today, plant breeding has raised this proportion—the 'harvest index'—to around 50 per cent for wheat, rice and corn, through 'dwarfing' the length of the straw. However, given plants' requirements for the supporting infrastructure of roots, leaves and stems, there is a limit to how high the harvest index can go; the estimate is around 60 per cent (Sinclair, 1998, quoted in Evans, 2009). One of the 'holy grails' of agricultural R&D is, therefore, to move beyond increasing the share of photosynthate that goes to seed, towards the more fundamental innovation of improving the efficiency of photosynthesis itself. As yet, however, this goal remains a distant prospect.

This does not, however, imply that further yield increases are not possible. For example, two centuries back, Thomas Robert Malthus wrote:

> I think I may fairly make two postulata. First, that food is necessary to the existence of man. Secondly, that the passion between the sexes is necessary and will remain nearly in its present state. These two laws, ever since we have had any knowledge of mankind, appear to have been fixed laws of our nature, and, as we have not hitherto seen any alteration in them, we have no right to conclude that they will ever cease to be what they now are, without an immediate act of power in that Being who first arranged the system of the universe,

and for the advantage of his creatures, still executes, according to
fixed laws, all its various operations. . . .

Assuming then my postulata as granted, I say, that the power
of population is indefinitely greater than the power in the earth to
produce subsistence for man. Population, when unchecked, increases
in a geometrical ratio. (Malthus, 1830)

Thus, though Malthus did predict that population would expand in
geometric progression and outstrip food production, and that crises
will set in to restore the system to equilibrium, it did not happen
exactly that way because of the use of technology. Food production
kept pace with population growth. In the real world, crop yields
rarely reach their yield potential, because of such constraints as
water, nutrients, imperfect adaptation to local environments, and
pests, diseases and weeds (Evans, 1998). Thus, if these constraints
can be tackled, yield will increase over time.

In this regard, the potential of GM crops to 'feed the world' has
attracted considerable attention. Indeed, GM crops have shown
their potential to deliver improvements in crop yields. The first
generation of GM technologies focused on improving the resistance
of crops to so-called 'biotic stresses', such as weeds and pests, either
through building defences against pests into the plant itself (as with
crop varieties that contain a gene from the *Bacillus thuringiensis*
microbe, which produces a toxin to protect plants against pests
such as corn borers), or through allowing plants to work with
herbicides (as with 'Roundup Ready' crop strains). Meanwhile,
'biotic stresses' (like too much or not enough water, extremes of
temperature, salinized or acidified soils) are becoming the focus of
R&D on the next generation of GM crops. Research has also shown
that plants can be engineered to over-express the gene that allows
roots to absorb more nitrogen, thus allowing crops to produce
the same yield with a 50 per cent or even two-thirds reduction in
the amount of nitrogen fertilizer needed (Biello, 2008, quoted in
Evans, 2009). These advances have the potential to deliver a double
win against the objectives identified earlier, improving both the
resilience of crops to climate change and land degradation and
their sustainability (in particular by making them more efficient
in their use of water).

At the same time, however, there are risks associated with GM
technologies. The danger of cross-pollination from GM plants has
been documented in a number of cases and could be a serious

threat to the environment and biodiversity. The adverse impact of GM crops on friendly insects, soil microbes and water are not to be overlooked:

> Widespread use of herbicide resistant plants can lead to excessive use of herbicides and consequently to the development of resistance to herbicides in weeds. As transgenic plants containing pharmaceuticals, new pesticide genes and other properties, are being grown in experimental fields, the issue of bio-containment of transgenes gains special importance. (Byravan, 2010)

The other dangers of GM crops such as collateral damage to health have also been highlighted. The experience of the Green Revolution in the world provides considerable grounds for caution on this front—as early as 1993, excessive application of new insecticides and herbicides meant that 700 pests, 200 pathogens and 30 weeds had already developed resistance to agri-chemicals (Thacker, 1993, quoted in Evans, 2009).

The polarity of views has another interesting dimension. M. S. Swaminathan, the 'Father of Economic Ecology', is on record as saying that genetically modified organisms (GMOs) should not be grown in or marketed to the developed world, where they are not necessary, but should be created for the developing world, to meet the food needs of large populations living in poverty and to allow such nations to develop a reliable food surplus that they can sell to the developed world (Sexton, 2009).

On the balance, it seems that while GM crops may have an important contribution to make on the resilience and sustainability fronts (discussed later), at this point in time, the track record to date of GM technologies does not inspire confidence that they will make a significant contribution to raising the yield potentials of the world's main cereal crops—wheat, rice and maize—and that too without impairing human health and causing collateral damage to environment and other crops.

Adopting Ecologically Integrated Approaches

If improving the resilience and resource use of individual crops through biotechnology is one avenue for exploration, another is to achieve the same results through working with whole food-growing *systems*, rather than just with individual crops, in particular through integrating natural biological and ecological approaches—such as soil regeneration, predation and parasitism—into food production. One example of this is integrated pest management (IPM). The hallmark approach of IPM is to control pests through the influence of natural predators and parasites that prey on them, thereby reducing the need for pesticides. But IPM does not provide one-glove-fits-all solutions: it is highly space- and time-specific. Given the huge diversity among the countries of Asia-Pacific in every dimension of agriculture, from soil conditions, agro-ecological environment and climatic conditions to agricultural practices, a staggering array and amount of research is needed to establish which strategies will work for which of the different crops and in what different contexts, and an incredible amount of expertise is also required on the part of the millions of small and marginal farmers implementing the approach on the ground. The concept of IPM is nearly half a century old, but usable IPM programmes for many important crop pests are still lacking, even in developed countries. There is much that remains to be done. Second, integrated soil fertility management (ISFM) provides another approach that combines the use of both inorganic fertilizers and organic approaches such as composts, manures and nitrogen-fixing plants in order to increase yields at the same time as rebuilding depleted soils, improving moisture retention and protecting the natural resource base. Experts whose research on soil is based on this approach argue that, taken together, organic and inorganic approaches to increasing soil fertility offer the prospect of a virtuous spiral: organic methods increase the efficiency of fertilizer, while fertilizer helps increase the returns on organic methods through positive interactions on the biological, chemical and physical properties of soil. Like IPM, ISFM is highly place-, space- and time-specific. Thus, like IPM, ISFM practices typically need to be adapted from place to place, being cognizant of seasonality and the size of farm-holdings, and relying heavily on farmers' knowledge as well as access to inputs. Third, a related approach to soil quality is 'minimum tillage' (also called 'conservation tillage'). Although the plough is a traditional feature of agriculture in most Asian countries, there is increasing recognition that tilling the land as little as possible can have benefits for the soil, through minimizing energy and pesticide use and reducing erosion. Under minimum tillage systems, crop residues are left on top of the soil as 'mulch', and new seeds are simply sown into undisturbed soil.

Weeds are controlled by herbicides rather than ploughing, which reduces soil erosion and improves the soil's capacity to sequester CO_2. However, implementation of the practice of minimum tillage is concentrated in a handful of countries: most countries in the region have no more than 2 million hectares under no-till systems (Brown, 2006, quoted in Evans, 2009).

BOX 7.4 Types of Agriculture

- **Sustainable Agriculture:** This refers to the ability of a farm to produce food indefinitely, without causing severe or irreversible damage to ecosystem health (UNCCD, 2008). Two key issues are biophysical (the long-term effects of various practices on soil properties and processes essential for crop productivity) and socio-economic (the long-term ability of farmers to obtain inputs and manage resources such as labour). Sustainable agriculture integrates three main goals: environmental stewardship, farm profitability and prosperous farming communities.
- **Urban/Peri-urban Agriculture:** This is the practice of cultivating, processing and distributing food in, or around (peri-urban), a village, town or city (Bailkey and Nasr, 2000). Urban farming is generally practised for income-earning or food-producing activities, though in some communities the main impetus is recreation and relaxation. Urban agriculture contributes to food security and food safety in two ways: first, it increases the amount of food available to people living in cities, and, second, it allows fresh vegetables and fruits and meat products to be made available to urban consumers.
- **Organic Agriculture:** This is a production system that sustains the health of soils, ecosystems and people (Organic World Foundation, 2008). It relies on ecological processes, biodiversity and cycles adapted to local conditions, rather than the use of inputs with adverse effects. Organic agriculture combines tradition, innovation and science to benefit the shared environment and promote fair relationships and a good quality of life for all involved.
- **Conservation Agriculture:** Conservation agriculture is a concept of resource-saving agricultural crop production that strives to achieve acceptable profits together with high and sustained production levels while concurrently conserving the environment (FAO, 2008a). The first key principle in conservation agriculture is practising minimum mechanical soil disturbance, which is essential to maintaining

---

(BOX 7.4 *Continued*)

(BOX 7.4 *Continued*)

minerals within the soil, stopping erosion and preventing water loss from occurring within the soil. The second key principle in conservation agriculture is much like the first principle in dealing with protecting the soil. The principle of managing the topsoil to create a permanent organic soil cover can allow for growth of organisms within the soil structure. This growth will break down the mulch that is left on the soil surface. The breaking down of this mulch will produce a high organic matter level, which will act as a fertilizer for the soil surface. The third and final principle that is exercised by the FAO is the practice of crop rotation with more than two crop species. This process will not allow pests such as insects and weeds to be set into a rotation with specific crops. Rotational crops will act as a natural insecticide and herbicide against specific crops.

- **Precision Agriculture:** Precision farming is a new technology that allows farmers to look at their fields more site-specifically than before and apply inputs in a manner more specific than a blanket application. This technology saves money while holding or enhancing yield output of the field. Environmental pollution can also be reduced using this method (Kansas State University, 2009). Precision agriculture uses ICT to cover the three aspects of production, namely, *(a)* for data collection of information input through options as Global Positioning System (GPS) satellite data, grid soil sampling, yield monitoring, remote sensing, etc; *(b)* for data analysis or processing through Geographic Information System (GIS) and decision technologies as process models, artificial intelligence systems, and expert systems; and *(c)* and for application of information by farmers.
- **Industrial Agriculture:** Industrial agriculture is a form of modern farming that refers to the industrialized production of livestock, poultry, fish and crops (Agri-vision, 2007). The methods of industrial agriculture are techno-scientific, economic and political. They include innovation in agricultural machinery and farming methods, genetic technology, techniques for achieving economies of scale in production, the creation of new markets for consumption, the application of patent protection to genetic information, and global trade. These methods are widespread in developed nations and increasingly prevalent worldwide. Most of the meat, dairy, eggs, fruits and vegetables available in supermarkets are produced using these methods of industrial agriculture.

(BOX 7.4 *Continued*)

(BOX 7.4 *Continued*)

- **Biodynamic Agriculture/Ecological Agriculture:** Biodynamic agriculture, a method of organic farming that has its basis in a spiritual worldview (anthroposophy, first propounded by Rudolf Steiner), treats farms as unified and individual organisms, emphasizing balancing the holistic development and interrelationship of the soil, plants and animals as a closed, self-nourishing system (Biodynamic Farming, 2009). Regarded by some proponents as the first modern ecological farming system, biodynamic farming includes organic agriculture's emphasis on manures and composts and exclusion of the use of artificial chemicals on soil and plants. Methods unique to the biodynamic approach include the use of fermented herbal and mineral preparations as compost additives and field sprays, and the use of an astronomical sowing and planting calendar.
- **Community-supported Agriculture:** Community-supported Agriculture, or CSA, is a direct marketing alternative for small-scale growers (Adam, 2006). In a CSA, the farmer grows food for a group of shareholders (or subscribers) who pledge to buy a portion of the farm's crop that season. This arrangement gives growers up-front cash to finance their operation and higher prices for produce, since the middleman has been eliminated. Besides receiving a weekly box or bag of fresh, high-quality produce, shareholders also know that they're directly supporting a local farm.
- **Slash and Burn (Swidden) Agriculture:** Slash and burn consists of cutting and burning of forests or woodlands to create fields for agriculture or pasture for livestock, or for a variety of other purposes. It is sometimes part of shifting cultivation agriculture, and of transhumance livestock herding. Burning removes the vegetation and may release a pulse of nutrients to fertilize the soil. Ash also increases the pH of the soil, a process which makes certain nutrients (especially phosphorus) more available in the short term. Burning also temporarily drives off soil micro-organisms, pests and established plants long enough for crops to be planted in their ashes. Before artificial fertilizers were available, fire was one of the most widespread methods of fertilization.

Source: Compiled by Eric Roeder, China.

Because heavily ploughed soil releases carbon dioxide and methane as once buried organic matter is exposed, techniques such as minimum tillage can play a significant role in tackling climate change—an important point given that organic matter contained in soil is the single largest pool of carbon. Indeed, although the figure is frequently being revised upwards with new discoveries, over 2,700 Gt of carbon is stored in soils worldwide, which is well above the combined total of atmosphere (780 Gt) or biomass (575 Gt), most of which is wood. In future, attention needs to be focused not only on minimizing emissions from soil, but also on using soil actively to sequester carbon.

Livestock management is an area that warrants serious consideration. In the developing countries of Asia, livestock rearing is a multifunctional activity (FAO, 2009b). Apart from contributing to food and income generation directly, livestock acts as a store of wealth, collateral for credit and an essential safety net at times of crisis. More integrated approaches to livestock management are thus needed. They can, *inter alia*, produce dramatic improvements in the sustainability of agriculture, both through reducing the grain-intensiveness of meat production and through easing the pressure of grazing on rangelands, apart from protecting small farmers from the fallout of shocks. The incorporation of soya meal into feed rations, which produces enormous increases (in some cases a doubling) in the efficiency with which grain is converted into animal protein, is a case in point. In India—home to the world's largest dairy industry—cattle are fed on roughage such as wheat and rice straw or corn stalks. China is successfully using a similar model in the eastern provinces of Hebei, Shangdong, Henan and Anhui, which now produce more beef than the grazing provinces in the north-west.

Finally, aquaculture is also emerging rapidly as a high-potential element of the sustainable agriculture toolkit. Asian aquaculture production has expanded rapidly in recent years, and in principle this trend has the potential to deliver great efficiencies in protein production as compared to meat. This is especially important in the context of strong demand growth for protein in developing countries of Asia and the Pacific, especially China and India, where rising incomes and prosperity have allowed people to migrate towards a more animal protein-intensive diet. Globally, seafood currently represents 20 per cent of animal protein consumption, and this share is rising (Rabobank, 2008, quoted in Evans, 2009).

Adopting a New Paradigm in Agricultural Innovations

Perhaps the most striking feature of agricultural innovations in future will be a shift from the high external input-intensive model of the Green Revolution of the 1960s towards what we call the high tacit and explicit knowledge-intensive model of the second Green Revolution in the 21st century. The first Green Revolution had two distinctive features. First, while it achieved significant yield increases in the Asia-Pacific region through promotion of high external input agriculture (HEIA) (including irrigated water, chemical fertilizers, chemical pesticides and insecticides and energy use), it also brought with it several attendant problems. Second, the first Green Revolution was confined to the production of food through application of new technology in settled agriculture, what we call 'induced food production'; it did not extend itself to autonomous food production, or the production of food in micro-environments[7] and common property resources. Attention to autonomous food production is critical, for it provides food to the poor, adds variety and palate to their diet and provides much-needed protein otherwise absent from their diets.

Now, a second Green Revolution is needed, one that will not only increase yields even more than the first, but also move agriculture from high external input-intensive agriculture to high tacit and explicit knowledge-intensive food production in both induced and autonomous food production systems. The second Green

---

[7] 'A micro environment is a distinct small-scale environment which differs from its surroundings, presenting sharp gradients or contrasts in physical conditions internally and/or externally' (Chambers, 1990). Micro-environments include: home gardens (also known as homestead or household kitchen gardens, backyard or dooryard gardens); vegetable and horticulture patches (protected with wells, etc.); river banks and riverine strips; levees and natural terraces; valley bottoms (wade, fadama, vlei, etc.); wet and dry water courses; rainstreams (dividing and braiding, etc.); dry river beds (nallas, wadis, luggas, etc.); drainage lines; alluvial pans; artificial terraces; silt trap fields (depositional fields, gully fields, etc.); raised fields and ditches or ponds (especially in wetlands); water-harvesting structures in their many forms; hedges and wind breaks; clumps, groves or lines of trees or bushes; pockets of fertile soil (termitaria, former livestock pens, etc.); corners or strips sheltered by aspects of slope, configuration, etc.; plots protected from livestock; flood recession zones, small floodplains; springs and patches of high groundwater and seepage; strips and pockets of impeded drainage; lake basins; ponds including fish ponds; and animal wallow (e.g., for buffaloes).

Revolution must integrate traditional knowledge and technology with advances in modern-day science and agricultural engineering including plant genetics, plant pathology and information technology, and encompass ecologically integrated approaches, like intergraded pest and soil fertility management, minimum tillage and drip irrigation, all discussed earlier for both forms of food production.

A high tacit and explicit knowledge-intensive food production system also commends itself on grounds of resilience and equity, as it will attempt to return the 'power to produce' to farmers rather than investing the whole of it in corporate boardrooms. Under this new approach, with the application of R&D, science, innovation, local technology and local knowledge, food production will be highly intensive in output terms, but will be less demanding in terms of fertilizers, pesticides, water, energy and soil stress.

However, within knowledge-intensive approaches, there are two schools of thought. The first is the more high-tech, relying on life sciences and biotechnology to deliver increased yields, crop resilience and sustainability, in which knowledge is initially heavily concentrated at the 'top' of the process—in the laboratories of biotechnology companies and seed companies. Knowledge then moves downwards to farmers who apply the technologies in the field. The biotech companies who own the patents to engineered crop strains (and may enforce them through technologies to ensure that crops do not produce new seed by use, among others, of 'terminator technology') are also in a powerful economic position if farmers depend on their seeds for future cropping.

The second of the knowledge-intensive approaches is focused more on the entire systems of crop production rather than on individual crops, and on more ecologically integrated approaches, in which research is heavily adapted to local context. Here the line between 'laboratory' and 'field' becomes much more permeable, and the approach is more participative. It has been said that the promise of ecologically integrated approaches is to 'replace investment in chemicals and their associated pest-surveillance systems by investment in people' (Kenmore, 1996).

While both approaches can help to deliver resilience and sustainability, the latter commends itself in the Asia-Pacific context as it scores better on the equity front. Ecologically integrated approaches ultimately distribute power and autonomy outwards to individual farmers, while life sciences approaches empower seed companies, introducing in the process the potential for a dependency relationship among farmers. If we test agricultural innovations both on their technical merits and in terms of who will benefit and how these innovations will change distributions of power, the ecological approaches are better suited for Asia-Pacific conditions, where approximately 85 per cent of the farmers are small and marginal farmers.

In this context, it is important to bear in mind that the intellectual property rights (IPR) regimes under the Agreement on Trade-Related Intellectual Property Rights (TRIPS) of the World Trade Organization (WTO) need also to be re-examined for the second Green Revolution to succeed. Since much of the tacit knowledge will need to be made widely available for dissemination and intensive use, the IPR needs to be robust enough to guarantee that the community or poor and small farmers who have been using this knowledge for generations do not lose out and become eventually liable to pay royalty under patent laws, as almost happened in the case of neem and turmeric (see Box 7.5).

BOX 7.5   A Farmer's Perspective on TRIPS

I farm a 2.6 hectare area where I grow rice, corn, various vegetables, fruit trees and root crops. I am one of the 34 practising rice breeders in Magsasaka at Siyentipiko Para sa Pag-unladng Agrikultura (MASIPAG), and my farm is my laboratory. It is also here where I do on-the-job coaching for other farmers who want to learn how to do systematic breeding. It was during a meeting in MASIPAG in 1998 that I first came to hear about the Trade-Related Aspects of Intellectual Property Rights (TRIPS). After three months of educating and consulting with other grassroots members in Negros, we organized a mass mobilization against TRIPS in which 7,000 farmers and their support groups participated. As a farmer-breeder, the impact of putting these intellectual property systems in place can be summarized in four points:

- Privatization of genetic resources: TRIPS enforces the private ownership of genetic resources. This will restrict access to seeds for planting and breeding materials, a factor which is sure to affect crop improvement, both in the big institutions and on farms. TRIPS means monopoly control and ownership which is contrary to the

---

(BOX 7.5 *Continued*)

(BOX 7.5 *Continued*)

free sharing that we farmers have been practising for generations. Scientists will not be willing to exchange materials freely anymore and farmer-breeders like me will lose the most.

- Promotion of the wrong agricultural agenda: TRIPS will push agricultural research in the wrong direction: towards corporate agendas in public research, high-value export crops rather than poor people's food crops, and uniformity in the field rather than diversity. In addition, our governments' research priorities are currently shifting to modern biotechnology at the expense of research and development for sustainable agriculture, which is more useful to the majority of our farmers who are small.
- Restriction to saving, exchanging and selling of seeds: Taking care of the seed is essential for small farmers to survive. But now with TRIPs, the act of saving, exchanging and selling seeds is being prohibited. For example, in the proposed Philippine Bill on Plant Variety Protection, it states that farmers are only allowed to save, exchange and sell seeds if it is for non-commercial purposes and done in their own landholdings. But the reality is, 1.2 million farming families in the country are landless. This favours big resource-rich farms while putting aside the interest of resource-poor farmers.
- Undermining farmers' rights: TRIPS tramples on our inherent rights as farmers, which have been established for thousands of years. How can someone suddenly claim ownership over genetic resources? And make farmers pay royalties on them! We Filipino farmers have been prey to this. . . . scientists took the credit for the *burdagol* rice variety, which was in fact, developed by a farmer. Although there was no intellectual property right involved in this case, we can draw from this experience: how much worse it would be when the TRIPS regime is established.

We will not submit ourselves to such a regime, and continue to uphold our rights as farmers to do whatever is necessary to protect, conserve and improve our seeds which belong to all of us collectively and to no one privately.

Source: Excerpted from *Masipag News & Views*, Saturday, 22 September 2001.

Change in Agricultural Paradigm

Paradigms are basic belief systems and respond to three basic questions concerning the nature of reality (ontology), the nature of the relationship between the knower and the known (epistemology) and the methodology involved. It is less recognized but well entrenched that a paradigmatic shift in agriculture is warranted which the governments should facilitate. Some of the elements of the paradigmatic shift foreshadowed earlier need to be summarized here.

There is a whole gamut of indigenous knowledge systems lying in people's domain. That should be tapped and used for the benefit of agricultural production. The 'one glove fits all' solution to raising agricultural productivity by using NKP and irrigated water may need to change, at least in some cases, to accommodate centuries of wisdom to operate on the fields. From this, it follows that the method of agricultural extension service should be change from 'command and control' system towards more decentralized approaches which implies that agricultural extension has to migrate from 'lecturing' mode to the praxis of 'listening and learning' to 'transfer of technology'.

Then there are multiple perspectives of stakeholders of farming systems and they are often at variance with the outsiders' perspective. While outsiders may seek a growth in grain production, other stakeholders like the farmers' household may seek a growth in total output from their crop fields: grain, fodder, fuel, grass, legumes, fish and so on. A policy environment is needed to promote this shift from focusing only on grain output to total output from land. And countries in the region will have to move from closed systems of agriculture to more open systems that accommodate multiple perspectives, from narrow approaches to more holistic approaches.

The question that is gaining ground is: what is agricultural development all about? Agriculture is not only about raising income for the farmers but about providing a basis of life: it is a way of life. This aspect of agriculture has to be highlighted so that a system that sustains hundreds of millions of lives is not converted into a commercial machine. The powers that be need to move from 'closed' mindsets to 'open' mindsets. A series of indicators are used to monitor and work progress in agriculture and the fulfillment of those yardsticks creates an illusion that agricultural development has occurred. Such emphasis is on physical targets like total grain output or rate of agricultural growth and not on people who live on agriculture who are the real people for whom agricultural development must take place. If we misquote Max-Neef et al., the best 'agricultural development process will be that which allows the greatest improvement in the people's quality of life.... Human needs (including food) must be understood as a system; that is all human needs are inter-related and interactive'.

Questions are also being raised on who owns agricultural development and how sustainable is the process? Policies pursued so far seems to indicate that agricultural development belongs to many stakeholders and that is why state policies have a multi-pronged approach. While this process has yielded substantial results, it will not be able to do meet growing demand for food in the face of inevitable demographic changes and consumption habits discussed in Chapter 1. Agriculture must be seen as owned by the farmers, especially small farmers and policy option should be supportive of their aspirations and goals.

Conditions for Success in Smallholder Production

If small farms are indeed the answer to food security in Asia-Pacific, we need to know what conditions need to be created for smallholder farming to flourish. The answer to this complex question can usefully be organized into a few broad areas.

First, smallholder farmers interface with five different types of capital: *physical capital* (roads, bridges, buildings, plants, machinery, infrastructure and so on); *natural capital* (water, fisheries, forests, land, weather, biodiversity and so on); *human capital* (health and nutrition, education, skills, competencies); *social capital* (kinship groups, associations, mutual trust institutions); and *financial capital* (savings, credit and debt—formal and informal—remittances, pension, wages) (see Figure 7.1).

Land

Of the various kinds of capital that small farmers depend on, the most obvious is natural resources capital, and within natural resources capital, land is by far the most important.

A range of factors often undermines small farmers' access to land. One is that as land gets divided through inheritance, so farms become smaller—declining average farm sizes can be seen in many parts of Asia. With the reduction in farm sizes, the endowment entitlements of farmers to food also get reduced. Insecure property

FIGURE 7.1   A Framework for Forms of Capital Accessed by Small Farmers

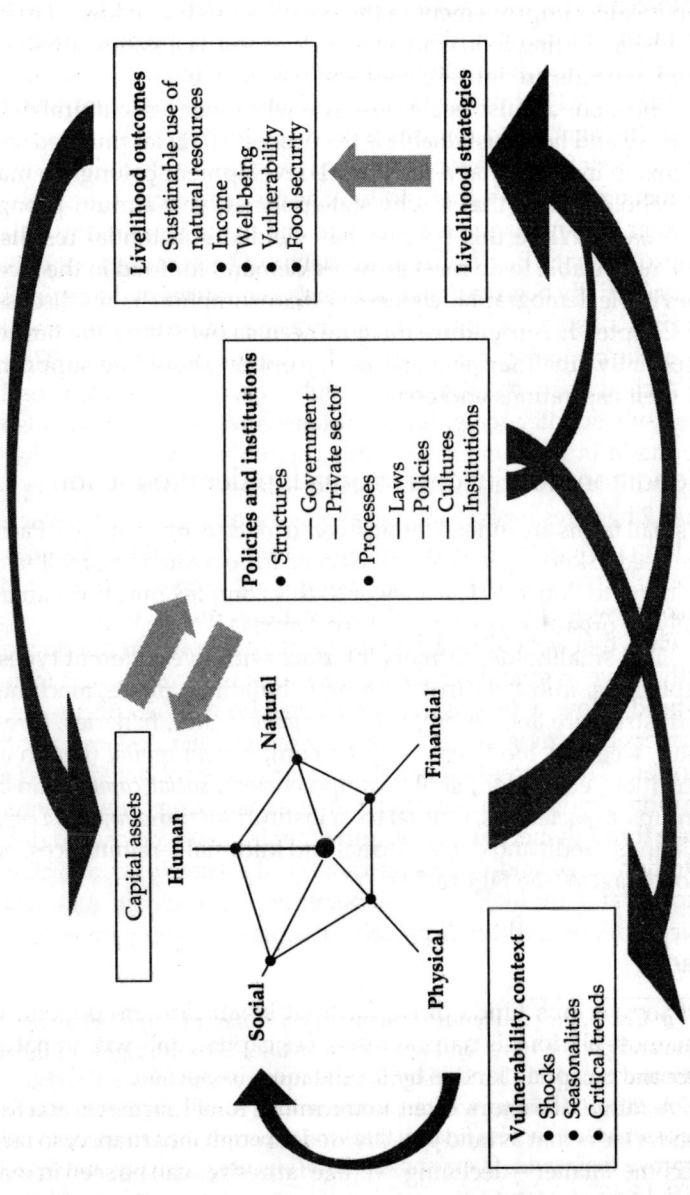

Livelihood outcomes
- Sustainable use of natural resources
- Income
- Well-being
- Vulnerability
- Food security

Livelihood strategies

Policies and institutions
- Structures
  — Government
  — Private sector
- Processes
  — Laws
  — Policies
  — Cultures
  — Institutions

Capital assets

Human

Natural

Financial

Social

Physical

Vulnerability context
- Shocks
- Seasonalities
- Critical trends

Source: Department of International Development of the United Kingdom.

rights[8] and illegal seizures of land coupled with weaknesses in law enforcement or local government are also often factors that reduce farmers' endowment entitlements to food. Women in particular often have unequal right or access to land or have insecure tenure (UNDP, 2011), thus jeopardizing their endowment entitlements even further.[9] Land reforms to address these problems can have major benefits, because long-term security in farm-holding for farmers encourages them to look at the long-term sustainability of their land management practices and make investments in land improvements. But land reforms have not been carried out or have only partially been carried out in many countries of the region like India, Pakistan and Bangladesh. In some countries, the effect has been the opposite, such as in the Philippines (Ombion, 2007).

A choice between large-scale, high-tech, centralized approaches vis-à-vis smaller-scale, labour-intensive approaches has also to be made in the context of farms themselves in the Asia-Pacific region. Most farms in Asia-Pacific—about 85 per cent—are less than 2 hectares in size, and the average area is getting smaller due to both population growth and the onslaught of natural processes like soil erosion. There are two competing lines of thought on the links between land access and agricultural productivity.

First, that agricultural development policy should not purpose-fully strive to guarantee land access to poor families, and instead should allow a process of land consolidation to occur—large farmers becoming larger by buying out smaller, less competitive farmers. The contention here is that large, consolidated farms can be more productive than small farms, capitalizing on economies of scale through mechanization and other improved technologies. In addition, large landholders can more easily absorb price and climate shocks, and are likely to have better access to credit. As a result, purveyors of credit will perceive them as less risky investments.

---

[8] In China, the land tenure system is very different. Land belongs to the state and people have land use contracts till 2028, with rights of inherit-ance and transfer. This is a significant improvement over what obtained before 2000. 'However without secure tenure, rural residents do not have the asset base to access finance that would permit then to move to cities, improve their land or expand off-farm business' (Huang and Rozelle, 2009: 16–17).

[9] See Chapter 3 for details.

Credit will then allow even more investment in yield-enhancing technology, kick-starting a cycle of ever higher productivity. Additionally, some argue that small farms in Asia-Pacific should be substituted by large farms like the Brazilian model (see Box 7.6). For example, it has been argued that

...the remedy to high food prices is to increase supply. The most realistic way is to replicate the Brazilian model of large, technologically sophisticated agro-companies that supply the world market. There are still many areas of the world that have good land that could be used far more productively if it were properly managed by large companies. To contain the rise in food prices we need more globalization, not less. (Collier, 2008, quoted in Evans, 2009)

BOX 7.6   Brazilian Adventure

Brazil's farms are sustainable, thanks to abundant land and water. They are many times the size of even American ones. Farmers buy inputs and sell crops on a scale that makes sense only if there are world markets for them. And they depend typically on new technology. Brazil's progress has been underpinned by the state agricultural research company and pushed forward by GM crops. Brazil represents a clear alternative to the growing belief that in farming small and organic are beautiful.

Source: Economist, 28 August 2010, p. 9.

The second line of thought is the opposite: that productivity is generally higher on small farms. This is mainly due to two factors: greater input use and the 'incentive' effect. Per unit area, small farmers generally invest more in providing inputs like fertilizers, water and labour into their agricultural operations, which leads to higher yields. Because small farms use only household labour to run the farm operations, while larger farms generally need to hire outside labour, household members have a greater stake in increased agricultural productivity than do hired workers. Thus, the impact of household labour is generally higher than that of hired labour. It is also argued that because the immediate challenge in the Asia-Pacific region with respect to food security is as much to increase food availability—a longer-term goal—as to improve poor people's economic and social access to food (ESCAP, 2008a), bigger farms may not be the answer. Additionally, because small farmers are more likely to practise low external input agriculture (LEIA), as opposed to the high external input agriculture pursued in large farms, small farms are more conducive to environmentally sustainable development.

There actually is a range of small farms that may employ quite different strategies. Three distinct livelihood strategies for small farmers (Darwood et al., 2006, as quoted in Evans, 2009) have been identified as under:

- 'Hanging in', where activities are undertaken by small farmers in order to maintain their livelihood at a 'survival' level.
- 'Stepping up', where investments are made by small farmers in existing activities in order to increase their returns.
- 'Stepping out', where small farmers are engaged in existing activities to accumulate assets as a basis for investment in alternative, higher-return livelihood activities.

Thus, while some small-holder farms may not get beyond subsistence levels, it is also possible for some of them to become a viable livelihood strategy, either in themselves, or as part of a broader portfolio. At their best, small farms can also deliver wider local economic benefits. Moreover, there is also evidence that small farms can become significant export earners, given adequate functions to aggregate their output. Vietnam, for example, has gone from being a food-deficit country to being a major food exporter, largely as a result of improvements in smallholder farming. As a result, Vietnam's experience of rising food prices has also been more positive than that of many other countries: the effect of costlier food in rural areas has been largely offset by increased incomes (Von Brown et al., 2008, quoted in Evans, 2009).

Both lines of argument have mountains of evidence supporting their claims. However, the conditions surrounding agriculture (like whether small farmers have access to affordable credit, availability of input subsidies, or access to appropriate technology together with extension services) determine the strategy that is likely to produce the greatest gains in agricultural productivity. It needs noting, however, that the purpose of facilitating access to land for the small farmer is not just to increase agricultural productivity, but also to reduce food insecurity among the rural poor through providing a critical livelihood resource. For this reason, many governments in Asia—even those that simultaneously promote some degree of land consolidation, particularly for growing export crops—have employed a variety of land reform strategies to increase poor farming families' access to land, from straightforward expropriation and redistribution of rich farmers' land, as in China, to the creation of 'land ceiling' laws like in the states of Kerala and West Bengal, India, under which individuals were allowed to own only a fixed amount of land, with the excess parcels purchased by the state and redistributed to poorer families; or the 'willing seller, willing buyer' strategy where the government buys land at market rates from large land owners and then assists landless families with grants and loans to purchase the land so acquired. Obviously, the ideological environment determines which form of land redistribution is exercised. While in some situations, powerful rural elites make true land reform politically impossible, such as in Pakistan and certain states in India, in others the glacial pace of land reform is leading to rising resentment among the rural landless, such as the extremist movements in eastern India.

A more moderate approach to improving land access is to concentrate on improving the legal administrative framework around land ownership and tenure. Often, rural families have only customary, not legal, title to the land they have used for generations. Without official title, powerful interests can forcibly dispossess families of their land. Improvement of the legal and administrative framework can prevent such expulsions from occurring, and also provide a basis for legal challenge if they do occur. Similarly, legal protections can protect sharecroppers and tenants (renters of land) from exploitative terms of tenancy as also eviction. One example is West Bengal, where the state government recorded the names of all those who were sharecroppers and tenants on others' land, prescribed the rent that sharecroppers and tenants were required to pay to the land owner, and enjoined that they could not be evicted at will by the land owner. This focus on linking poor families to a responsive and efficient legal and administrative structure has generally been one of the most favoured approaches to improving land access for the rural poor.

Water

Water resources are another form of natural resource capital, access to which is critical for small famers to succeed. The availability of water is central to ensuring food security. Irrigation can make the difference between one harvest and two, between half a year of food insecurity and no food insecurity. At present, only about one-third of all farms in Asia are irrigated. And most of the irrigation is under the largely inefficient flood-or-furrow system, though it is known that water productivity can be increased substantially, for example, by using overhead sprinkler irrigation (which can reduce water use by 30 per cent below flood-or-furrow systems) or drip irrigation (which typically halves water use) (Brown, 2006). Water use efficiency can increase very substantially, but, at present, very few countries in the Asia-Pacific region use drip irrigation extensively (less than 1 per cent in China and India) (Postel and Vickers, 2004, as quoted in Evans, 2009).

In some countries, including India and China, vast amounts of untapped groundwater exist. And there is a peculiar anomaly that pervades the Asian region: while surface water belongs to the state and its appropriation is subject to regulatory regimes, groundwater is private property and a free good. People are free to draw groundwater at will in most countries of Asia. As a matter of fact, lack of property rights and/or pricing mechanisms, together with inefficient subsidies for water use and/or free energy for water extraction, like in Punjab and Haryana, imply that farmers and indeed other water users lack clear incentives to use water efficiently. Small farmers and/or poor people are often the ones who lose out from unsustainable water use, especially given the fact that corruption is frequently associated with water use and irrigation. Thus, groundwater depletion along with unsustainable levels of water extraction from rivers and lakes throughout the Asia-Pacific region are problems that require immediate action, even without factoring in the impacts of climate change.

In addition, climate change is likely to intensify water problems in the coming decades, in two ways: by reducing total rainfall and by causing rainfall to become more erratic and unpredictable, with a higher frequency of extreme weather events like droughts and floods. The rainfall pattern over the last several years in many countries in Asia like China, India, Pakistan and Vietnam illustrates both of these problems: an overall drop and higher year-to-year variability. Though the debate is still far from settled, climatologists argue that a strong link exists between these trends and climate change. The increasing unpredictability of rainfall distribution *within* a growing season is also of concern. Many food crises have occurred when harvests failed despite total annual rainfall being more than sufficient, but poorly distributed over the crop cycle.

Against this background of a challenging environment of falling groundwater tables and increasingly unreliable rainfall, improved water management is crucial. Here also, effective and equitable governance mechanisms are essential. Public works and rural infrastructure programmes should concentrate on soil and water conservation activities that are appropriate and effective in each local context. As suggested by the UN's International Fund for Agricultural Development (IFAD), a 'Blue Revolution' is needed to focus policy attention on the availability and efficient use of water. Water 'harvesting' and storage systems should be made increasingly efficient, especially since the prospects of increasing water scarcity loom excessively large and since climate change is expected to lead to greater variability in precipitation and availability of water. Many developing countries in Asia-Pacific have limited capacity in tackling these issues. Thus, significant investment and an efficient regulatory framework will be required to build the infrastructure needed to harvest groundwater on a sustainable basis. Access of poor families to more water can occur only if efficiency of current use through drip irrigation and other technologies is improved.

The dramatic improvement in agricultural productivity in Gujarat, India, through effective use of water harvesting and storage systems has some lessons to offer (Bhalla and Singh, 2009).

Further, small farmers need access to markets. For example, when food prices were on the rise in 2007 and 2008, farmers did not have adequate access to markets, which is one reason why not many farmers in Asia-Pacific could reap the benefits of higher prices of food. Where institutions existed, farmers have reaped benefits, like in Vietnam. The most obvious and tangible need here is for infrastructure such as rural roads, but other kinds of infrastructure, such as communication networks that allow farmers access to up-to-date market and price information, are essential too. Improving the operation of markets themselves, and the ways in which smallholders access them, is also important.

**BOX 7.7   Good Management of Water:
The Case of Gujarat, India**

Gujarat started by planning large dam projects such as the Sardar Sarovar Project (SSP) to achieve a breakthrough in agriculture. To this day, its progress remains limited. Only a small portion of the potential command area has been covered with irrigation facilities. A lot of land continues to be rain-fed.

That's why Gujarat has embarked on a major exercise to conserve water and use it more efficiently in the fields. The most important turning point in the state's agriculture has been the innovative management of its groundwater resources. The state has adopted a combination of rainwater harvesting—that traps water that would otherwise drain away—and micro-irrigation—that supplies each drop of water more efficiently and directly to the plant. The movement has been a roaring success and stories abound of conversion of barren lands into fertile farms, rising yields and falling costs of cultivation across the state.

Between 2001 and 2006, the state government ordered the building of check dams wherever possible. The slogan was that rainwater in the fields should remain there and the water in rivulets should remain there too. There was little sense in letting all this water drain into the sea. The strategy worked. And farmers began to see a rise in water tables year after year. The state government also streamlined the supply of electricity to water pumps. Because they were getting subsidized power, farmers had little incentive to save on its use or keep pumps in good order to lower power consumption. As a result, much power was wasted. Also, power theft was widely prevalent. Further, farmers faced the problem of low-voltage power that helped nobody. To tackle the situation the state government ensured uninterrupted supply of *quality* power to farms for three-phase pumps for at least four hours a day, but only at night. This guaranteed that farmers could use the pumps only for a limited time and had to make the most of it. During the day, industry got quality power. The scope for power theft reduced. The Government of Gujarat is also studying a policy adopted in Raipur, where electricity tariffs for pumps of more than 5 horsepower invite the higher industry-level tariffs, and smaller pumps enjoy the subsidized agricultural tariffs, to encourage farmers to draw less water. It is also aimed at ensuring that all farms get irrigated water; not just those owned by rich farmers.

The Government of Gujarat was still painfully aware that the real need was to save water and use it even more efficiently. That's when the government formed the Gujarat Green Revolution Company (GGRC) to provide micro-irrigation. The new company adopted a

(BOX 7.7 *Continued*)

(BOX 7.7 *Continued*)

twin strategy. First, it made the subsidy for micro-irrigation available to all farmers, not just the poor ones. The initial investment to install the plumbing for micro-irrigation could be prohibitive. Even after the subsidy, it would come to a big sum and poor farmers would hesitate to make that investment. But for the richer ones, the subsidy made it a compelling proposition and they jumped in. This, in turn, showed the way for poorer farmers who followed.

Second, GGRC tightened norms for the subsidy scheme ensuring that companies did not sell pipes and move on to clinch more sales. It insisted that micro-irrigation technology providers also offer extension services. To ensure compliance, it introduced a series of norms—like how many agronomists must be employed for a given expanse of land, how many field visits the experts must make and even the price at which the systems could be sold. This led to widespread use of micro-irrigation.

To make greater inroads, the Government of Gujarat may order that power connections will be granted only if a farm has micro-irrigation facilities. This is so because drip, sprinkling and spraying systems that come under the definition of micro-irrigation deliver water very close to the plant or even to the roots, avoiding delivery of water where it is not needed, thereby reducing the growth of weeds. They do not allow water to seep too deep into earth.

Micro-irrigation is spreading fast across Gujarat. Back in the Dolpur farm of the Patel brothers, a modern drip irrigation system is at work. They have had to pay the full price for it because they chose to go in for micro-irrigation before the subsidy scheme was set in motion. The brothers bring scientific knowledge of soil, water and weather to their farming practices. They have even built a small soil and water analysis laboratory. 'We know that while one well would normally irrigate three to four acres, using newer automated techniques, we could irrigate 15–20 acres,' says Jitesh Patel. The modern systems have made sure that the four hours of power are plenty for their large farm. 'We sleep better, have saved on labour and also on water,' he says adding, quite proudly, 'We now enjoy a higher status in society.'

Source: Bhalla and Singh (2009).

A related question has to do with which markets small farmers are able to access, and in particular whether they can successfully break into markets for higher value added products such as fruits, vegetables, fish, nuts, spices and flowers, which now account for more than half of all developing country agricultural exports (outstripping more traditional 'cash crops' such as tea and coffee). In all of these cases, the advance of globalization means that small farmers are increasingly finding that their sales avenue is less through traditional markets and more through larger purchasers such as multinational food companies and supermarkets (which account for rapidly growing market shares in many developing countries). For example, Walmart and Reliance Industries in India are typically into the sale of these commodities, and will want to source larger volumes of produce that are far in excess of what many small farmers can produce, manage and market. Additionally, these larger corporations often require standardization and quality assurance, product safety and traceability as also stability in future supplies, which are problematic for small farmers. However, these small farmers can reap the benefits of globalization by selling directly to the larger corporation, if an intermediary can discharge the functions of aggregating produce (and playing a quality assurance role). This can lead to a win–win situation of ensuring that the standards required by large corporations are met while tackling the problems of aggregation and redressing what would otherwise be a very imbalanced negotiation. In the past, this aggregating role was often played by parastatal public-sector bodies such as marketing boards, but in many developing countries these were rolled back or abolished under the structural reform programmes mandated by international financial institutions in the 1980s and 1990s. While in many cases, these bodies were corrupt and inefficient, nevertheless they provided an important service, which is why many countries now find themselves facing a gap where parastatals used to be. The options for plugging this gap will vary from the public sector through cooperatives and public/private partnerships all the way down to private companies. In all cases, the *form* that the aggregating function takes is less important than ensuring that the *function* is delivered. Thus, there is a need for private companies (such as in Vietnam), community-based organizations (such as self-help groups in India), grassroots non-governmental organizations and farmers' organizations (such as Indigenous Multi-Purpose Cooperative [IMPCI], Namitpitan Bulo Farmers' Associations, Inc. [NBFAI], Bado Dangwa Federation of Association and Cooperatives [BDFCO] and Pide Aguid Fidilisan Multi-Purpose Cooperative [PAFMPCI] in the Philippines), that are capable of better fulfilling the role that the erstwhile quasi-government agencies played. Added to these, small farmers also need access to financial capital or credit. When they lack access to credit on reasonable terms, they typically become more vulnerable in a range of ways to financing their operations through lenders who levy extortionist interest. This reduces the ability of small farmers to invest in new technology, and innovation is diminished (eroding their capacity to increase food production). Their capacity to cope with volatility in prices is reduced (diminishing their ability to maintain stable food supply for their families), and by default, they become susceptible to predatory forms of lending, often resulting in loss of the very land for which they borrow,[10] and even death (as for example in India, where suicide among heavily indebted farmers became such a major problem and a salient election issue that the Government of India had to resort to the loan waiver scheme for farmers, aggregating ₹65,000 crores[11]). Here also, part of the picture is the rollback of agricultural services that used to be provided by developing country governments during the heyday of aid investment in agriculture, following the onset of the era of liberalization, privatization and globalization. In some cases, retail outlets of both national and international companies such as seed and fertilizer companies have stepped in to fill the gap, providing small farmers with inputs, finance and extension services, and sometimes also agreeing to contracts for small farmers' produce.[12] While on the face of it this may look fine, there is always the danger of conflict of interest, and small farmers become vulnerable to the adverse consequences of moral hazards as well as to the ignorance of people at the retail outlets. For example, in India farmers were advised by retail outlets of pesticides dealers to use a particular pesticide, which was not only inappropriate but also raised the cost of cultivation, contributing to making agriculture unviable (Mukherjee, 2001).

---

[10] 'Farmers are losing their land either voluntarily or involuntarily to debt and some are getting hopelessly poor' in Thailand. See Kanchanakak (2010).

[11] Or ₹650 billion. One crore is 10 million and ₹45 = US$1 approximately.

[12] See Mukherjee (2004) for examples in India.

At their best, improved arrangements to increase access to capital, water, extension services and marketing can put small farmers relatively on a more level playing field with larger players, and help them to achieve significant increases in productivity. It is a matter of public history in many parts of South Asia that the power imbalance between small farmers and large companies means that such arrangements can be predatory too, for example through extortionate interest rates or by leaving small farmers to bear all of the risk of crop failure (together with the attendant vulnerability to falling into debt).

As with access to markets, small farmers can improve their relative position if they can aggregate themselves into larger units, which will (in turn) be able to access credit, inputs and other services on more preferential terms. Organized groups of small farmers can also improve their political power relative to large companies and land owners, allowing them to pursue lobbying strategies with a greater chance of success. However, in the Asia-Pacific region, small farmers are too numerous to be able to organize themselves. For example, in Thailand, 24 million or 40 per cent of the population are farmers (Kanchanakak, 2010), and in India, there are 760 million farmers according to the 2001 census. For them to be organized, they need a facilitator. SorKorPor (or the Farmers' Federations Association for Development Thailand) is currently strengthening its organizational processes, information and database systems and members' capabilities in agriculture and self-help activities; Aliansi Petani Indonesia (Alliance of Peasants in Indonesia) is another example. But their reach is limited in comparison to the need, though they are big in absolute terms.

Finally though no less importantly, smallholders depend on access to knowledge. Innovations in farming methods will be fundamental to enabling farmers to deliver rising yields, at the same time as using more resilient and sustainable practices. During the 20th century Green Revolution, however, smallholder farmers struggled to access more capital-intensive innovations, and consequently often lost out in the so-called 'innovators' rent'—a point that re-emphasizes the need for access to credit. However, credit is just one part of the story. There is not only the need to invest in research in the laboratories on better seeds, better ways to tackle pests and insects and knowledge-intensive techniques of farming, but also to propagate new knowledge and highly knowledge-intensive techniques in smallholder farming sectors. The need for extension services for this purpose cannot be overemphasized. Thus, governments in the Asia-Pacific region—or the private sector—need to invest in extension services that can bring research findings to farmers and share best practices with them. This is particularly true because in many developing countries in Asia and the Pacific, these services have dwindled significantly in the wake of globalization, liberalization and privatization, a situation made even worse during the long period of neglect towards investment in agriculture. The need is huge for significant investment for these extension services to be rebuilt and rehabilitated. Furthermore, the extension services have also to be modernized in that they should not only help transfer knowledge from the laboratory to the fields and help farmers to augment productivity by adapting and adopting new technology, but also demonstrate to small farmers how they can raise their income and collaborate with one another, and even with the private sector and agricultural research. That is, extension services must migrate from the delivery mode of prescribing technological practices, to the empowerment mode of building capacity among farmers to take advantage of available opportunities, both technical and economic. For this migration in approach to be effective requires retraining and reskilling extension workers to mobilize small farmers, manage farm and non-farm incomes and so on.

Finally, smallholders need access to risk management mechanisms. Social protection systems can be used effectively by small farmers and producers in this regard. These issues will be discussed in detail in the next section.

Sharply Focused Social Protection: The General Case

If food supply is one half of the story, the other half is all about access to food: who gets to buy it and how it is traded; who cannot get food even if there is no shortage of buying power; and who is physically prevented from accessing food due to a variety of reasons. High food prices, lower wages, social barriers and age-related disabilities hurt poor consumers hardest, whether

BOX 7.8   Technical Options for Adaptation of Rice Crop to Support Food Security

- *Selection of appropriate planting date to manage risks of temperature variations*: Germination and emergence of rice seedlings are more likely to be governed by maximum and minimum temperature. The selection of the appropriate date for planting the rice crop holds the key. As temperature varies from month to month, it is possible to select the right date for crop establishment in such a way that the reproductive and grain-filling phases of rice fall into those months with a relatively favourable temperature. This would minimize the negative effect of temperature increase on rice yield. Efforts to collect and disseminate the information on month-to-month variation in temperature regimes in major rice-growing areas, therefore, are essential for helping rice production to adapt to climate changes.

- *Selection and development of appropriate rice varieties to manage risks associated with temperatures, salinity, droughts and floods*: Rice varieties have different abilities to tolerate high temperature, salinity, drought and floods. Rice varieties with a high level of salinity tolerance have been utilized to expedite the recovery of rice production in areas damaged by the tsunami. The selection of appropriate rice varieties is, therefore, another technical option for adaptation to global climate changes. Also, the development of rice varieties that have not only high-yielding potential, but also a good degree of tolerance to high temperature, salinity, drought and flood, would be very helpful under the environment of global warming.

- *Optimization of high CO_2 concentration for higher yield*: The high CO_2 concentration present in the atmosphere under global warming could be harnessed to increase the productivity of the rice crop. The grain yield of IR8, for example, was significantly increased with carbon dioxide enrichment before and after heading. C4 plants, such as maize and sorghum, are more productive than C3 rice and wheat, because C4 plants are 30 to 35 per cent more efficient in photosynthesis, especially when the level of CO_2 concentration in the atmosphere is high. Cloned genes from maize may be required to regulate the production of enzymes responsible for C4 synthesis to alter the photosynthesis of rice from C3 to C4 pathway.

Sources: Nguyen (2004); Peng et al. (2004).

or not urban or rural social protection systems exist to overcome problems of access. Over the centuries, many western countries developed mechanisms that reduced the risk of food insecurity, including transport networks that allowed food to be shipped easily, sophisticated storage facilities, agricultural technologies that produced massive crop surpluses and so on. The development of these mechanisms was fuelled by tremendous increases in wealth, a result of conquest, rapid scientific innovation and, originally, sheer geographical luck.

During the first half of the 20th century, experimentation with various other mechanisms for preventing food insecurity, including developing grain reserves, bans on food exports out of deficit areas and price controls, were attempted. These interventionist policies were motivated by evidence that markets were unable to guarantee food security, especially in poor rural areas with pronounced seasonality in agricultural production. Against the backdrop of theses early initiatives, modern governments have experimented with a wide array of initiatives towards providing social protection, which have proved successful in tackling food insecurity. Figure 7.2 illustrates some of the good practices, arranged into the categories of 'emergency assistance', 'the social protection safety net' and 'agricultural livelihoods development'. Emergency assistance measures are targeted at people who are suffering from food insecurity due to natural or man-made calamities

FIGURE 7.2   Intervention Framework for Social Protection

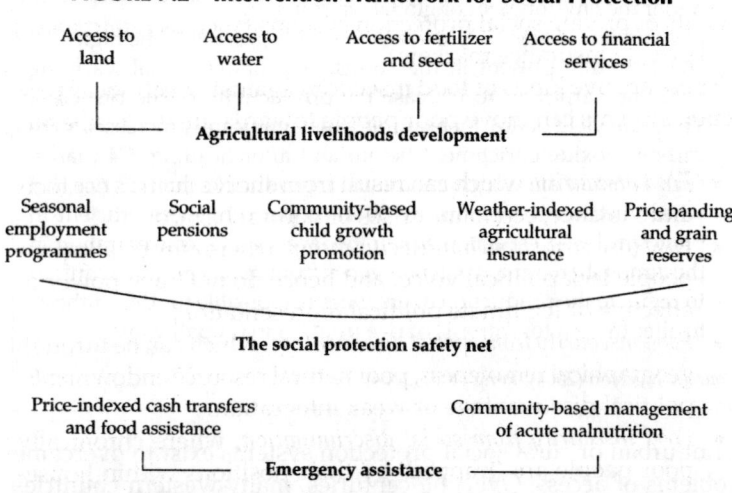

Source: Adapted from Devereux et al. (2008).

and need immediate help. The social protection safety net attempts to prevent families from falling into food-insecure situations in the first place, through a mix of employment, nutrition, price control and other policies. Finally, agricultural livelihoods development initiatives focus on improving productivity through better access to key inputs for food production. These initiatives are long-run measures working towards a future where rural households have high and stable enough incomes that ensure food security through both endowment and exchange entitlements, and where the social protection safety net will rarely need to come into play. Taken together, the group of ideas shown in Figure 7.2 represents an intervention framework for fighting food insecurity.

Within the broad term 'social protection' as currently understood, there is an enormous diversity of mechanisms and approaches apart from those shown in Figure 7.2. In acute crises, emergency safety nets such as food aid or humanitarian relief are a form of social protection. But social protection approaches are also used in much longer-term contexts: for example, cash transfers (which may be unconditional, or linked to conditions such as school attendance, working or accessing health care), asset transfers, vouchers, pensions, and transfers of inputs such as fertilizers.

Social protection systems have lately gained significant attention. Part of the reason for their growing salience is an increasing recognition that the chronically poor tend not to have access to the benefits of private social protection systems (such as remittances) or to private insurance markets.

There are five kinds of food insecurity against which social protection systems can move poor people towards greater resilience:

- *Food insecurity*, which can result from shocks such as conflicts and violence, economic crises or natural hazards.
- *Food insecurity from limited citizenship*, where chronically poor people lack political voice, and hence do not have power or effective or legitimate political representation.
- *Food insecurity from spatial disadvantage*, which can be through geographical remoteness, poor natural resource endowments, political disadvantage or weak integration.
- *Food insecurity from social discrimination*, where chronically poor people are 'trapped' by their positions within households, communities and countries.

- *Food insecurity from poor work opportunities,* either where employment opportunities are limited, or where the employment available is exploitative or of low return.

These factors need to be taken into account in developing a social protection system. The success of existing social protection approaches, where they have been put in place, shows that, if well designed, they can achieve their aim of building resilience to shocks at individual, household and community levels, though there is a long way to go. Despite calls for universal access to social protection systems by 2020, the fact remains that most people in Asia-Pacific lack access to any form of social protection. For example, social protection even in South-East Asia has been low, though gradual steps towards universality through social insurance or publicly funded schemes are in evidence. In the Republic of Korea, a comprehensive set of welfare policy instruments has emerged, including unemployment benefits and unconditional cash transfer programmes, with the idea of social rights underpinning a new rationale for public assistance. Similarly in Thailand, short-term safety net and social fund programmes were introduced including the landmark programme '30 Baht' Health Plan (this has since been made free), which provided the basis for the current universal scheme. But coverage across the sub-region is still highly variable (Cook, 2009), and systematic comparative data are lacking. A common agenda or set of instruments around 'social protection' driven by international experience has also been slow to emerge.

The foregoing is particularly important because the existing instruments of social protection are often gender-insensitive and/or inefficient and lacking in range, reach and depth or some combination of these. For example, the widow pension scheme in India provides only ₹250 per month,[13] if the widows get it at all, and the scheme covers a very small percentage of widows,[14]

---

[13] This is less than US$6 per month. See Chapter 3, Box 3.4 for further details.

[14] There are similar examples in other regions as well. For example, China's basic health insurance covers only 30 per cent of its 1.3 billion population. The system is being revamped and the amount that each person covered will get as subsidy is about $17 per year starting in 2010.

and is riddled with informal taxation. And even these inefficient instruments come with a high cost: subsidies, outright grants (like pensions) and price support can create havoc with government finances, while price control, though beneficial to food consumers, concomitantly carries the unintended effect of reducing farmers' incentive to produce more food.

What international action is required to build up the capacity of social protection systems? First, a note of realism is important. Apart from the sheer number of people who have access to social protection, there is also the point that the international development sector's enthusiasm for social protection systems is relatively new. Much more evidence is needed about which forms work best in which contexts; as the *Chronic Poverty Report* puts it, the period up to 2010 'must be treated as a genuinely experimental phase'.

Second, significant investment—of time and political will as well as money—is needed to reform and increase the capacity of international humanitarian aid systems. And then again, more integrated ways of working are also needed among the development actors working for food security. In addition to the need for better coordination within the international humanitarian community, better integration is needed between humanitarian relief agencies and the international community focused on development more broadly. At present, humanitarian relief and development assistance are often seen as largely discrete spheres by practitioners in both areas. Social protection, on the other hand, blurs the line between the two.

In the context of fragile states with limited capacity where poverty reduction and political stability are threatened by a whole range of factors including climate change, it is essential that the international community move from rapid reaction to investment in longer stability without tripping itself up. The issue transcends the work of one agency. There is an urgent need to work out the operational implications of a more seamless approach to social protection.

But above all, moving to this more integrated approach requires the international community to work with developing country governments at both national and local levels. Social protection systems at their best can have a transformational political impact, both through progressive social change and through building a social compact in which the state acts to reduce people's risks in return for their commitment to the state. If, on the other hand, the international community allow itself to fall into the very real trap of supplanting states' responsibilities, then that impedes this process of evolution towards greater state accountability and legitimacy.

The international community can help to answer and allay these concerns, through, for example, providing predictable financing (where their record falls far short of their rhetoric). But resource transfer and technical assistance on their own are not enough. The international community also needs to think intelligently about *influence:* about their own position as *de facto* political players, the nature of drivers of change in the wider society that open up opportunities for pro-poor change, and how the international community can partner with other progressive political actors in promoting social protection provision.

Social Protection: Agriculture Per Se

As we have mentioned in Chapter 6, improved agriculture is one of the best forms of social protection, as it provides livelihoods and employment, especially for women in and a key pathway out of poverty[15] and food insecurity. During economic down-turn, agriculture provides enormous respite. Investment in agriculture, therefore, increases food availability, economic access to food (especially for women), and provide good social protection. Making agriculture more viable will also reduce poverty and food insecurity, both through improved availability of and access to, food. Strategies for revitalizing agriculture may help the rural poor in several ways including, by connecting them to cities and markets; improving delivery of health and education; diversifying agriculture; introducing insurance against crop failures and price declines; and promoting social mobilization to influence agricultural policy. Strategies which seek to facilitate the rural poor moving out of agriculture, for example, by training them to enter labour markets and promoting the rural non-farm sector, hold great promise (ESCAP, 2008b).

---

[15] World Development Report (2008).

Social Protection: Employment

As mentioned in Chapter 6, employment is a social protection measure and is crucial for food security. Generally, managing the interdependence between consumption and income is the key to ensuring food security[16] as the need to have access to income and other resources to access food can not be over-emphasized. Employment opportunities during shocks are even more important as it is crucial to enable people to purchase food and maintain assets. Guaranteed employment for a minimum period, during disasters such as public employment guarantee schemes like food-for-work or cash-for-work programmes, acts as insurance, for example, as they help the poor protect their valuable productive assets caused by unexpected income loss, unemployment, or other adverse effects.[17] Food security outcome of employment for women is better because women have a higher propensity to allocate incremental income to food, especially their children. Along with employment, measures to reduce and ultimately eliminate wage discrimination are good social protection measures to ensure that all sections of population benefit equally from the employment opportunities, leaving none at greater risk of food insecurity. Elimination of wage discrimination will additionally help the already-disadvantaged subpopulations often relegated to do the most insecure and '3-D jobs', generally in the informal sector without enforcement of labour laws, pension funds and workmen's compensation.

Social Protection: *Ex Ante* Management of Shocks

We have seen in Chapter 6 that large sections of population in the Asia-Pacific Region are vulnerable to several shocks (man-made or nature-induced) causing food insecurity. Where food insecurity may have already existed and is compounded by sudden adverse shocks, much can be gained from improved *ex ante* risk management and insurance and insurance-like responses. Devising counter cyclical risk management policies and programs, prior to the onset of natural disasters, is a problem that governments of the Asia-Pacific region must tackle. Flexible targeting, financing,

---

[16] Sen (1997).
[17] Jalan and Ravallion (1999) and Barrett et al. (2004).

and implementation arrangements should be the hallmark of such counter cyclical risk management policies and programs. Shocks can be of two kinds: *idiosyncratic* and *covariate.* Though the distinction is important but, as we saw in Chapter 6, the two can merge into one another and a growing body of empirical evidence suggests that idiosyncratic risks are the dominant causes of food insecurity in this region. (The main sources of idiosyncratic and covariate risks are shown in Table 2.6 in Chapter 2.) And a majority of the people who are food-insecure are the ones who suffer from idiosyncratic shocks caused by water-borne diseases caused by lack of access to potable drinking-water and other health facilities; transitory shortfalls in access to and utilization of food and damage to assets and income. It is, thus, important that governments of the region tackle idiosyncratic shocks to prevent an overwhelming majority of food-insecure individuals or households from suffering especially when they fall outside 'humanitarian emergencies' associated with covariate shocks. The need for effective, ubiquitous and continuing *de jure* insurance programs, whether through financial innovations such as micro insurance, index insurance schemes or community-based insurance programmes, is huge.

Public distribution system for providing food to the poor and vulnerable groups has to be made more effective and effective. These systems need to be universalized for acutely disadvantaged groups like the old, infirm, disabled, widows and ethnic minorities. In this regard, community-managed alternative public distribution system (APDS),[18] for food security and other attendant benefits, need to be strongly supported, including the creation of an appropriate regulatory environment that creates an environment conducive to the spread of APDS on a large scale.

Social Protection, Small Farmers and Food Production

Social protection to small farmers is especially important, and has beneficial effects on agricultural production leading to improvements in food security. Various instruments can alleviate their

---

[18] See the Chapter on community-based responses.

liquidity constraints, create demand for their products, create multiplier effects throughout the local economy, deal with the vagaries
of seasonality, and help risk taking by small-holder. Devising these
instruments is a big challenge.

The Challenge of Alleviating Liquidity Constraints of Small Farmers: Cash and Food Transfers and their Multiplier Effects

As discussed earlier, lack of access to liquidity during agricultural
operations is a major hindrance for farming, forcing smallholders
to purchase only a part of the required inputs, eventually leading to
lower endowment entitlement to food. Cash transfer programmes
have been found to prevent adoption of damaging coping strategies (such as sale of assets and removing children from school);
relieve, at least partially, the liquidity constraints allowing small
farmers to accumulate productive assets and boost household
investments in farming as well as non-farm micro-enterprises,
thereby unleashing untapped productive and income generating
potential, all contributing to betterment in food security. A small
percentage of transfers are even invested in fertilizer and seeds,
leading to higher food production. Thus, the governments of the
region must meet the challenge of making available cash to small
farmers in time.

Food transfers to small farmers also have a bearing on the farming sector in Asia, as they increase labour supply for agricultural
work, wage work and business activities Such transfers also help
local production, labour markets and consumption patterns when
food for food transfer is locally sourced or from elsewhere within
the region. Local purchases of food for school meals have been
found to stimulate food production by augmenting demand, not
only for staple crops but also for vegetables, meat, eggs and dairy
products. Such transfers should be well-targeted and well-timed
as a social protection measure to reach households and small farmers who are already priced out of the market, to stem depletion of
productive assets, jeopardizing long-term food security.

Cash transfers are likely to have more positive secondary and
multiplier effects than food transfer, because of the 'multiplier-
accelerator effect', which apply equally to transfers to economically

inactive groups and to transfers to small farmers. But, the synergies with agriculture are likely to be higher in the case of latter, as discussed in Chapter 6. To have large distributional impacts of economic multipliers a number of factors, including the openness and structure of the local economy and its linkages with urban centers and other large markets have to be increased. The expenditure patterns of different groups receiving cash transfers may have to be influenced to be pro-farming. Given that there is sound evidence in support for localized multiplier effects of social transfers, considerable attention needs to be focused on this aspect of a food security agenda.

Small Farmers, Timing and Seasonality

Given the seasonal character of *agricultural production*, facilitating access to inputs for smallholders who face seasonal cash constraints cannot be over emphasized is a critical function of any state policy. Subsidies like the proverbial fertilizer subsidies, and power subsidy for irrigation as also free seed distribution have had their fair share of criticism as seen in Chapter 6 but they have increased food production in many countries in the Asia-Pacific Region including bigger countries China and India and relatively smaller ones like Thailand and Vietnam, helping both household and national food securities. The case for providing resources to small farms, in different ways including through microfinance provisioning, must ne on the national agenda of the countries of Asia.

Household food security of small farmers has traditionally been undermined by seasonality of fluctuations in food and asset prices, which raises the cost of food while at once reducing the market value of assets sold to buy food, at 'distress prices'. For allocation of productive resources efficiently, in the face of uncertainty in commodity markets, and preventing adoption of risk-reducing strategies which may lead producers, consumers and traders to engage in diversification into lower value crops but more stable production, not using purchased inputs, and not trading in remote locations, needs to be prevented. There are advocates who disfavour 'interventionist' measures in favour of market-based solutions, but because market-based solutions can not help those who are out of the market, state intervention to stabilize price volatility is a must. Hence, countries of the region, especially large countries like China and India will have to intervene in grain markets to ensure price stabilization for the benefit of small farmers and eventually of the consumers. Like seasonality in agricultural production there is also a *seasonality in labour market*. This means agricultural labour remain unemployed for considerable periods of time in a year, called the lean season. 'Employment-based safety net', food- or cash-for-work programme (like the National Rural Employment Guarantee Programme, in India), provide smallholders a supplementary source of food or income for consumption smoothing purposes during lean season. And quite apart, under such programmes the assets constructed by the public works activities do boost agricultural production by enhancing market access and soil fertility. The governments of the region need to devise such programmes so that availability of jobs in public works programme may not induce smallholders to divert their own labour away from vital own-farm activities. One way out is to restrict such programme during periods of high agricultural activity. Governments have to manage the trade-off between social protection for immediate consumption needs and longer-term returns to agriculture. Thus reducing the vulnerability of small farmers to seasonal variations (in labour demand, agricultural production, food availability and prices and asset loss) through efficient social protection interventions needs to be a major focus of public policy in the Asia-Pacific region.

Predictability and Management of Risks and Shocks

We have discussed in Chapter 6 that farmers face idiosyncratic risks and covariate risks of every description ranging from food-fuel inflation seen in 2008–09 to civil strife and fire. While there are various preventive socio-economic measures ranging from macroeconomic changes to specific taxes, subsidies and incentives that can be place, there is a need to instal certain social protection interventions to help small farmers affected by both kinds of risks under reference. And the synergies between social protection and agricultural policies is very powerful in the context of risk reduction.

Predictable and regular social protection, including insurance covers, perceived by the farming community as a dependable source of income, boosts productivity by enhancing risk-taking behaviour of small farmers: risk-averse small farmers will be more than willing to increase investment in agricultural productive activities, which are otherwise risky but more productive. Predictable social protection acts as an insurance against shocks. In the 'Employment Guarantee Scheme'[19] in the state of Maharashtra in India, providing low-waged unskilled manual labour in rural areas for anyone willing to work, released scarce resources hitherto held as precautionary savings to more productive investments: farmers in Maharashtra started planting more of higher-yielding as against drought-tolerant crops. Similarly effective post harvest technology including crop storage systems reduce risks associated with price fluctuations and reduce post-harvest losses and help farmers get better prices for their produce, if they are net sellers of food. Thus effective post harvest technology reduces risk aversion and increase their investment capacity of small farmers. Governments must devise social pretction programmes that have the required insurance effects such the ones discussed above. Two kinds of social protection measures especially commend themselves in this regard for their reach and implementability: improvements in agriculture and employment generation programmes and *ex ante* management of shocks though insurance mechanisms of various kinds. This is a challenging prospect but it will become more so because of rising cost of inputs, along with demographic changes and weakening or vanishing traditional forms of social capital.

Generally, various kinds of insurance against idiosyncratic risks are available, such as life insurance, health insurance and draft animal insurance, but they are very expensive for those who are most likely to suffer these risks, namely, poor farmers, especially the small ones. Insurance against covariate risks, like crop insurance, is available, but access to these remains relatively rare in the smallholder sector, though it can help smallholders to reduce their exposure to both commodity prices and weather-related risks. Insurance against covariate risks are not generally available as the risks are too large, barring a few publicly funded schemes (like crop insurance in India and China, which is almost 100 per cent subsidized), and where available they are prohibitively expensive.

---

[19] This scheme has now become the Mahatma Gandhi National Rural Employment Guarantee Scheme applicable all over India.

Pure crop insurance where it is not subsidized by 100 per cent for smallholders has failed for a number of reasons: high transaction costs, moral hazard, adverse selection, covariate risk and delayed payouts, all of which make private crop insurance economically unviable for insurers and inaccessible or unresponsive to client needs.

Index insurance has been used increasingly in the past decade, particularly in countries that do not have a history of traditional crop insurance.[20] Index insurance can be used to mitigate a wide variety of weather risks, from inadequate rainfall to catastrophic losses from typhoons or earthquakes. Index insurance is insurance where claims paid are based on the value of an index, normally weather-related, rather than actual losses. Typically, there is a trigger and a limit which define the range of values over which claims are paid. The trigger is the point at which payments begin. Once the trigger is reached, the payment increases incrementally with the value of index up to a limit. Once the limit is reached there is a full payout. The claim amount is normally calculated by multiplying the payment rate by the amount of cover purchased. For example, the payment rate per hectare is multiplied by the number of hectares insured. In most index insurance programmes, everyone insured within a specific area is paid at the same rate and there is no need for the insured to lodge a claim, as it is automatically triggered according to the value of the index.

The most common index is volumetric rainfall, although a wide variety of weather-related indexes have been used. Indices can also be constructed from aggregate statistics such as area yields (area yield index). An example of this is county yields in the United States and district or village yields in India. In these programmes, the trigger is normally set just below the long-term (trend-adjusted) average county or district yield. A few contracts have used the normalized difference vegetation index (NVDI), which measures the level of photosynthesis of ground vegetation, which is a proxy for vegetation condition and growth.

Key advantages of index insurance programmes over traditional crop insurance include: no loss assessment costs, availability of data

---

[20] Purchasers of index insurance range from governments to small-scale farmers.

for setting rates, speed of payouts, and the fact that loss assessment is normally on the basis of objective, publicly available data. Moral hazard is also reduced as the behaviour of the insured cannot affect the index measure. For many developing countries, farm-level loss assessment, required in traditional crop insurance, is impractical due to the size of the farms, which is often less than 1 hectare. In some cases, loss adjustment costs have proved to be higher than the risk premium. Speed of payout is particularly important when payments are used for disaster response, for example, in drought-related disaster that confronts small farmers. Normally, aid is distributed once a drought is imminent, but with index insurance the lack of rainfall could trigger payments during the growing season, making the disaster response more effective and often cheaper.

The main disadvantage of index insurance is basis risk, which occurs when the claims paid do not match actual losses. Contract design and index selection are key factors in minimizing basis risks. Careful design of the insurance contract can reduce the amount of basis risk and improve the affordability of the product. Thresholds, limits and insured values can be set at values that make the premium affordable. Designing the contract involves identifying and quantifying the risk to be insured, deciding on the how the index will be measured and monitored, and pricing the risk. Pricing requires the ability to model the probability of payments, and this depends on the quality and length of the data inputs. Crop growth models are often used to determine the most critical growth periods for different crop types.

In order to reduce spatial basis risks, insurers often stipulate a maximum distance from the measurement instrument and the insured location. A distance of 20 kilometres between the weather station and the insured farm is often used for rainfall index products, although this is fairly arbitrary and most countries will not have sufficient weather station coverage. A contract can have different risk layers, an example of which is livestock insurance in Mongolia.

There are several impediments to implementing any programme of index-based insurance. First, in general at least 30 years of historical data are required to develop a probability distribution for an index and to show the future trends in weather patterns. While this can be improved upon in future, complex techniques will be needed to simulate historical index distributions and to include climate change predictions. The lack of measurement infrastructure also needs to be addressed if index insurance is to be expanded significantly. Satellite data may be a cheaper alternative to weather stations. Second, there is also a lack of regulatory and legal standards. The International Association of Insurance Supervisors (IAIS) has yet to produce guidance on how insurance laws, regulations and practices may need to be adapted to include index insurance. Regulators should be actively involved in developing the market for index insurance. Governments in the Asia-Pacific region will need to drive regulatory changes, infrastructure needs and data collection. They can also subsidize the design and development costs of products.

In a larger sense, many of the tasks associated with climate change adaptation also come under the broad heading of risk management (see Box below on 'Balance Sheet of Selected Food Distribution Programmes in South Asia'). The range of actions that may be necessary to adapt to a changing climate is immense, and

---

BOX 7.9   Flood Insurance in the Mekong Delta

The Vietnam Bank of Agricultural and Rural Development (VBARD) was offered index-based flood insurance by Bao Minh Insurance Corporation, a domestic insurance company, which in turn obtained reinsurance from Paris Re. The VBARD required the insurance to cover loan defaults when farmers suffered crop damage due to flooding. The index was the water level at the Tan Chau gauging station on the upper Mekong River. Flood modelling has shown that flooding in the Dong Thap province is highly correlated with water levels measured at the station. The flood event index is the maximum three-day moving average of daily water level at the Tan Chau station during the period of cover. The trigger level is 2.8 metres, and claims are paid for each centimetre of water above that level with maximum payout at 3.5 metres. Historical analysis suggests that the trigger will be reached every 1 in 7 years. The product was commercially priced and no subsidies were involved when calculating the price. The insurance has competition from the government, which provides recapitalization for VBARD in the case of portfolio and business interruption losses.

Source: Hellmuth et al. (2009).

BOX 7.10   Development of Index Insurance in India

Rainfall insurance was first introduced in 2003 by ICICI Lombard General Insurance Company and BASIX, a rural credit provider. The programme covered 230 peanut and castor bean farmers within one province during the monsoon season. In 2004, the programme was expanded to three more weather stations and to include cotton farmers. In order to reduce basis risk, the contract design was refined to include a three-phase payout that weighted the importance of rainfall during specific stages of crop growth. BASIX also began to train loan agents to sell the product. The product was further refined so that the start date changed depending on the start of the monsoon period. Cover did not start until at least 50 mm of rain had fallen. This insured that soil moisture conditions were sufficient for planting. The success of the BASIX programme encouraged other insurers to offer similar insurance products. Index crop insurance is now available in 13 states, insures around two million farmers, and 2.7 million hectares. In 2009–10, the sum insured was US$875 million with an associated premium of US$77 million. The majority of sales are through a network of credit institutions. A study in 2007 found that the main factors affecting rural households' willingness to buy index insurance were: understanding of the insurance product, participation in a pre-existing network, perceptions of future weather risk and access to other risk management strategies. The growth of index insurance has led to a large investment in meteorological infrastructure by both the government and by private companies. India currently has around 3,500 stations collecting rainfall data, all with good historical data. However, it is estimated that the country needs at least 5,000 automatic weather stations and 20,000 automatic rain gauges for a successful, comprehensive weather index programme.

Source: United Nations Department of Economic and Social Affairs (2007).

will differ dramatically across geographical contexts, from different crop strains to mechanisms for harvesting and storing rainwater for use during dry spells or droughts. Furthermore, according to some experts there is a disjuncture in time between adaptation and climate change: while climate change usually occurs over a very long run, adaptation measures generally are much more short-term measures. This reinforces the importance of access to knowledge and institutional arrangements for dissemination of knowledge for ensuring that innovative ways of adapting to climate change are disseminated as widely as possible and that newer ways of adapting are constantly devised and formulated.

There is, consequently, a move towards 'weather-indexed' insurance where the insurance is against a local index, say, rainfall shortage or days of hailstorm or snow or frost, which are correlated with harvest outcomes and farmers get compensated if the index reaches a 'trigger' level, irrespective of crop losses. Quick disbursement of payments allows farmers to smoothen their consumption following a poor harvest, without adopting strategies such as selling productive assets. Because 'weather-indexed' insurance covers households and farms are credit-worthier, investments in productive assets and higher-yielding crops are also promoted However, high costs and adoption on a commercial basis remain the challenges of 'weather-indexed' insurance. Additionally, the premium for such insurance may be too high for smallholders. This is the basis for arguing that financing for 'weather-indexed' insurance should partake the nature of a social protection measure.

Social Protection and Climate Change

We have discussed in Chapter 6 how climate change have affected and will affect large number of people in the Asia-Pacific region, including the fact that a rise in water stress is o likely to affect 185 million to 1 billion people in South and South-East Asia. The socio-economic impacts will be significant because of its geographic exposure, reliance on climate sensitive sectors, low incomes, and weak adaptive capacity. We also saw that Asia-Pacific contains some 1.7 billion hectares of arid, semi-arid, and dry sub-humid land, with degraded areas including the expanding deserts in several countries, the steeply eroded mountain slopes and the deforested and overgrazed highlands in some parts of the Region. Since dry-land agriculture constitutes a large proportion of food production in the Asia-Pacific region, the projected decline in agricultural productivity and other natural resource-based livelihoods will trigger, *inter alia*, relatively large income losses and food insecurity. In the absence of public policy actions, climate change will expose households in Asia-Pacific region toward more and more frequent covariate shocks as also greater uncertainty about them, posing significant problems for people in adopting a coping mechanism that works.

Managing climatic shocks have traditionally been the responsibility of households and community-based institutions with little external support. This has to change as large and repetitive covariate climatic events could overwhelm the risk coping capacity of many community-based institutions. There are unresolved issues surrounding the design of external support for adaptation measures to climate change relating to governance and design adaptation measures for ensuring pro-poor local climate action. But these can no longer wait. Governmental efforts to enhance community resilience need to feature more prominently in the response to climate change induced food insecurity. Governments must ensure that adaptation responses cover a wide range of sectors and draw on a variety of approaches and disciplines. And the responses must be at national or international level (such as development of more heat and rain tolerant seeds), and at the very local (like adoption water saving techniques). Thus, governments need to identify the most effective instruments of social protection to support community-based adaptation to climate change as also implement them at the right level. Arguably, while community-based adaptation could be the most important element of the adaptation response at the local level, social protection could contribute more to national-level responses.

In this connection, governments have to put in place catastrophe safety nets. Though there are benefits and challenges of catastrophe safety nets, they commend themselves because even a system of subsidized insurance is preferred to post-disaster aid, and because the reinsurance market in the region is not yet prepared to commit sufficient and affordable capital to markets serving the poor. In this regard, governments must take care that catastrophe safety nets are closely coupled with a risk management programs like a vulnerability assessment.

Food Reserves and Stocks

First, all countries ought to be able to agree on the basis of recent experience that, whatever their stance on the degree of openness or protection that they favour, volatility of food prices is a common enemy. Sudden swings in markets and changes in trading conditions can create political contexts in which political constituencies find themselves suddenly disadvantaged, and are as such fertile ground for outbursts of violence, civil unrest or conflict, which a range of countries have already experienced as a result of the food price spike during 2009. Therefore, policy makers should examine options for creating buffers in the international trade system, to make it more resilient to shocks and stresses.

The 'traditional' approach has been for governments to hold food stocks at the national level, and with adequate overseeing this approach can deliver benefits. The Chinese government, for example, holds significantly higher grain stocks than many other countries, and argues that this has helped it to mitigate the impact of global turbulence on Chinese consumers. In many other countries, however, government-held grain stocks have been much less effective, due to a range of reasons including inefficiency of the public sector, politicization or capture by interest groups, or outright corruption.

However, stocks can still be held at levels other than the national, and by agencies other than central governments. In many cases, the local or community level can work well; aid agencies such as Oxfam have for many years invested in community grain banks that protect farmers and consumers alike against market fluctuations through buying just after harvest (when prices are low) and then selling stocks during lean seasons, at a price below market cost but with sufficient margin to cover management costs. The experience of the Deccan Development Society is instructive in this regard. Similarly, region-level stocks can also make sense. For example, the East Asia Emergency Rice Reserve (subsequently renamed ASEAN Plus Three Emergency Rice Reserve Programme), which was started as an improvement over the ASEAN Food Security Reserve, had a pilot programme beginning in 2004 and ending in 2010, that provided disaster assistance and malnutrition programmes in Cambodia, Indonesia and Myanmar (Hermida, 2007). With the rice stocks increasing substantially to a total of 787,000 metric tons, the Rice Reserve Programme is expected to be more effective. Discussions are on to make the reserve under reference permanent, with an ASEAN+3 meeting held in October 2010 to finalize an agreement (Rahil, 2010). The South Asian Association of Regional Cooperation (SAARC) has also established the SAARC Food Bank. The most obvious approach to the problem might be to create a physical, public, regionally managed grain reserve. A framework to establish such a reserve under the name and style of the 'Asia-Pacific Grain Security System' could be as follows (while national governments may adopt appropriate policies[21] to ensure grain security in their own countries):[22]

1. Establish an agriculture-related international social service structure to provide technology, information, consulting and marketing, and transportation services across the region. This will implicitly require that a regional food distribution system be developed, supported by a regional transport system in line with what is ultimately envisaged in the Asian Highway Network[23] and the Trans-Asian Railway. This needs to be established to ensure that people, even in isolated and alienated parts of the region, have access to food-grains.

2. The recent spurt in globalization has meant that global rules play an increasingly important role in the lives of people everywhere, including small farmers in countries of Asia and the Pacific. These farmers have the potential to feed not only themselves, but many others. Without international rules that

---

[21] Including, among other things, measures for enhancement of general economic growth, expansion of employment and decent rewards for work; diversification of production (all to enhance employment and livelihood access); enhancement of medical and health care (to improve health status); availability of potable water and sanitation (to allow better food absorption); arrangement of special access to food on the part of poor people, including deprived mothers, the aged, disabled and small children; spread of basic education and literacy especially among women (to bring down fertility rates, *inter alia*); strengthening democracy and the news media (to generate political incentives in preventing hunger); and reduction in gender-based inequalities (to improve equitable intra-familial access to food). See Sen (1997).

[22] Some of these measures were suggested for China by Ding (2004).

[23] This will include upgrading the 17 per cent of the road in the Asian Highway Network which is classified as below standard to at least class III level.

provide justice for the farming community, especially for small farmers, the prospects of long-term solutions to grain security are dim. International agricultural trade rules in Asia and the Pacific must be altered to ensure that the livelihoods of small farmers are strengthened, which means ensuring that farmers can sell their crops on their local markets and have opportunities to export to other markets. The countries of the Asia-Pacific region must agree to remove protectionism and market segmentation across countries in the region. A unified grain market for the region should be developed, in which all farmers and all entities should be encouraged to enter.

3. Set up an Asia-Pacific International Grain Bank (APIGB) of appropriate size and spread headed by a board nominated by member countries, to allow grain surplus countries to sell their grains to it and allow deficit countries to access grain from it at times of distress. Run, *inter alia*, public distribution systems (PDS) for indigent people. Some examples of PDS in the Asia-Pacific region have important lessons to offer in this regard (Box 7.12).

The broad features of the Asia-Pacific International Grain Bank shall be as follows:

- The APIGB would buy grains which are culturally accept-able,[24] from grain-surplus countries of the region at remu-nerative prices to be determined by its board every year, and sell these grains at affordable prices to the grain-deficit countries of the region, which may stand in need of food-grain.
- The grains bought up by the APIGB can be stored at places where they were purchased for being moved to deficit countries when the contingency arises. This will obviate the need to move large contingents of food stocks across the region, noting that in some cases it is easier to move grains internationally than nationally.

---

[24] Fortunately, rice is eaten throughout the Asia-Pacific region, which will obviate the need to hold a wide variety of grains.

BOX 7.11 Balance Sheet of Selected Food Distribution Programmes in South Asia

Country	Programme	Coverage and Targeting	Benefits	Drawbacks
India	Fair price shops (1970s through the present)	State-wise Extensive: urban and rural areas in the state of Kerala.	Food and agriculture prices became more stable than world market prices. The programme provided price support and improved equity among grain producers.	The programme was partly successful in terms of its spread and in terms of achieving its basic objectives. The programme was untargeted.
Pakistan	Subsidy & ration (1970s): Ration books were used and food was distributed through ration shops. Subsidy & ration (1943–87): Food was distributed through ration shops with quotas being imposed according to supply availability and location.	Narrow coverage; about one-third of the population mostly urban, one shop per 2,000 people.	Rationing was positively co-related with child nutritional status.Initially poor people benefited by obtaining food at subsidized prices.	Bogus ration cards were issued which were diverted to the free market at considerable profit. The subsidies were (at least in part) utilized by corrupt officials and mill owners. The floor quality declined due to which many people did not obtain ration: 16 per cent of the targeted population availed of this subsidy.
Sri Lanka	Coupon system (after 1979): Coupon issued on the basis of income and family size.	Covered an estimated half of lower-income population, urban and rural.	The programme was well targeted.	The contribution of food stamps to the household budget was only 11 per cent of the total household budget in the poorest quintile and 7.2 per cent in the highest quintile.

Source: Mahbub Ul Haq Human Development Centre (2002), with modifications.

BOX 7.12   Sen on Democracy and Famines

'Elections and the possibility of public criticisms make the penalty of famines affect the rulers as well—not just the starving victims. Even the provision of health care can be substantially influenced by the nature of the political process and deeply disrupted by military and sectarian governance. Much does depend, however, on the enterprise with which democratic freedoms are used by the citizens.'

Source: Sen (1997).

- If there are national granaries which store the buffer stocks of national governments,[25] the APIGB could establish a contractual relationship with them.
- A credit system and features of a grain market made up of spot transactions and futures transactions should be incorporated in the APIGB to ensure regional grain security.
- Repayment of grain loans to the APIGB by countries which would avail of credit facilities would be either in convertible currency or grains borrowed.
- Initial financing for the APIGB should be provided out of the accumulated reserves.

4. Establish Sub-regional Grain Insecurity Early Warning Systems (for example, for South Asia, South-East Asia and Central Asia), to warn member states of the concerned sub-regions of any potential risks emanating from natural and/or man-made disasters and of other impending dangers, such as volatility of international grain prices, and help them to ward off potential crises in grain security. The Sub-regional Grain Insecurity Early Warning Systems would be linked with other related early warning systems like the tsunami early warning systems being set up, because disasters are a major cause of grain insecurity.

---

[25] Such as the warehouses of the State Trading Corporation of India in different parts of India.

5. Enter into a binding international agreement among countries of the region to cooperate so that 'Green Box' support policies, within the framework of the World Trade Organization, across the region are strengthened through such measures as joint ventures to increase investments in productivity-enhancing technologies, agricultural infrastructure, marketing infrastructure and quality control. These could, together with national measures, usher in a 'Second Green Revolution' for Asia and the Pacific.

6. Agree on a Food Safety Protocol for GM foods and establish a reference group for testing the safety of GM food and crops to be cultivated in the region, recommending their release or otherwise to member countries. This is particularly important because people and grains move across borders without going through the formal process, so that GM crops may be transferred from one country to another without the host country even being aware of such transfers.[26]

7. Intellectual property rights may be reworked so that the rights of the farmers are not trampled on, and their intellectual property rights are protected, as already discussed earlier in this chapter. This may even require renegotiation of the TRIPS agreement.

Eliminating the Seven Gender-based Food Inequalities

Eliminating gender-based food inequalities is too important a topic to be over-emphasized. We have discussed these in detail in an earlier chapter. Suffice it to say here that gender-based inequalities leading to food insecurities must be treated as a development issues. Governments should, therefore, adopt a multi-sectoral programme foe ensuring food security for women, comprising, *inter alia*, social security (including *de jure* and *de facto* insurance of different dimensions

---

[26] For example, contamination and cross-breeding of GM papaya have been detected recently in Kamphaeng Phet Thailand. See Wongruang (2005).

especially for women farmers), affirmative actions (including reservation of seats for women in all legislatures as a fair outcome and realization of the benefits of law), changing the legal architecture affecting women and making right to food, education, health care and information for women, justiciable rights, with women themselves being considered as potentially active agents of major social change rather than as solicitors of social equity. Governments ought also to make guaranteed employment for specified number of days a legal right for women especially for women having depend-ents with disabilities, elderly, widows, migrant women, women categorized as internally displaced persons and women-headed-households and for women facing discrimination of any kind.

Good Governance

Establishing good governance is an essential prerequisite for ensuring food security, both to ensure the right kind of incentives and to manage the different kinds of interventions that are necessary for food security. An essential part of good governance is the preponderance of democratic systems. Apart from the fact that famines do not occur in democratic countries with a relatively free press and active opposition parties, because the political executives will have incentives to prevent food insecurity to garner votes in the next round of elections, democracies also provide platforms for diverse interest groups, who seek policy actions from the government that promote food security, to lobby on issues having bearing on food security in the last analysis. For example, the 'Right to Food Campaign' in India is mobilizing public opinion for a bill that will eventually become the Right to Food Act, which would not have been possible in a non-democratic setup.

International Actions

International actions could include improved technical assistance to weaker economies on negotiating long-term land supply agreements. An area for immediate action by the international community should be a rapid, clear-headed assessment of whether, and in what circumstances, large-scale investment projects for land acquisition in low-income countries by overseas purchasers looking to boost the security of their food supply will actually deliver development benefits. Such deals are already becoming a significant feature of the global food landscape. For example, China (which has 7 per cent of the world's arable land but feeds 20 per cent of its population), which does not have enough arable land to produce adequate food, has entered into long-term food purchase agreements, land leases or land purchases in countries with surplus land. China is reported to have acquired the owner-ship or leasehold of 1.24 million hectares of land in the Philippines and 700,000 hectares in Laos; and the United Arab Emirates is said to have acquired 900,000 hectares in Pakistan. South Korea is also reported to have acquired 690,000 hectares in Sudan. China's Ministry of Agriculture has proposed acquiring offshore land as a central policy objective similar to the country's existing stance on purchasing rights to overseas energy resources. China was 'negotiating deals . . . to buy more than 2 million hectares of land in countries as far flung as Mexico, Tanzania and Australia. The United Arab Emirates is seeking some 800,000 hectares in Pakistan alone, while Saudi Arabia is negotiating for 1.6 million hectares in Indonesia' (Montero, 2008). More recently Daewoo, a South Korean conglomerate, has signed a lease agreement with Madagascar to acquire 1.3 million hectares there (fully half of that country's arable land). Often these agreements or purchases are not by governments but by companies, like in the case of a South Korean company buying land in Madagascar or Gulf countries buying such lands in Indonesia. Such purchases displace small and marginal farmers and have the potential of rendering them into landless agricultural labourers. They are also germane to social problems in the countries where land is purchased.

These partnerships could help to deliver capital investment in infrastructure, technology and productivity gains in low-income countries while also driving poverty reduction through rural growth. But there is no guarantee that they will do so, and there are real risks that the benefits could be highly concentrated among a few land owners in the investment-receiving countries, without achieving wider wins for poor people—replicating problems that are well chronicled in other commodity sectors such as oil and min-ing. The case of the South Korea–Madagascar land deal appears to exemplify this situation: reports in the *Financial Times* suggest that

Daewoo expected to pay nothing for the deal, with the benefits for Madagascar limited instead to employment creation.

The problem of limited developing country governmental capacity to negotiate such investment deals with the corporate giants of developed countries in an environment of adverse bargaining power for the recipients, also increases the potential danger for poor deals for the recipient countries, jeopardizing the interests of poor people even further.

Consequently, the international community should urgently undertake a review of third-country investment programmes in land acquisition for food production and the circumstances in which they can deliver development wins, which would bring valuable additional analytical capacity to bear on an important immediate-term issue. The basic premise should be 'Buy food in exchange for technology transfer.'

There is a need to remedy the situation by allowing countries which do not have access to adequate arable land to guarantee food security of their people by investing in low-income countries with surplus land through an 'exchange agreement'. Such agreements would enjoin that the countries needing to jack up their food availability would bring into the host country capital, infrastructure, agricultural technicians and cutting-edge technology in agricultural engineering to produce much more food in the host countries than the latter could do by themselves. The donor country in return will be assured a supply of food equivalent to an agreed proportion (say half) of the increases in food production caused by injection of 'new men, material, capital and technology' during the payback period, and will have priority in buying food after the payback period is over. Producer countries that lack the capability to negotiate these complex and innovative exchanges need to get assistance from international agencies like FAO and IFAD to develop these capabilities, pretty much in line with what some of these agencies did in building up national capacities for negotiating WTO accession. These institutions should also ensure that producer countries are aware that this support is available. All these measures will of course be contingent upon developed country agricultural trade liberalization and expansion of food aid in cash (to be used to purchase food in developing countries), thus investing in the agricultural sectors of developing countries at the same time.

Another international action could be to establish an Asia-Pacific International Food Agency (APIFA).[27] There is need for an integrated response to calls for emergency food assistance, instead of several piecemeal efforts being made by different sub-regional groupings like SAARC and ASEAN in the form of grain banks and emergency reserves. The core mission of the Asia-Pacific International Food Agency would be to coordinate collective action in future food crises, through a response system-based on strategic food reserves in member countries, similar to what is being done by the International Energy Agency in the wake of the oil crisis. The APIFA could be an affiliate of UNESCAP or FAO in the region. It would not act as a mechanism for price support for producers, a role best performed by national governments, or a permanent system for managing food aid, which role lies in the domain of other agencies like WFP.

A third action could be providing protection against protectionism. Under the triple crisis of food, fuel and finance, countries are scrambling towards protectionism, more so in the agricultural sector where export restrictions were introduced by nearly 20 countries (FAO, 2008b), including in the Asia-Pacific region by India, Vietnam, Pakistan and Thailand. (Thailand even considered developing a cartel of rice-exporting countries along the lines of OPEC.) Trust in the world markets is on the wane, in as much as some countries are embarking on policies that border on autarchy, despite knowing that food self-sufficiency and food security are different things and that self-reliance, not self-sufficiency, in food is a robust insurance against the covariate shocks to which agriculture is prone. For trade in agriculture to command support, importing countries' legitimate concerns of security of food supply need to be addressed while ensuring that food-exporting countries do not hurt domestic consumers by complying with binding commitments to meet the needs of importing countries. The hibernating Doha round of trade negotiations should be resurrected to explore the potential for a new set of WTO rules on controls on export and imports of food, to concurrently manage the seemingly conflicting

---

[27] A similar idea by way of an Asian Grain Security System was proposed in Mukherjee (2008), and in Chapter III of ESCAP et al. (2005), which is germane to the concept of the Asia-Pacific International Food Agency discussed here. See also Hayes and Choudhary (2010).

needs of food-exporting and food-importing countries in times of crisis and rebuild the international food policy architecture (Hayes and Choudhary, 2009).

From the outset, it ought to be manifest that a wide range of solutions is needed. Thus, for instance, separate farming systems are required for irrigated areas, for arid and semi-arid areas, for lowland in humid areas and for uplands in humid areas. The 'one jacket fits all' system that many countries were pursuing under the aegis of the Green Revolution is unlikely to yield results.

An area which warrants immediate attention is the development of five codes.[28]

1. A *drought code* has to be developed to minimize the adverse impact of drought, laying down both what measures are to be taken once it is known that droughts are likely to set in, and the measures to be taken once drought sets in. The drought code should include adaptation of crop life–saving technologies and contingency plans to change the cropping pattern according to moisture availability. It could also lay down how to implement alternative strategies depending on weather conditions. A seed reserve of the substitute crops could be maintained as a measure of crop security. In this connection, the famine codes in India are a good example, however imperfect they may have been. They are a collation of guidelines developed by the various 'famine commissions' established to investigate the spectacular failures of the governments of the time in reacting to several catastrophic famines that devastated India following the imposition of British rule in 1858. The codes were not fully implemented until the early decades of the 20th century, but are in many ways the precursor of anti-hunger interventions as we know them today, emphasizing large public works employment programmes like NREGS, free relief to those unable to work, and the relaxation of taxes during times of impending famine. The codes were also innovative in that they recognized the seasonal pattern of food crises: when food prices rose more than 40 per cent above 'normal' for the time of year,

---

[28] The first three codes were briefly mentioned in Swaminathan (2008).

it was termed as a 'scarcity rate' and was seen as a sign of potential famine, and therefore, time to activate the codes. The famine codes also represented an acceptance that the state had a moral and legal duty to protect poor and powerless people against the worst consequences of their poverty and vulnerability. By so doing, they introduced the idea that people receiving assistance had legal entitlements, and were not merely unfortunate beneficiaries.

2. A *flood code*, both to prevent excessive distress and damage and to promote post-flood rehabilitation plans for food production, needs to be in place. For example, the code could include a 'flood strategy' to cover the entire river basin area and promote the coordinated development and management of actions regarding water, land and related resources. It should be noted that mitigating the risk of flooding in upstream or headwater areas involves a wide range of innovative, small-scale solutions; on the other hand, in lowland flooding, warning periods and the duration of flood events are longer and large-scale measures have to be taken. The flood code must promote transnational efforts to restore rivers' natural flood zones in order to reactivate the ability of natural wetlands and floodplains to retain water and alleviate flood impacts. It should devise ways and means to deal with all flooding-related problems: flooding, rising groundwater tables, sewage network disruption, erosion, mass deposition, landslides, ice flows, pollution, etc. The code may lay down protocol for regular flood forecasting and warning as a prerequisite for successful mitigation of flood damage. A scheme for planning for specific preparedness to alert, rescue and operate safety measures must be implemented at all levels, including the public, by maintaining regular basic information and continuous on-going training actions. With appropriate and timely information and preparedness, everyone who may be vulnerable to flood events should be able to take—if possible—their own precautions, and thus seriously limit flood damages. The code could lay down the parameters for provision of food aid to those affected by flood, support to animal property of the affected population, provision of supply of seeds for quick-growing varieties of alternative crops, and rehabilitation of agricultural land

BOX 7.13 Strategies to Eliminate Hunger

Track 1: Strengthen productivity and incomes	Linkages-maximizing synergy	Track 2: Provide direct access to food
Diversification and growth of the economy.	Democratic governance Vibrant civil Society	Mother and infant feeding
Low-cost, simple technology (water management, use of green manures, crop rotation, agro-forestry, poly-culture)	Strong 'fourth estate'	Supplementary nutrition to children (such as midday meals in schools) and pregnant women
	Local food procurement for safety nets	Unemployment and pension benefits
Rural infrastructure (roads, electricity, etc.)	Support to rural organizations	Food-for-work and food-for-education programmes
Provision for improved irrigation and soil nutrition	Primary health care and reproductive health services	Targeted conditional cash transfers
Natural resource management (including forestry and fisheries)	Prevention and treatment of HIV/AIDS	Food banks and food distribution system for indigent people (safety nets)
Market and private-sector development	Asset redistribution (including land reforms)	Emergency rations
Food safety and quality	Education, especially for girls and women	
Agricultural research, extension and training	Potable drinking-water	
Support rice for agricultural produce	Enabling legal framework (right to food/fair wages, etc.)	

Sources: FAO (2004); Mukherjee (2004).

damaged by flood waters. It may include a compensation system to support the victims of flood disasters to restore their livelihoods and their living conditions within the shortest possible time. Insurance solutions at the private or public level or subsidence by the state, which reinforces solidarity, should be furthered.

3. A *good weather code* is needed to maximize benefits from good weather for agriculture, such as raising nurseries of appropriate plants so that in years of excessive rainfall, extensive tree-planting and sand-dune stabilization drives can be undertaken in arid areas. The code may include provision for R&D for developing and discovering seeds that are flood-resistant and drought-resistant.

References

Abdulai, A., Christopher B. Barrett and John Hoddinott. 2004. 'Does Food Aid Really Have Disincentive Effects? New Evidence from Sub-Saharan Africa'. *World Development*, vol. 30, no. 10, pp. 1689–1704.

Adam, Katherine L. 2006. 'Community Supported Agriculture'. National Centre for Appropriate Technology, NCAT Agriculture Specialist, ATTRA Publication IP289.

Agri-vision. 2007. 'Trend of Industrial Agriculture'. Agriculture Guide, 28 November. http://www.agri-vision.net/trend-industrial-agriculture (accessed in March 2009).

Associated Press. 2010. 'Asia Needs More Farm Investments to Feed Hungry'. 7 July. http://www.businessweek.com/ap/financialnews/D9GQ5G4G0.htm (accessed on 13 September 2010).

Bailkey, M., and J. Nasr. 2000. 'From Brownfields to Greenfields: Producing Food in North American Cities'. *Community Food Security News*, Fall 1999/Winter 2000, p. 6.

Barrett, Christopher B., Christine M. Moser, Oloro V. McHugh and Joeli Barison. 2004. 'Better Technology, Better Plots or Better Farmers? Identifying Changes in Productivity and Risk among Malay Rice Farmers'. *American Journal of Agricultural Economics*, vol. 86, no. 4, pp. 869–88.

Bhalla, G. S., and Gurmail Singh. 2009. 'Economic Liberalisation and Indian Agriculture: A Statewise Analysis'. *Economic and Political Weekly*, vol. 44, no. 52, 26 December, pp. 34–44.

Biello, D. 2008. 'Oceanic Dead Zones Continue to Spread'. *Scientific American*, 15 August.

Biodynamic Farming. 2009. 'What Is Biodynamic Farming?' http://www.biodynamic-farming.com/?p=3 (accessed in March 2009).

Bradsher, Keith. 2011. 'U.N. Food Agency Issues Warning on China Drought'. *International Herald Tribune*, 8 February.

Brown, Lester. 2005. *Outgrowing the Earth: The Food Security Challenge in an Age of Falling Water Tables and Rising Temperatures*. London: Earthscan.
————. 2006. *Plan B 2.0: Rescuing a Planet under Stress and a Civilization in Trouble*. New York: Norton.
Byravan, Sujatha. 2010. 'The Inter-academy Report on Genetically Engineered Crops: Is It Making a Farce of Science?' *Economic and Political Weekly*, vol. 45, no. 43, 23 October, pp. 14–15.
Chambers, Robert. 1990. *Microenvironments Unobserved*. Gatekeeper Series 22, International Institute for Environment and Development, London.
Chronic Poverty Research Centre. 2009. 'The Chronic Poverty Report 2008–09: Escaping Poverty Traps'. London: Overseas Development Institute.
Collier, P. 2008. 'Food Shortages Think Big'. *Times*, 15 April.
Cook, Sarah. 2009. *Social Protection in East and South East Asia: A Regional Review*. Sussex: Institute of Development Studies.
Darwood, A., R. S. Wheeler, I. MacAuslan, C. P. Buckley, Jonathan Kydd and E. W. Chirawa. 2006. 'Promoting Agriculture for Social Protection or Social Protection for Agriculture: Policy and Research Issues'. Discussion Paper, Future Agricultures Consortium. http://www/futures-agricultures.org.
Devereux, S. 2002. 'Can Social Safety Nets Reduce Chronic Poverty ?' *Development Policy Review*, vol. 20, no. 5, pp. 657–75.
Devereux, S., and S. Coll-Black. 2007. 'Review of Evidence and Evidence Gaps on the Effectiveness and Impacts of DFID-Supported Pilot Social Transfer Schemes'. *DFID Social Transfers Evaluation*. Brighton: Institute of Development Studies.
Devereux, Stephen, Bapu Vatila and Samuel Hauenstein Swan. 2008. *Seasons of Hunger*. London: Pluto Press.
DFID. 2000. *Sustainable Livelihoods Guidance Sheets*. Department for International Develoment. www.livelihood.org/info/info_guidancesheets.htm as quoted in 'The Sustainable Livelihoods Approach', Input Paper for the Integrated Training Course of NCCR North-South Aeschiried, Switzerland (9–20. September 2002), compiled by M. Kollmair and St. Gamper, Development Study Group, University of Zurich (IP6), July 2002. http://www.nccr-pakistan. org/publications_pdf/General/SLA_Gamper_Kollmair.pdf (accessed on 7 March 2012).
Diouf, J. 2008. 'The Challenges of Climate Change and Bioenergy'. Address to the High-level Conference on World Food Security, Rome, 3 June.
Ding, Shengjun. 2004. 'Grain Key to China's Success in Achieving National Stability'. *Jakarta Post*, 19 February, p. 7.
Dorward, A. 2006. 'Markets and Pro-poor Agricultural Growth: Insights from Livelihood and Informal Rural Economy Models in Malawi'. *Agricultural Economics*, vol. 35, no. 3, pp. 157–69.
ESCAP (Economic and Social Commission for Asia and the Pacific). 2008a. *Towards Sustainable Agriculture and Food Security in the Asia Pacific Region*. Bangkok: United Nations.
————. 2008b. *Economic and Social Survey of Asia and the Pacific 2008*. New York: ESCAP.
ESCAP, ADB and UNDP. 2005. *A Future within Reach: Reshaping Institutions in a Region of Disparities to Meet the Millennium Development Goals in Asia and the Pacific*. Bangkok: United Nations.

ESCAP, ADB and UNDP. 2009. *Achieving Millennium Development Goals in an Era of Global Uncertainty.* Bangkok: United Nations.

Evans, A. 1998. *Rising Food Prices: Drivers and Implication for Development.* London: Chatham House.

———. 2009. *The Feeding of Nine Billion.* London: Chatham House.

FAO (Food and Agriculture Organization). 2004. *The State of Food Insecurity in the World 2004.* Rome: FAO.

———. 2008a. 'What Is Conservation Agriculture?' Agriculture and Consumer Protection Department, Conservation Agriculture. http://www.fao.org/ag/ca/1a.html (accessed in March 2009).

———. 2008b. *Food Outlook,* November 2008. www.fao.org/docrep/011/ai474e/ai474e13.htm (accessed on 6 April 2008).

———. 2009a. 'Investment'. Paper prepared for 'How to Feed the World 2050', High-Level Expert Forum, Rome 12–13 October.

———. 2009b. *The State of Food and Agriculture: Livestock in Balance.* Rome: FAO.

———. 2010. *United against Hunger.* Rome: FAO.

Hayes, Kristy, and Biplove Choudhary. 2010. *Rebuilding International Food Policy Architecture.* Colombo: UNDP Regional Centre.

Hellmuth M. E., D. E. Osgood, U. Hess, A. Moorhead and H. Bhojwani (eds). 2009. *Index Insurance and Climate Risk: Prospects for Development and Disaster Management.* Climate and Society No. 2. International Research Institute for Climate and Society (IRI), Columbia University, New York, USA.

Hermida, Jet. 2007. 'Emergency or Expediency? A Study of Emergency Rice Reserve Scheme in Asia'. AsiaDHRRA, 16 December. http://asiadhrra.org/wordpress/2007/12/16/emergency-or-expediency-a-study-of-emergency-rice-reserve-schemes-in-asia/ (accessed on 27 January 2012).

Huang, Jikun, and Scott Rozelle. 2009. 'Agricultural Development and Nutrition: The Policies behind China's Success'. Occasional Paper No. 19, World Food Programme, Beijing.

ILO (International Labour Organization). 2004. *A Fair Globalization: Creating Opportunities for All.* Geneva: ILO.

IPCC (Inter-governmental Panel on Climate Change). 2007. *Climate Change 2007: Mitigation.* Contribution of Working Group III to the Fourth Assessment Report of the Inter-governmental Panel on Climate Change. http://www.ipcc.ch/pdf/assessment-report/ar4/wg3/ar4-wg3-frontmatter.pdf (accessed on 15 July 2008).

Jalan, Jyotsna, and Martin Ravallion. 1999. 'Are the Poor Less Well Insured? Evidence on Vulnerability to Income Risk in Rural China'. Policy Research Working Paper 1863, World Bank, Washington D.C.

Kanchanakak, Pornpimol. 2010. 'The Story of a Girl Who Isn't and a People Who Aren't'. *Nation,* 5 August, p. 12A.

Kansas State University. 2009. 'Precision Agriculture: Tomorrow's Future Here Today'. Biological and Agricultural Engineering Department. Kansas State University. http://www.bae.ksu.edu/precisionag/ (accessed in March 2009).

Kenmore, P. E. 1996. 'Integrated Pest Management in Rice', in G. Persley (ed.), *Biotechnology and Integrated Pest Management.* Wallingford: CABI.

Mahbub Ul Haq Human Development Centre. 2002. *Human Development Report in South Asia.* Karachi: Oxford University Press.

Malthus, Thomas Robert. 1830 [1798]. *An Essay on the Principle of Population* with *A Summary View* (Introduction by Antony Flew). Penguin Classics.

McCord, A. 2005. 'Win-Win or Lose-Lose? An Examination of the Use of Public Works as a Social Protection Instrument in Situations of Chronic Poverty'. Paper for conference on Social Protection for Chronic Poverty, Manchester, Institute for Development Policy and Management.

Montero, David. 2008 'Wealthy Countries Seek Land in Cambodia, Madagascar and Brazil'. *Christian Science Monitor*. http://www.csmonitor.com/2008/1222/p01s06-wosc.html.

Mukherjee, Amitava. 2001. *The Voices of the Silent Majority: Community Perspectives on Air Pollution, Crop Yield and Rural Livelihood*. Report for the T. H. Huxley School at the Imperial College for Science, Technology and Medicines, London. London: School of Oriental and African Studies, University of London and Aldershot: Ashgate.

———. 2004. *Hunger Theory and Perspectives*. Aldershot: Ashgate.

———. 2008. 'Food Insecurity: A Growing Threat in Asia'. January 2008. http://www.unapcaem.org/publication/FoodInsecurityInAsia.pdf (accessed on 27 January 2012).

Nguyen, V. N. 2004. 'FAO Programme on Hybrid Rice Development and Use for Food Security and Livelihood Improvement'. Paper presented at the Concluding Workshop of the IRRI-ADB funded project 'Sustaining Food Security in Asia through the Development of Hybrid Rice Technology', IRRI, Los Baños, Philippines, 7–9 December.

Ombion, Karl G. 2007. 'Philippines's Land Reform Program a Failure, Study Shows'. PinoyPress, 3 December. http://www.pinoypress.net/2007/12/03/philippiness-land-reform-program-a-failure-study-shows/ (accessed on 1 September 2009).

Organic World Foundation. 2008. 'Making a Difference, Inspiring Action: That is What Organic Agriculture Is'. http://www.organicworldfoundation.org/organic_agriculture.html (accessed in March 2009).

Peng, S., J. Huang, J. E. Sheehy, R. C. Laza, R. M. Visperas, X. Zhong, G. S. Centeno, G. S. Khush and K. G. Cassman. 2004. 'Rice Yield Decline with Higher Night Temperature from Global Warming', in E. D. Redona, A. P. Castro and G. P. Llanto (eds), *Rice Integrated Crop Management: Towards a Rice Check system in the Philippines*. Nueva Ecija, Philippines: PhilRice.

Postel, S., and A. Vickers. 2004. 'Boosting Water Productivity', in Worldwatch Institute, *State of the World, 2004*. New York: W. W. Norton.

Rabobank. 2008. *The Boom Beyond Commodities: A New Era Shaping Global Food and Agribusiness*. London: Rabobank.

Rahil, Siti. 2010. 'Japan, East Asian Nations Closer to Pact on Emergency Rice Reserve'. *Japanese Times*, 8 May.

Sen, Amartya. 1981. *Poverty and Famines: An Essay on Entitlement and Deprivation*. Oxford: Clarendon.

———. 1997. *Hunger in the Contemporary World*. London: STICERD and London School of Economics.

Sexton, Kay. 2009. 'Genetically Modified Crops: A Danger or an Agricultural Right?' 16 April. http://redgreenandblue.org/2009/04/16/genetically-modified-crops-a-danger-or-an-agricultural-right/#comments (accessed on 23 July 2010).

Sinclair, T. R. 1998. 'Options for Sustaining and Increasing the Limiting Yield Plateaus of Grain Crops'. Paper for the 1998 Symposium on World Food Security, Kyoto, for USDA, Washington, D.C.

Studdert, L., K. Soekirman and J. Habicht. 2004. 'Community-based School Feeding during Indonesia's Economic Crisis: Implementation, Benefits and Sustainability', *Food and Nutrition Bulletin*, vol. 25, no. 2, pp. 156–65.

Swaminathan, M. S. 2008. 'Challenge of a Hunger Free Asia', in UNDP/FAO, *Combating Hunger: A Seven Point Agenda*. Colombo and Bangkok: UNDP/FAO.

Takle, Eugene, and Don Hofstrand. 2008. 'Global Warming—Agriculture's Impact on Greenhouse Gas Emissions'. AgDM Newsletter Article, April. http://www.extension.iastate.edu/agdm/articles/others/TakApr08.html (accessed on 2 February 2011).

Taylor, J., and A. Yunez-Naude. 2002. 'Farm/Non-farm Linkages and Agricultural Supply Response in Mexico: A Village-wide Modelling Perspective', in B. Davis, T. Reardon, K. Stamoulis and P. Winters (eds), *Promoting Farm/Non-farm Linkages for Rural Development: Case Studies from Africa and Latin America*. Rome: FAO.

Thacker, J. R. M. 1993. 'Transgenic Crop Plants and Pest Control'. *Science Progress*, no. 77.

UNCCD (United Nations Convention to Combat Desertification). 2008. 'Combating Land Degradation for Sustainable Agriculture'. World Day to Combat Desertification, 17 June. http://www.unccd.int/publicinfo/june17/2008/docs/2008WDCDthemeconceptnote.pdf (accessed in March 2009).

———. 2009. 'Combating Desertification in Asia: Fact Sheet 12'. http://www.unccd.int/publicinfo/factsheets/ showFS.php?number=12

UNDP (United Nations Development Programme). 2007. 'Climate Change and the MDGs in Asia Pacific: Challenges and Opportunities'. *Inside Asia Pacific*, vol. 2, no. 2, July 2007. http://www.undprcc.lk/rcc_web_bulletin/Issue2/country_Bangladesh.shtml (accessed on 22 October 2008).

———. 2011. *Power, Voice and Rights*. Colombo: UNDP, and New Delhi: Macmillan.

UNESCAP (United Nations Economic and Social Commission for Asia and the Pacific). 2009. *Towards Food Security and Sustainable Agriculture in Asia Pacific*. Bangkok: United Nations.

United Nations. 2008. 'Comprehensive Framework for Action'. High-level Task Force on the Global Food Crisis 2008, United Nations, July. www.un.org/issues/food/taskforce/docs.html (accessed on 3 September 2008).

United Nations Department of Economic and Social Affairs. 2007. *Innovation Briefs*, No. 2, March 2007.

Von Brown, J., Akhter Ahmed, Kwadwo Asenso-Okyere, Shenggen Fan, Ashok Gulati, John Hoddinott, Rajul Pandya-Lorch, Mark W. Rosegrant, Marie Ruel, Maximo Torero, Teunis van Rheenen and Klaus von Grebmer. 2008. 'High Food Prices: The What, Who and How of Proposed Policy Actions'. International Food Policy Institute, Washington, D.C.

WHO (World Health Organization). 2008. 'Obesity and Overweight'. Briefing Paper. http://www.who.int/dietphysical/media/en/gsfs_obesity.pdf.

Wongruang, Piyapron. 2005. 'Test Seems to Show GM Papaya Rampant'. *Bangkok Post*, 2 June, p. 4.

World Bank. 2000/2001. *World Development Report 2000/2001: Attacking Poverty.* Washington D.C.: World Bank and Oxford University Press.

———. 2008a. *World Development Report, 2008: Agriculture for Development.* Washington, D.C.: World Bank.

———. 2008b. 'Double Jeopardy: Responding to High Food and Fuel Prices'. Working Paper presented at G-8 Hokkaido-Toyako Summit, World Bank, Washington, D.C., 2 July.

Index

About the Author

Amitava Mukherjee is a Senior Expert in the Macro-economic Policy and Development Division, United Nations, Economic and Social Commission for Asia and the Pacific, Bangkok (UNESCAP). Prior to this, he was the Head, Asian and Pacific Center for Agricultural Engineering and Machinery, in Beijing of the United Nations and Head, Centre for Alleviation of Poverty through Sustainable Agriculture, Bogor, Indonesia. He is also the Executive Director (on leave), Development Tracks RTC, a knowledge enterprise in Delhi.

Dr Mukherjee studied and did his research at the Ranchi University, Ranchi, India; Rice University, Houston, USA, the University of South Florida, Tampa, USA and the London School of Economics, London. He held visiting appointments at the Department of Public Policy at the Graduate School of Business, Stanford University, and the Department of Economics, Stanford University, Stanford, USA; School of Oriental and African Studies, University of London, London and at the London School of Economics and Political Science, London. Dr Mukherjee also held the Chair in Economics and Planning, and co-ordinated the faculty in the Centre for Micro-Planning and Regional Studies at the Lal Bahadur Shastri National Academy of Administration, Mussoorie, India.

He has written and lectured extensively on poverty, food security and participatory community development both in India and abroad; has published a large number of professional reports, articles/papers and over 35 books. Two of his latest publications are: *Hunger: Theory, Perspectives and Reality* (2004) and *Voices of the Poor* (2007).

He is the recipient of several awards and honours, including the Bihar Governor's Cup for the Best Scholar, Vivekananda Centenary Gold Medal, Hem Nalini De Memorial Gold Medal,

Rotary Foundation International Fellowship, Henry Lee Fellowship and the Commonwealth Award. In recognition of his services to rural development in China he was made an Honorary Citizen, Huain County, China.